Membrane Processing

The Society of Dairy Technology (SDT) has joined with Wiley-Blackwell to produce a series of technical dairy-related handbooks providing an invaluable resource for all those involved in the dairy industry, from practitioners to technologists, working in both traditional and modern large-scale dairy operations. For information regarding the SDT, please contact Maurice Walton, Executive Director, Society of Dairy Technology, PO Box 12, Appleby in Westmorland, CA16 6YJ, UK. email: execdirector@sdt.org

Other volumes in the Society of Dairy Technology book series:

Probiotic Dairy Products (ISBN 978 1 4051 2124 8)
Fermented Milks (ISBN 978 0 6320 6458 8)
Brined Cheeses (ISBN 978 1 4051 2460 7)
Structure of Dairy Products (ISBN 978 1 4051 2975 6)
Cleaning-in-Place (ISBN 978 1 4051 5503 8)
Milk Processing and Quality Management (ISBN 978 1 4051 4530 5)
Dairy Fats (ISBN 978 1 4051 5090 3)
Dairy Powders and Concentrated Products (ISBN 978 1 4051 5764 3)
Technology of Cheesemaking, Second Edition (ISBN 978 1 4051 8298 0)
Processed Cheese and Analogues (ISBN 978 1 4051 8642 1)

Membrane Processing

Dairy and Beverage Applications

Edited by

A.Y. Tamime

Consultant in Dairy Science and Technology, Ayr, UK

WILEY-BLACKWELL

A John Wiley & Sons, Ltd., Publication

SDT

This edition first published 2013 © 2013 by Blackwell Publishing Ltd.

Blackwell Publishing was acquired by John Wiley & Sons in February 2007. Blackwell's publishing program has been merged with Wiley's global Scientific, Technical and Medical business to form Wiley-Blackwell.

Registered office: John Wiley & Sons, Ltd, The Atrium, Southern Gate, Chichester, West Sussex, PO19 8SQ, UK

Editorial offices: 9600 Garsington Road, Oxford, OX4 2DQ, UK
The Atrium, Southern Gate, Chichester, West Sussex, PO19 8SQ, UK
2121 State Avenue, Ames, Iowa 50014-8300, USA

For details of our global editorial offices, for customer services and for information about how to apply for permission to reuse the copyright material in this book please see our website at www.wiley.com/wiley-blackwell.

Library of Congress Cataloging-in-Publication Data has been applied for

9781444333374 (hardback)

A catalogue record for this book is available from the British Library.

Wiley also publishes its books in a variety of electronic formats. Some content that appears in print may not be available in electronic books.

Cover image: Adapted with permission from Tetra Pak
Cover design by www.hisandhersdesign.co.uk

Set in 10/12.5pt Times by Laserwords Private Limited, Chennai, India.

1 2013

Contents

Preface to the Technical Series

For more than 60 years, the Society of Dairy Technology (SDT) has sought to provide education and training in the dairy field, disseminating knowledge and fostering personal development through symposia, conferences, residential courses, publications, and its journal, the *International Journal of Dairy Technology* (previously published as the *Journal of the Society of Dairy Technology*).

In recent years, there have been significant advances in our understanding of milk systems, probably the most complex natural food available to man. At the same time, improvements in process technology have been accompanied by massive changes in the scale of many milk processing operations, and the manufacture a wide range of dairy and other related products.

The Society has embarked on a project with Wiley-Blackwell to produce a Technical Series of dairy-related books to provide an invaluable source of information for practicing dairy scientists and technologists, covering the range from small enterprises to modern large-scale operation. This eleventh volume in this series, on *Membrane Processing – Dairy and Beverages Applications*, provides a timely and comprehensive update of the principles and practices involved in this technology. The commercial exploitation of membrane technology is developed further in its applications to the processing of milk and milk products, plus a wide range of non-dairy beverages, where membranes provide novel opportunities for separation, fractionation and concentration.

Andrew Wilbey
Chairman of the Publications Committee, SDT

Preface

In the last two decades, there have been significant developments in membrane filtration processes for the dairy and beverage industries. The filtration systems can be classified into four main groups: reverse osmosis (RO), nanofiltration (NF), ultrafiltration (UF) and microfiltration (MF). The primary objective of this book is to assess critically the pool of scientific knowledge available to the dairy and beverages industry, as a tool for process and product innovation, quality improvement and safety.

Appraisals of the key technical aspects of membrane processing stages to control and maintain the consistency of dairy and beverages products are also included. Although this has produced some overlap in the coverage of membrane processes, I have felt justified in allowing this overlap because it emphasises the prime importance of processing in the preparation of milks and beverages, to achieve the desired quality and consistency of these products for the end-user.

Generally, the book is divided into three main parts. Part I, which consists of five chapters, reviews the principals, developments and designs of membrane processes that are mainly used in commercial dairy and beverage applications. Successful applications of membrane processes in the food industry and requirements pertaining to ensure hygienic conditions in the equipment are reviewed in Chapter 4, and Chapter 5 details the aspects of food industrial effluent treatment, recovery of detergent from cleaning-in-place (CIP) systems and reuse of wastewater.

The application of membrane technology in the manufacture of different dairy products could be briefly summarised as follows: (a) MF for the removal of bacteria and spores from skimmed milk for the production of an extended shelf-life product; (b) UF is used to concentrate the fat and protein components in milk, which can be used for standardisation of the protein content prior to cheesemaking and also to concentrate the fermentate for the production of concentrated yoghurt (labneh); in addition, a wide range of cheeses are commercially manufactured from ultrafiltrated milk (e.g. Quarg, Feta); (c) reduction of salts in milk is achieved using NF, electrodialysis (ED) has been used for whey; (d) combined applications of MF, UF and NF have been used to fractionate milk components and, in some instances, the composition of cow's milk has been modified to be similar to mare's milk; (e) RO is widely used as a lower energy alternative to evaporation where low degrees of concentration are needed.

Part II of the book provides information on the applications of membrane processes in the manufacture of dairy products. This ranges from on-farm concentration of milk as a pre-treatment for cheesemaking to fractionation of milk and whey to provide a wide range of ingredients for food and other applications.

Part III considers membrane applications during the manufacture of fruit juices, beer and cider, wine and vinegar. These include concentration, deacidification and dealcoholisation processes.

Dairy and beverage processors are encouraged to use this book as a reference in the application of membranes, both to aid the creation of novel products and to improve their process economics.

A.Y. Tamime

Contributors

Editor

Dr. A.Y. Tamime
Dairy Science & Technology Consultant
24 Queens Terrace
Ayr KA7 1DX
Scotland – United Kingdom
Tel. +44 (0)1292 265498
Fax +44 (0)1292 265498
Mobile +44 (0)7980 278950
E-mail: adnan@tamime.fsnet.co.uk

Contributors

Dr M. Ashokkumar
University of Melbourne
Particulate Fluids Processing
Centre School of Chemistry
Victoria 3010
Australia
Tel. +61 (0)3 8344 7090
Fax: +61 (0)3 9347–5180
E-mail: masho@unimelb.edu.au

Dr A. Cassano
Istituto per la Tecnologia delle Membrane (ITM-CNR)
Università della Calabria
via P. Bucci, 17/C
87030 Rende (CS)
Italy
Tel. + 39 0984 492067 or 492014
Fax + 39 0984 402103
E-mail: a.cassano@itm.cnr.it

Dr-Ing T. Demmer
Goethering 56
D-85570 Markt Schwaben
Germany
Tel. +49 8121 40891
E-mail: thd042@googlemail.com

Dr M. Duke
Victoria University
Institute for Sustainability and Innovation
PO Box 14428
Victoria 8001
Australia
Tel. +61 (0)3 9919 7682
Fax: +61 (0)3 9919 7696
E-mail: mikel.duke@vu.edu.au

Dr G. Gesan-Guiziou
Director of Research INRA
UMR1253 Science and Technologie du Lait et de L'oeuf
INRA – Agrocampus Ouest
65 Rue de Saint Brieuc
35042 Rennes Cedex
France
Tel. +33 (0)2 23 48 53 25
Fax: +33 (0)2 23 48 53 50
E-mail: Genevieve.Gesan-Guiziou@rennes.inra.fr

Dr K. Grainger
c/o Tarsus Hotel
Daventry Road
Southam
Warks CV47 1NW
England – United Kingdom
Mobile +44 (0)7956 004855
E-mail: keith@keithgrainger.com

Dr S. Gray
Victoria University
Institute for Sustainability and Innovation
PO Box 14428

Victoria 8001
Australia
Tel. +61 (0)3 9919 8097
Fax: +61 (0)3 9919 7696
E-mail: Stephen.gray@vu.edu.au

Dr A. Hausmann
Victoria University
Institute for Sustainability and Innovation
PO Box 14428
Victoria 8001
Australia
Tel. +61 (0)3 9919 7690
Fax: +61 (0)3 9919 7696
E-mail: angela.hausmann@live.vu.edu.au

Dr S.E. Kentish
University of Melbourne
Particulate Fluids Processing Centre
Department of Chemical and Biomolecular Engineering
Victoria 3010
Australia
Tel. +61 (0)3 8344 6682
Fax: +61 (0)3 8344 4153
E-mail: sandraek@unimelb.edu.au

Ms L.L.A. Koh
University of Melbourne
Department of Chemical and Biomolecular Engineering
Victoria 3010
Australia
Tel. +61 3 8344 6682
Fax: +61 3 8344 4153
E-mail: l.koh5@pgrad.unimelb.edu.au

Dr F. López
Universitat Rovira i Virgili
Facultat d'Enologia
Departament d'Enginyeria Química
av. Països Catalans 26
43007-Tarragona
Spain
Tel. 34 977 558 503
Fax: 34 977 559 621
E-mail: francisco.lopez@urv.es

Dr N. Milne
Victoria University
Institute for Sustainability and Innovation
PO Box 14428
Victoria 8001
Australia
Tel. +61 (0)3 9919 7646
Fax: +61 (0)3 9919 7696
E-mail: Nicholas.milne@vu.edu.au

Dr V.V. Mistry
South Dakota State University
Dairy Science Department
Brookings SD 57007
United States of America
Tel. + 1 (0)605 688 5731
Fax +1 (0)605 688 6276
E-mail: Vikram.Mistry@sdstate.edu

Dr L. Ramchandran
Victoria University
Institute for Sustainability and Innovation
PO Box 14428
Victoria 8001
Australia
Tel. +61 (0)3 9919 7684
Fax: +61 (0)3 9919 7696
E-mail: Lata.ramchandran@vu.edu.au

Dr S. O'Mahony
University College Cork
School of Food & Nutritional Sciences
Cork
Republic of Ireland
Tel. + 353 (0)21 4903625
E-mail: sa.omahony@ucc.ie

Dr B. Ozer
Abant Izzet Baysal University
Faculty of Engineering and Architecture
Department of Food Engineering
14280 Bolu
Turkey
E-mail: adabarbaros@gmail.com

Dr K. Smith
Dairy Processing Technologist
Wisconsin Centre for Dairy Research
1605 Linden Drive
Madison, WI 53706–1565
United States of America
Tel. +1 (0) 608 265 9605
E-mail: smith@cdr.wisc.edu

Dr S.J. Tuohy
2e Technical Development Ltd.
Hollybrook House
Corrin
Fermoy
Co. Cork
Republic of Ireland
Tel. : + 353 (0)25 31144
Mobile: + 353 (0)87 2657706
E-mail: 2etechdev@gmail.com

Dr T. Vasiljevic
Victoria University
School of Biomedical and Health Sciences
Werribee campus
Victoria 8001
Australia
Tel. +61 (0)3 9919 8062
Fax: +61 (0)3 9919 8284
E-mail: Todor.Vasiljevic@vu.edu.au

1 Development of Membrane Processes

K. Smith

1.1 Historical background

The ability of membranes to separate water from solutes has been known since 1748, when Abbé Nolet experimented with the movement of water through a semi-permeable membrane. Depending on the reference, either Abbé Nolet or Dutorchet coined the word *osmosis* to describe the process. Throughout the 18th and 19th centuries, membranes were used exclusively for laboratory applications, and often consisted of sausage casings made from animal intestines or the bladders of pigs, cattle or fish.

The first synthetic membranes were produced by Fick in 1855, and appear to have been made of nitrocellulose. Membranes based on cellulose were known as collodion and had the advantages of reproducible characteristics compared with the previously used animal-based membranes. Bechhold further advanced the process for manufacturing collodion membranes when he developed methods for controlling pore size and measuring pore diameters in 1907. He is generally credited with first using the term *ultrafiltration* (UF). In addition, Richard Zsigmondy at the University of Göttingen, Germany, patented a membrane filter in 1918 that was referred to as a cold ultrafilter. His work becomes the basis of the membrane filters produced by Sartorius GmbH.

Collodion membranes produced by the Sartorius GmgH of Germany became commercially available in 1927. The primary use of membranes until the 1940s was the removal of micro-organisms and particles from liquids and gases and research applications. There was a critical need to test drinking water in Europe for microbial content following the Second World War, and membranes were developed that could rapidly filter water and capture any micro-organisms on the membrane surface, where they could quickly be enumerated to determine the safety of the water for human consumption.

In addition to the separation of relatively large particles from water, there was interest in developing membranes that could desalinate sea or brackish water. The term *reverse osmosis* (RO) had been coined in 1931 when a patent was issued for desalting water; however, the available membranes could not withstand the pressures required.

Although many improvements were made in the following years, including the use of other polymers for constructing membranes, membranes were limited to laboratory and small specialised industrial applications. Factors limiting the use of membranes included a lack of reliability, being too slow, not sufficiently selective and cost.

Membrane Processing: Dairy and Beverage Applications, First Edition. A. Y. Tamime.

A breakthrough came in the early 1960s when Sourirajan and Loeb developed a process for making high-flux, defect-free membranes capable of desalinating water. Researchers at the time believed the best approach to improving flux would be to reduce the thickness and thereby the resistance to flow of the membrane. Sourirajan and Loeb attempted to produce such membranes by taking existing cellulose acetate membranes and heating them while submersed in water in a process known as annealing. They expected the membrane pores would increase in size by such a process, but instead the pores became smaller and the membrane more dense. When they attempted the same process with cellulose acetate UF membranes, they discovered not only did the pores become smaller but the ability of the membrane to reject salt increased, as did flux. The flux improvement was such that the membranes could be a practical way to desalinate water.

The annealing process of Sourirajan and Loeb had created an anisotropic or asymmetric membrane. Anisotropic membranes have different behaviour depending on which side of the membrane is used for the separation. Although this type of membrane had been seen over 100 years earlier with natural membranes, it had not been reproduced with the synthetic variety.

The key to the anisotropic membrane of Sourirajan and Loeb was the thin 'skin' on one surface of the membrane. The skin typically was approximately 0.1–0.2 μm thick and had a dense structure whereas the remainder of the membrane had a very porous open structure. The thickness of the membrane essentially determined the flux and so by reducing the effective separating distance from 100–200 μm to 0.1–0.2 μm the rate of liquid crossing the membrane dramatically improved, but because of the small pores in the skin the rejection of salt remained high.

Many changes in the production of membranes occurred during the 1960s, 1970s and 1980s. By continuing the work of Sourirajan and Loeb, others were able to develop additional methods for producing membranes. Initial membrane modules were plate-and-frame (Danish Sugar Corporation) or hollow fibre (Amicon) designs, but membranes in formats, such as spiral-wound and tubular (Abcor), were introduced shortly afterwards. The thickness of the separating layer was further reduced to less than 0.1 μm. Large plants using RO, UF and microfiltration (MF) were operating around the world by 1980.

Cellulose acetate remained the material of choice until the mid-1970s, when methods of producing composite membranes for water desalination were developed. By combining polysulphone and polyamide, composite membranes had the advantage of high salt rejection combined with good water flux and increased resistance to temperature and chemicals. Nanofiltration (NF) or 'loose RO' membranes became available in the mid-1980s. The NF membranes operated at lower pressures than RO systems, and were able to permeate monovalent ions. They found immediate application in producing ultrapure water by permeating trace salts from water produced by RO.

In addition, membranes made from inorganic materials, such as zirconium and titanium dioxide, became commercially available in the mid-1980s. Membranes made from these materials are referred to as mineral or ceramic and are available in tubular form for UF and MF. Union Carbide (USA) and Societé de Fabrication d'Elements (France) used carbon tubes covered with zirconium oxide for their inorganic membranes. Later Ceravèr (France) used a ceramic base with aluminium oxide. Chemical and temperature resistance were the significant advantages of ceramic membranes. It was originally thought that such membranes had an unlimited life, but subsequent experience has shown this is not the case.

Advancements in membrane composition and design along with operation of membrane systems have continued. A wide variety of membrane polymers and designs have been adapted for RO, NF, UF and MF, resulting in many commercial applications. The feasibility of membrane-based applications depends chiefly on the ability of the filtration process to economically produce an acceptable product. Membrane pore size distribution, selectivity, operating conditions, membrane life, capital and operating costs become important economic considerations. These parameters are in turn influenced by many factors, such as the membrane polymer, element configuration and system design.

1.2 Basic principles of membrane separations

Membrane filtration is a pressure-driven separation process using semi-permeable membranes. The size of membrane pores and the pressure used indicate whether the term RO, NF, UF or MF is used for a given separation. RO and NF systems use the highest pressures and membranes with the smallest pores, whereas MF has the lowest operating pressures and membranes with the largest pores. UF is intermediate in pressure used and membrane pore size.

1.2.1 *Depth versus screen filters*

In the past, filtration processes relied on depth filters. This type of filter has fibres or beads in a mesh-like structure. Particles in the feed solution become trapped or adsorbed within the filter network, which eventually clogs the filter, thereby resulting in replacement of the filter (Fig. 1.1). By contrast, screen type filters generally rely on pores, with the size and shape of the pores determining passage of particles. Pores are more rigid and uniform and have a more narrowly defined size than mesh openings in a depth filter. Components not able to pass through pores remain on the membrane surface and, therefore, do not typically become trapped within the membrane structure (Fig. 1.2). Because the fouling materials remain on the surface, internal fouling decreases and the membrane can be reused.

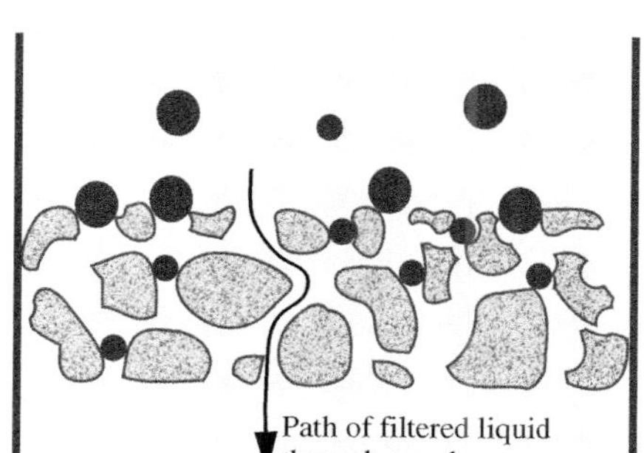

Fig. 1.1 Depth filter with particles entrapped within the membrane structure.

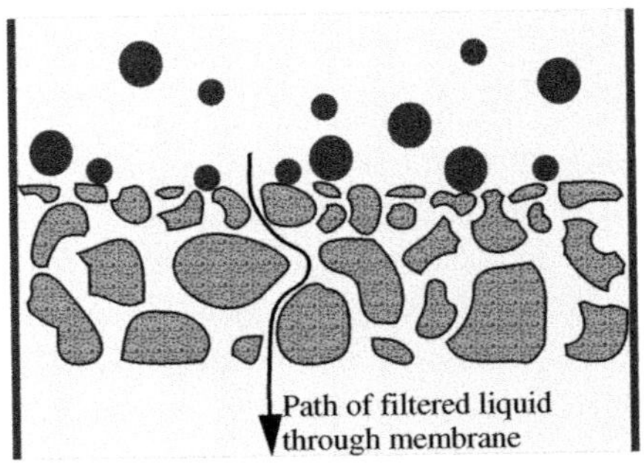

Fig. 1.2 Screen-type membrane separates particles at the membrane surface.

1.2.2 *Isotropic versus anisotropic membranes*

Membranes can have several types of internal structure. Terms, such as microporous, non-porous, isotropic and anisotropic, refer to the structure of the membrane. Typically, membranes are either isotropic or anisotropic. Microporous and non-porous refer to isotropic membrane structure. An isotropic membrane will have a relatively uniform structure (Fig. 1.3), i.e. the size of the pores is similar throughout the membrane. The membrane, therefore, does not have a top or bottom layer, rather the membrane properties are uniform in direction. Isotropic membranes generally act as depth filters and, therefore, retain particles within the internal structure resulting in plugging and reduced flux.

Microporous and non-porous membranes typically are isotropic. Microporous membrane structure can resemble a traditional filter; however, the microporous membrane has extremely small pores. Materials are rejected at the surface, trapped within the membrane or pass through pores unhindered, depending on particle size and size of the pores. A non-porous membrane will not have visible pores and materials move by diffusion through the membrane.

An anisotropic or asymmetric membrane has pores that differ in size depending on their location within the membrane (Fig. 1.4). Typically, anisotropic membranes will have a thin, dense skin supported by a thicker and a more porous substructure layer. The thin top layer provides high selectivity, whereas the porous bottom layer has good flux. Membranes used for commercial separations in the food industry are typically anisotropic.

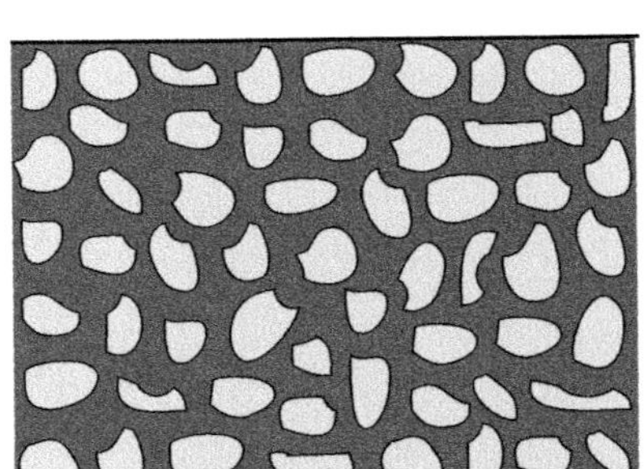

Fig. 1.3 Structure of an isotropic membrane.

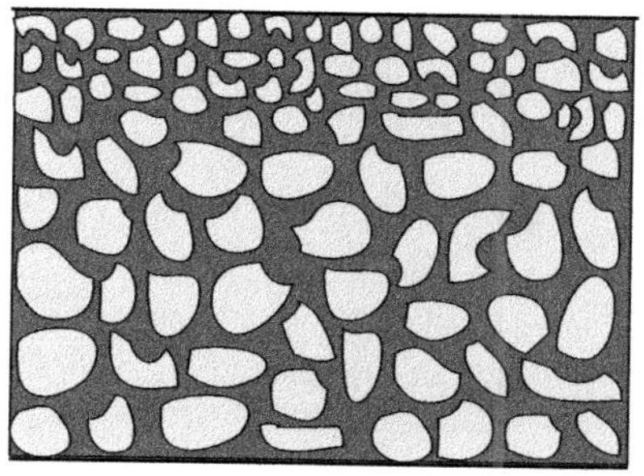

Fig. 1.4 Structure of an anisotropic membrane.

1.2.3 *Cross-flow filtration*

RO, NF, UF and MF systems all involve cross-flow filtration, which can be compared to the traditional method of perpendicular filtration. In traditional filtration (Fig. 1.5), the entire feed stream passes through the filtering media, i.e. the incoming stream flows perpendicularly to the filter with the filter retaining any trapped solids. The result is a filtered stream with solids trapped on and within the filter.

In cross-flow filtration (Fig. 1.6), the feed stream passes parallel to the membrane. Some of the incoming feed stream and particles will cross the membrane into the permeate section, whereas the other portion with the concentrated solids is the retentate stream. At any time only some of the water and particles will cross the membrane into the permeate stream, unlike traditional filtration where most particles are trapped after one pass through the filter. Because the feed stream flows parallel to the membrane rather

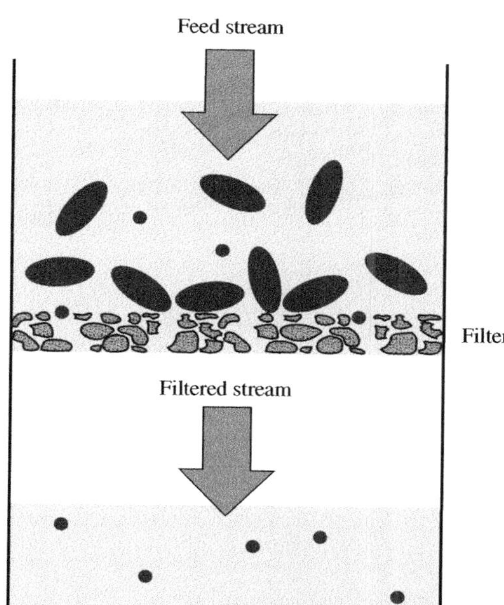

Fig. 1.5 Traditional filtration with perpendicular flow.

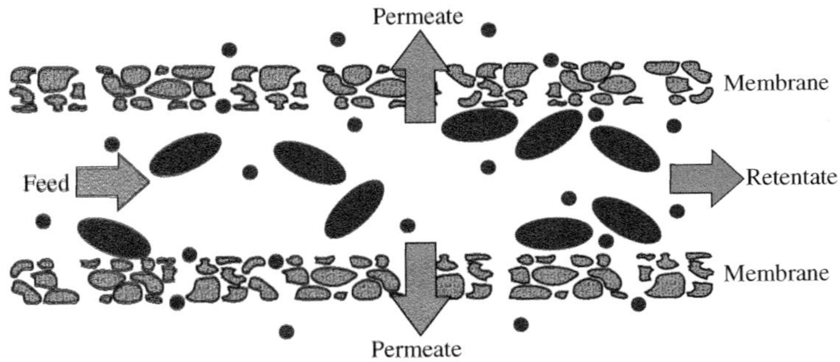

Fig. 1.6 Cross-flow filtration.

than perpendicular to it, cross-flow filtration is self-cleaning by comparison. Solutes and particles are continually swept along and away from the membrane surface by the retentate stream, thereby allowing longer operating times without cleaning than is possible with traditional filtration.

The affect of cross-flow permeate flow and thickness of the fouling or cake layer can be seen in Fig. 1.7. In perpendicular filtration, the flow of permeate is reduced as the thickness of the material on the surface of the filter, i.e. the thickness of the cake layer, increases over time. With cross-flow filtration, however, the thickness of the material on the membrane is limited by action of the feed stream sweeping across the surface of the

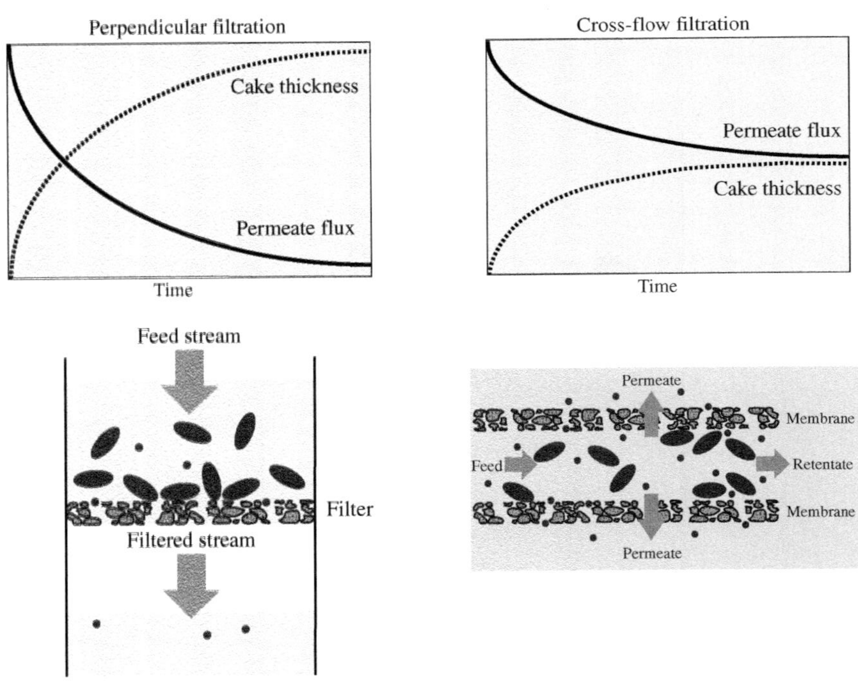

Fig. 1.7 Effect of perpendicular and cross-flow filtration on flux and cake thickness.

membrane. Because the thickness of the deposited material is limited, permeate flow is maintained at a higher level throughout filtering.

1.2.4 *Requirements of membrane processes*

The shared characteristics of membrane processes are pressure-driven processes using semi-permeable membranes. Pressure is used to reverse the direction of the osmosis process, while differences in membrane permeability determine separation of molecules. The process of osmosis is illustrated in Fig. 1.8. Solutions containing two different concentrations of dissolved materials are separated by a membrane that will allow only water to cross

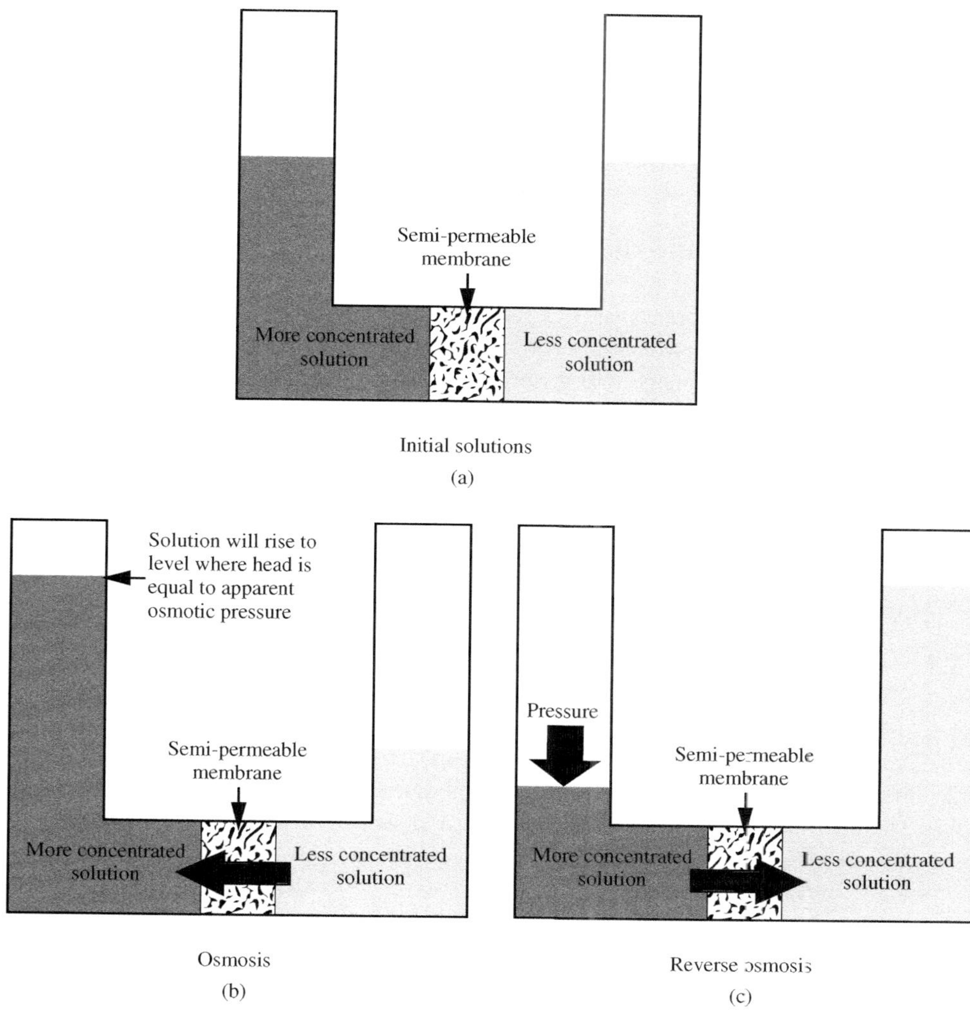

Fig. 1.8 The processes of osmosis and reverse osmosis.

(Fig. 1.8a). Nature will try to equalise the concentration of the two solutions. Since the dissolved material cannot cross the membrane, water must flow from the solution of lower concentration to the solution at the higher concentration (Fig. 1.8b). The flow of water will continue until the solutions are of equal concentration or no more water is available. The difference in the height of water in the corresponding tubes is a result of the movement of water from lower to higher concentration. The final water level in the more concentrated solution compared with the original level is equal to the apparent osmotic pressure.

In the process of 'reverse' osmosis, pressure is used to force water to flow in the opposite direction (Fig. 1.8c). Enough pressure must be applied to overcome the apparent osmotic pressure of the more concentrated solution before water can flow from the more concentrated to the less concentrated side. In doing so, the more concentrated side becomes even more concentrated through the loss of water. It is this ability to concentrate and separate that is taken advantage of in commercial membrane separations.

Another shared characteristic is the use of semi-permeable membranes. Membranes can be distinguished from filters by the size of the particulates that are separated. By convention, filters generally separate particulates that are greater than 1–10 μm in size, whereas membranes separate smaller particles. Semi-permeable refers to the ability to separate some particles from other particles.

1.3 Types of membrane separations

The classification of membranes as RO, NF, UF and MF is somewhat arbitrary, and has considerable overlap between categories. Generally, RO/NF membranes will retain molecules in the ionic size range, UF membranes will separate macromolecules, and MF will retain particles of micron size. Because RO, NF, UF and MF membranes differ in the size of molecules they separate, the osmotic pressure involved is considerably different between the processes. RO, which retains the smallest molecules, has the highest osmotic pressure to overcome and, therefore, requires the highest operating pressure. A range from 1.38 to 8.28 MPa is common for RO, 1.03 to 2.76 MPa for NF, 0.21 to 1.03 MPa for UF, and MF requires only from 0.07 to 0.69 MPa.

1.3.1 *Reverse osmosis*

RO membranes generally retain all compounds allowing only water to cross into the permeate. There are exceptions to this general statement and, at times, relatively large molecules may pass into the permeate. RO membranes can, therefore, either concentrate a feed stream (retentate stream) through removal of water or produce very pure water (permeate stream).

1.3.2 *Nanofiltration*

NF membranes are very similar to RO membranes with the exception that NF membranes will allow the passage of monovalent ions into the permeate. NF membranes are very effective at concentrating materials in the feed stream since only monovalent ions are removed into the permeate. The loss of monovalent ions into the permeate allows NF systems to operate at lower pressures than RO systems.

1.3.3 *Ultrafiltration*

UF membranes both fractionate and concentrate materials in the feed stream. Larger components, such as micro-organisms, lipids and proteins, typically are retained by UF membranes, whereas smaller molecules, such as sugars and minerals, pass into the permeate. UF systems operate at much lower pressures than RO and NF systems because of the permeation of sugars and minerals into the permeate stream.

1.3.4 *Microfiltration*

MF membranes also fractionate materials in the feed stream. Because MF membranes allow the passage of many larger components, such as smaller proteins, they are not as good at concentrating the feed stream compared with UF membranes. Operating pressures are the lowest of the four types of membranes and, in fact, MF membrane systems operated at higher pressures often lose their ability to permeate larger molecules and behave like UF membranes.

1.4 Theory of membrane transport

Although each of these membrane processes separates molecules of different sizes, the method of separation varies considerably. Generally, UF and MF membranes separate molecules based on size, shape and charge, whereas separation during RO and NF is based neither on size nor on shape.

1.4.1 *Transport models*

Two models, pore flow and solution diffusion, have been proposed for describing the movement of water and solutes into the permeate. The most significant difference between the two models is the size and permanence of membrane pores. In the pore flow model (Fig. 1.9), pressure-driven convective flow drives solutes through small pores or openings in the membrane. Solutes that are too large to move through the pores remain behind thereby resulting in a separation of solution components.

Separations in the solution diffusion model (Fig. 1.10) are due to differences in the solubility of solution components in the membrane and the rate at which the components will move across the membrane. Components flow from high to low pressure. Movement of individual components is not related and is determined by mobility, concentration and pressure gradients.

The size of the pores within a membrane probably determines the model that applies. A pore diameter of 5–10 Å probably represents the transition zone between the two models. It is very difficult to directly measure the size of pores within a membrane. Pore size often is inferred from an indirect technique, such as the size of molecules that will permeate the membrane. Given the limitations in determining membrane pore size, it is possible to

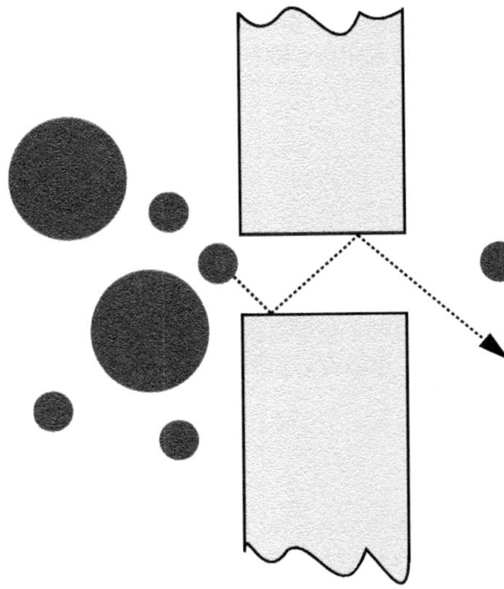

Fig. 1.9 Pore flow model.

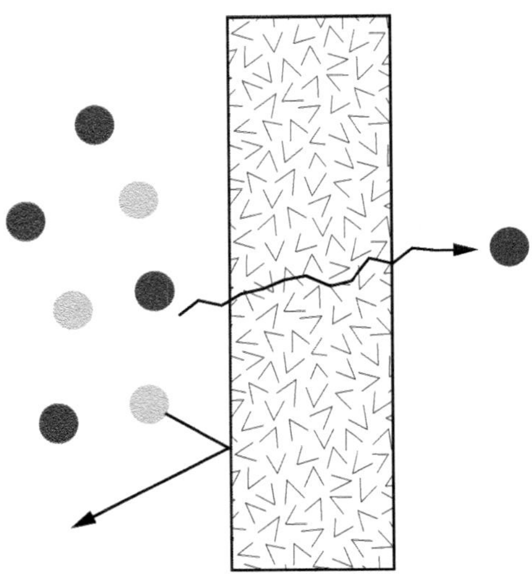

Fig. 1.10 Solution diffusion model.

assign RO, NF, UF and MF membranes into three general categories of separation models. RO membranes, which have no visible pores, are thought to follow the solution diffusion model. Separation in UF and MF membranes is probably based on the pore flow model. NF membranes are in the transition zone between the two models and, therefore, probably use a combination of the two models.

1.4.2 *Reverse osmosis/nanofiltration membranes*

Pores are not discernible in RO/NF membranes; however, if pores are present, they are thought to be 1–5 Å in diameter. It is the structure of water within the membrane that is important. The RO/NF membrane can be considered a water-swollen gel. Water is thought to move through the membrane based on diffusion and the ability of several water molecules to form a tetrahedral or ice-like structure through hydrogen bonding. One theory holds that water is absorbed into the voids of the RO/NF membrane where the water molecules form this tetrahedral structure. When pressure is applied, water on the retentate side of the membrane joins the structured water within the membrane, while water in the membrane on the permeate side is released into the permeate stream. Only those molecules or ions able to fit into this tetrahedral structure can cross the RO/NF membrane. Because ions are surrounded by a water shell, they do not readily fit into the ice-like structure and, therefore, do not cross into the permeate stream; however, small molecules, such as methanol, urea and lactic acid that exhibit hydrogen bonding, may be able to cross the membrane.

Based on this theory of RO/NF systems, water and solute passage across the membrane is not connected. Water flow into the permeate is based on pressure, whereas solute passage is based on the concentration gradient. An increase in pressure, therefore, will increase the flow of water across the membrane while solute passage remains unaffected and results in less solute in the permeate. Because each solute has its own electrochemical potential, the flux of each component is not related to other solutes. NF membranes are sometimes referred to as 'loose RO' membranes. In addition to the passage of compounds that resemble the structure of water, NF membranes also permit the passage of monovalent ions while rejecting polyvalent ions. Additional factors affecting the separating ability of both RO and NF membranes are discussed in Section 1.5.

1.4.3 *Ultrafiltration/microfiltration membranes*

Both UF and MF membranes separate compounds largely on size. Fundamentally, the membrane acts as a sieve by rejecting those molecules too large to fit through pores, while permitting smaller molecules to cross into the permeate stream. The change in pressure across the membrane is the driving force. Although molecular weight often is used to indicate the size of molecules retained by the membrane, other factors are important and are discussed in Section 1.5.

1.5 Factors affecting membrane separations

The methods of separation for RO and UF differ significantly; therefore, the factors affecting these processes also will differ. However, NF will more closely resemble RO, whereas MF is more similar to UF.

1.5.1 *Factors affecting reverse osmosis/nanofiltration separations*

Permeability of components during RO or NF is not based on size, but instead depends on the ability of a compound to cross the membrane using a pressure-driven solute diffusion process. Non-electrolytes and electrolytes, therefore, will be affected by different factors. Pressure, temperature, concentration of components and the type of compound affect RO/NF separations. Permeability of solution components during RO/NF also is affected by membrane composition. In general, as pressure increases the rate of water permeation increases. In turn, the concentration of dissolved solids in the permeate increases. A similar trend occurs with temperature during RO/NF. Increasing the temperature of the process increases the rate of permeation; however, the concentration of dissolved solids in the permeate also increases.

Permeation of components that typically do not cross a membrane is less affected by the presence and concentration of other solutes than components in the form of electrolytes. Conversely, electrolyte retention is affected by concentration and other solutes, with less retention of these components as their concentration increases. An example of the effect of membrane composition on permeability is illustrated by the membrane polymer cellulose acetate, which has a strong sorption of higher aliphatic alcohols and flavour compounds. Cellulose acetate is hydrophobic as are many of these compounds which, therefore, are drawn to the membrane as a result of being repelled by the water phase. Interactions between cellulose acetate and water also have a large affect on salt permeability during RO/NF. Increased interactions between cellulose acetate and water result in less water available for hydration of ions. This causes the salts to be less soluble and, therefore, more energy is required to hydrate these salts. Different amounts of energy needed to hydrate salts account for differences in permeability.

The amount of charge on the ions present in solution is very important to their permeability with NF membranes. NF membranes typically reject polyvalent ions while permitting the passage of most monovalent ions. The concentration of ions also is a factor with rejection of ions increasing as the concentration of ions increases in the solution.

1.5.2 *Factors affecting ultrafiltration/microfiltration separations*

Size, shape, deformability and hydrodynamic radius of a molecule are very important in determining whether a component is retained during UF/MF. Generally, linear, easily deformed molecules are less likely to be retained than highly structured, rigid molecules of equivalent molecular weight. Under pressure, the more flexible molecules can deform to fit through pores generally considered too small to allow their passage. Globular proteins, therefore, are used to define the molecular weight cut-off of a UF membrane since they are less likely to deform under pressure.

Membrane–molecule interactions may be more important than physical factors in determining the ability of a molecule to cross a UF membrane. An example of two materials having opposite effects is styrene and erythrocytes. Styrene, although small enough to cross UF membrane pores, will interact with pore walls and form agglomerates effectively blocking the pore. Erythrocytes, in contrast, are too large to pass through pores; however, their flexibility and lack of interactions with pore walls allow them to cross into the permeate.

Ionic strength and pH of the solution also influence the apparent size of a molecule. For example, a highly charged polyelectrolyte is more easily retained when in solution of lower ionic strength since these conditions give the molecule a larger effective radius and, therefore, limit passage through membrane pores.

The type of compound also is important. Acids, for example, are retained less readily than corresponding salts. Small organic molecules (alcohols, esters, aldehydes and ketones) and small non-ionic inorganic molecules, such as ammonia, are more likely to cross a membrane than ionic compounds.

Composition of the feed material also influences permeability. An example is small molecular weight proteins, which are more likely to cross a membrane when in a pure solution than when ultrafiltered in the presence of compounds such as larger molecular weight proteins.

1.5.3 *System parameters*

System parameters, such as operating conditions and membrane composition can influence passage of a molecule. Operating parameters including temperature, feed velocity, pressure and feed concentration can alter permeation of a molecule. For example, use of higher pressures and lower feed velocities can result in the rejection of proteins by a MF membrane that otherwise would permeate the protein. The affects of system parameters on separations will be discussed in a later section.

1.6 General characteristics of membrane processes

Certain characteristics of a particular membrane are especially important since they influence the economic viability and possible applications of the membrane system. Pore size and retention can be used as an indication of the ability of a membrane to retain a certain size molecule. Flux and membrane life, in contrast, affect the economics of processing with membranes.

1.6.1 *Retention and rejection*

The terms retention and rejection may be used interchangeably depending on whether the component is desired or undesired in the retentate stream. Retention (R) can be defined in several ways, and one common definition is:

$$R = \frac{c_f - c_p}{c_f} \tag{1.1}$$

where C_f is the concentration of a component in the feed stream and C_p is the concentration of a component in the permeate stream.

If a component freely permeates the membrane, R will be near zero, whereas a completely retained component has an R value of 1, and expressed as a percentage would be equivalent to 100% retention. Components having a rejection or retention value of zero will be found in the same concentration in the permeate and feed streams.

An alternative system that measures membrane performance according to permeability (P) is as follows:

$$P = 1 - R \tag{1.2}$$

or

$$P = 100 - R(percentage) \tag{1.3}$$

Although easily defined, rejection characteristics of a membrane are affected by many factors, and there is no uniform set of conditions used by manufacturers to determine this property. Pore diameter, chemical composition of the membrane and interactions between the membrane and feed material are major factors influencing rejection characteristics of a membrane. Rejection values also change depending on operating conditions and during operation of a membrane system.

Rejection is one method of classifying a membrane and, typically, RO and NF membranes are rated according to their ability to reject sodium chloride or other salts. Because RO membranes are used extensively for water desalination, the ability of the membrane to limit the passage of sodium chloride into the permeate is extremely important in the production of water for human consumption. Even small amounts of salt in drinking water can affect taste of the water; therefore, RO membranes often are rated on salt rejection.

1.6.2 *Pore size*

Pore size is used by some membrane manufacturers to indicate the separating ability of a membrane. Units of microns (μm) are often used. Reference to pore size is typical of membranes in the MF range or to specific categories of membranes, such as ceramic and track etched.

1.6.3 *Molecular weight cut-off*

Manufacturers often use molecular weight cut-off to indicate the separating ability of a given membrane. Membranes with the same molecular weight cut-off, however, may not have the same retention for a compound since manufacturers use a variety of methods to determine molecular weight cut-off. Generally for a given molecular weight cut-off, 80% of the molecules of that molecular weight will be rejected. Furthermore, molecular weight cut-off is typically used for membranes in the UF range.

1.6.4 *Flux*

Flux is the amount of permeate produced in a given time period, and the term generally is given as a volume or mass per unit membrane per unit time. A unit, such as L m^{-2} h^{-1}, is a possible volume flux unit, whereas kg m^{-2} h^{-1} is a mass flux unit.

Flux determines the area of membrane required to process a given amount of product to a certain concentration in a specific time period. The lower the flux for a given membrane the greater the membrane area required to process the same amount of product within a certain time than with a higher flux membrane. Flux, therefore, affects the economics of an operation, and is used as an indication of membrane fouling and cleaning adequacy. Factors influencing flux include pressure, feed velocity, temperature, viscosity and turbulence.

1.6.5 *Concentration factor*

Concentration factor (CF) expresses the degree of concentration for a feed and can be defined as:

$$CF = \frac{\textit{volume of original feed}}{\textit{volume of final feed}} \tag{1.4}$$

Concentration factor is often expressed as 1×, 2×, 3× etc. Volumes or weights may be used. An example would be 100 L of product that is processed to a final volume of 33 L:

$$100\ \text{L}/33\ \text{L} = 3\times$$

The ability to concentrate a feed material generally is determined by feed constituents, osmotic pressure and feed viscosity.

1.6.6 *Membrane life*

Although the cross-flow design allows reuse of a membrane, the operating life of a membrane is not indefinite. As membranes are used for processing, their characteristics flux and retention change, and it is the decreasing flux and retention with time that result in membrane replacement. The life of a membrane ends when membrane performance no longer meets specific performance criteria. In the case of a protein concentration operation, the loss of valuable protein into the permeate might be a criteria. In a water desalination system the criteria might be the permeation of excessive amounts of salt into the permeate, which is to be as drinking water.

Membrane life is influenced by many factors. Membrane composition is important, but often the processing and cleaning conditions have a greater influence. Composition, however, affects the resistance of a membrane to processing and cleaning treatments, and determines acceptable operating conditions.

1.7 Conclusion and future development

Although membranes have been known about for many decades, it has not been until relatively recently that technological improvements have permitted their widespread use by the food industry. Thus, the use of membrane processes for food applications can be considered a relatively new tool for the food processor. Knowledge about the theory of membrane systems operate will continue to increase as new applications for membrane systems are found.

Suggested literature

Baker, R.W. (2000) *Membrane Technology and Applications*, McGraw-Hill, New York.

Cheryan, M. (1998) *Ultrafiltration and Microfiltration Handbook*, Technomic Publishing Co., Pennsylvania.

Glimenius, R. (1985) Microfiltration – state of the art. *Desalination*, **53**, 363–372.

Gregor, H.P. & Gregor, C.D. (1978) Synthetic-membrane technology. *Scientific America*, **239**, 112–128.

Jonsson, G. (1980) Overview of theories for water and solute transport in UF/RO membranes. *Desalination*, **35**, 21–38.

Lonsdale, H.K. (1972) Theory and practice of reverse osmosis and ultrafiltration. *Industrial Processing with Membranes* (eds. R.E. Lacey & S. Loeb), pp. 123–178, John Wiley & Sons Inc., New York.

Lonsdale, H.K. (1982) The growth of membrane technology. *Journal of Membrane Technology*, **10**, 81–181.

Paulson, D.J., Wilson, R.L. & Spatz, D.D. (1984) Cross-flow membrane technology and its applications. *Food Technology*, **38**, 77–87, 111.

Pusch, W. (1990) Performance of RO membranes in correlation with membrane structure, transport mechanisms of matter and module design (fouling) – state of the art. *Desalination*, **77**, 35–54.

Strathmann, H. (1981) Membrane separation processes. *Journal of Membrane Science*, **9**, 121–189.

2 Principles of Membrane Filtration

A. Hausmann, M.C. Duke and T. Demmer

2.1 Introduction and definitions

This chapter provides a fundamental understanding of pressure-driven membrane operations, such as microfiltration (MF), ultrafiltration (UF), nanofiltration (NF) and reverse osmosis (RO), and the way these are used in the food industry. The three systems determining membrane performance are the membrane itself, the feed system and the operating conditions. All three influence each other and interact, resulting in a complex system of membrane operation. Each of these three systems will be discussed separately as well as pointing towards influencing correlations. Some novel membrane processes driven by other means than pressure will also be discussed briefly.

A membrane is a semi-permeable barrier, which means that at least one component of a fluid in contact with the membrane can pass through. The membrane enables selective separation of the compounds in a feed-stream, which is split into retentate (concentrate) and permeate (filtrate). The retentate contains the compounds that cannot pass through the membrane, the permeate is the product that passes through (Melin & Rautenbach, 2007).

2.1.1 *Membrane processes*

The many and various applications of membrane processes can be categorised into product concentration with the focus on the retentate, such as the pre-concentration of milk before cheese making, and solute removal for purification/clarification purposes as, for example, product sterilisation, demineralisation or desalination focusing on the permeate quality. Also possible are solute separation applications, where both permeate and retentate are considered valuable, as in fractionation of constituents: an example is the separation of milk proteins into a casein-enriched retentate and whey protein containing permeate, which is often referred to as 'ideal whey' (Lawrence *et al*., 2008). Membrane processes can be driven by differences in applied external pressure, osmotic pressure, vapour pressure and electric fields which are used for electrodialysis. This electro-membrane process works through an electric field introduced between the two membrane sides that allows for selective separation of ions in solution. The transport of ions across the membrane results in ion depletion on one side and ion concentration on the other. Electrodialysis can be utilised for several types of separation and concentration of salts, acids and bases from

Membrane Processing: Dairy and Beverage Applications, First Edition. A. Y. Tamime.

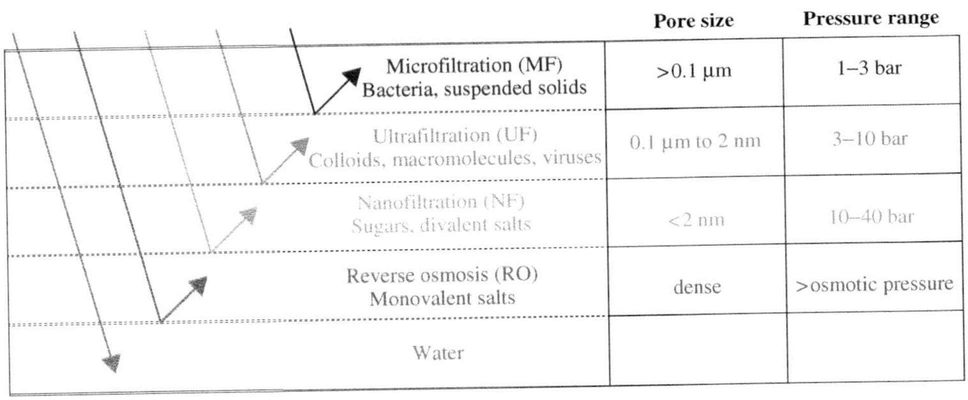

Fig. 2.1 Separation by the four pressure driven membrane processes. (Adapted from Koros *et al.* (1996) for pore sizes by International Union of Pure and Applied Chemistry (IUPAC) definitions, and Ayral (2008) for pressure ranges.)

aqueous solutions or for separation and concentration of monovalent ions from multiple charged components as well as the separation of compounds from uncharged molecules. Desalination is commonly carried out by electrodialysis. In the food industry it is, for example, used for desalination of cheese whey for infant formulas, fractionation of single whey proteins or demineralisation of sugar cane juice (Kariduraganavar *et al.*, 2006; Pouliot, 2008).

However, most membrane processes employed in the food industry to date are commonly driven by the application of pressure to the feed, while permeate is withdrawn from the system. As outlined in Fig. 2.1, the four pressure-driven membrane processes (MF, UF, NF and RO) differ by their average pore size, and can fractionate all major liquid food ingredients.

Based on the pore size differences, MF membranes retain suspended solids like bacteria or fat droplets. UF membranes are able to reject colloids, macromolecules, viruses and proteins. NF membranes can fractionate inorganic salts or diluted organic molecules, whereas RO membranes are mainly used for water desalination and the production of ultrapure water (Ayral, 2008).

2.1.2 *Definitions of membrane processes*

Flow regimes

There are two modes of operation for membrane processes, dead-end filtration and the cross-flow mode, with the ideal concepts shown in Fig. 2.2. In the case of dead-end filtration, the feed is pumped perpendicularly onto the membrane surface, whereas during cross-flow filtration it is pumped across the membrane surface tangentially rather than perpendicularly. Dead-end operation has to be run batch-wise to relieve the concentrated species, whereas the cross-flow mode allows for continuous process operation. For feed

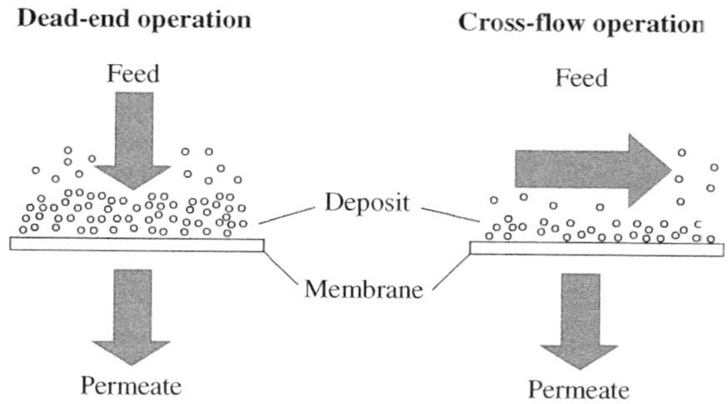

Fig. 2.2 Schematic of a dead-end filtration (left) and cross-flow filtration (right).

containing relatively high solids, cross-flow is advantageous in that it limits the build-up of solids on the membrane surface. Today the cross-flow mode is the standard operation for most filtration processes in the food industry (James *et al.*, 2003).

Membrane geometry

Membranes are available in flat sheet, hollow fibres or tubular geometries. Flat sheets are simply made with some kind of support (e.g. a porous backing or imbedded mesh) which supports the active (selective) layer. Hollow fibre and tubular membranes are similar in geometry as both appear tube-like. However, the difference comes from the tube diameters. Hollow fibre membranes have much smaller diameters than tubular membranes and, in general, the same material that forms the active layer also makes up the supporting layer. Both membranes are available as 'inside-out' membranes where the inner wall provides the active membrane layer or 'outside-in' where the outer wall acts as the membrane (Fig. 2.3). The remaining material in the membrane is more porous and, thus, acts merely to mechanically support the active layer and ideally offers no resistance to permeation.

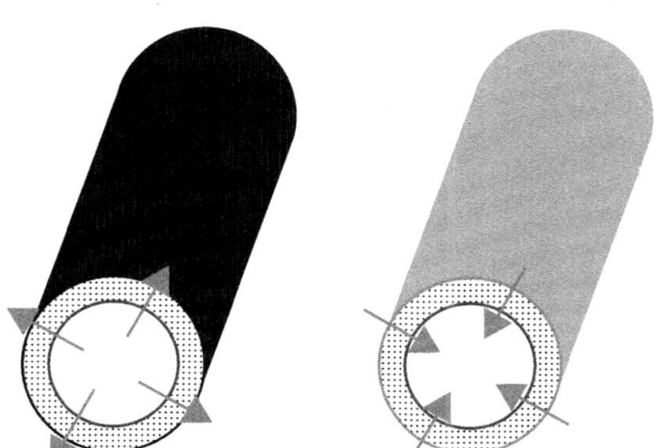

Fig. 2.3 Hollow fibre membranes for inside-out filtration with active layer on hollow fibre lumen (left) and for outside-in filtration with active layer on the outside (right).

Membrane module

For continuous membrane operations, membranes are packed into distinct units or membrane elements, known as the membrane module. The term 'module' stems from the modular design of membrane plants where identical membrane units are arranged and connected by pipelines. There are various module designs that meet the requirements of different membrane types. Flat sheet membranes can be arranged in spiral wound modules or plate on frame configuration as discussed in Chapter 3. All membrane module types consist of at least one entry for the feed to be separated and two exits for the filtered and retained components, as shown in Fig. 2.4. The area of membrane contained in these basic membrane units or modules is in the range of 1 up to 800 m^2.

Transmembrane pressure and flux

As mentioned above, typically the driving force for all four membrane processes (MF, UF, NF and RO) is the pressure across the membrane from feed to permeate side. For dead-end filtration this transmembrane pressure (TMP) is simply the pressure of the feed solution P_f minus the pressure on the permeate side of the membrane P_p:

$$\mathrm{TMP} = P_f - P_p \tag{2.1}$$

In cross-flow mode, as a consequence of flow, the pressure on the feed/retentate side declines while passing along the module from its inlet value P_f to the outlet pressure P_r (Piry *et al.*, 2008).

$$\Delta P = P_f - P_r \tag{2.2}$$

This pressure drop (ΔP) along the module is considered to be a source of energy loss in membrane systems, and the amount depends on factors influencing flow resistance, such as the cross-flow velocity, flow spacer design and channel width. In many module

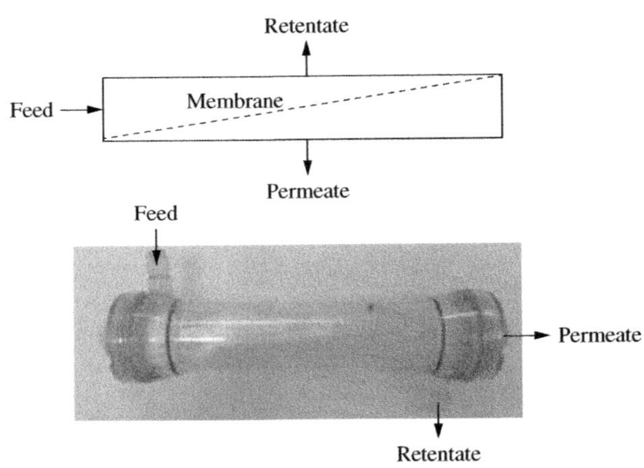

Fig. 2.4 Schematic membrane module (top) and an example of a hollow fibre module (bottom).

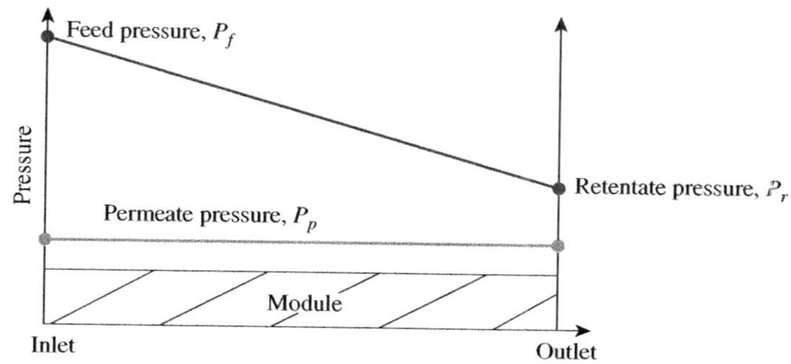

Fig. 2.5 Pressure relations along a membrane module in a cross-flow configuration.

configurations, the filtrate-sided pressure P_p of the module is almost constant due to the locally marginal flow velocity. To realise filtration over the entire length of the module, the permeate-sided pressure P_p needs to be below P_r which is the lowest pressure on the feed channel side, as shown in Fig. 2.5.

The TMP in cross-flow set-ups is, therefore, calculated as the average pressure applied over the length of the membrane module:

$$\mathrm{TMP}_{cross\text{-}flow} = \left(\frac{P_f + P_r}{2}\right) - P_p \tag{2.3}$$

Performance of a membrane process is assessed by measuring the filtrate throughput per unit membrane area and time that passes through the membrane; this flux is measured in $\frac{m^3}{m^2 \cdot s} = \frac{m}{s}$ or more commonly: $\frac{kg}{m^2 \cdot h}$. or $\frac{l}{m^2 \cdot h}$ as mostly seen in product sheets and other membrane media expressed simply as litre per square meter per hour (LMH) or (L m^{-2} h^{-1}).

Wall shear stress

The tangential force induced by the fluid flowing along the membrane in cross-flow mode is represented by the wall shear stress (τ_W). An approximation that allows for experimental determination was defined by Le Berre & Daufin (1996):

$$\tau_W = \Delta P \cdot \frac{d_{\text{hydraulic}}}{4 \cdot L} \tag{2.4}$$

where $d_{hydraulic}$ is the hydraulic diameter and L the membrane length.

Volume concentration ratio (VCR) and concentration factor (CF)

A measure for the volume reduction of the initial feed volume is the volume concentration ratio (VCR). It is defined as the quotient between initial feed volume V_f and retentate volume V_r (Cheryan, 1998).

$$\mathrm{VCR} = \frac{V_f}{V_r} \tag{2.5}$$

For concentration applications, the concentration factor (CF) is commonly used to express degree of concentration of a target compound. It is defined as the ratio of the concentration of a component i in the retentate ($c_{i,retentate}$) to the concentration of the same component in the feed ($c_{i,feed}$) (Koros *et al*., 1996).

$$\mathrm{CF} = \frac{c_{i,retentate}}{c_{i,feed}} \tag{2.6}$$

If the target compound is not retained completely but some gets lost into the permeate, the VCR needs to be increased above the CF to obtain the desired concentration. The relation between the retention factor, CF and VCR will be explained next.

Retention factor

The retention factor (Rf) of the various compounds in the feed stream reflects the membranes selectivity and can be calculated for each solute compound as the ratio between the concentration of that solute i in the bulk of the retentate ($c_{i,retenate}$) and its concentration in the permeate ($c_{i,permeate}$):

$$Rf = \frac{c_{i,retentate} - c_{i,permeate}}{c_{i,retentate}} = 1 - \frac{c_{i,permeate}}{c_{i,retentate}} \tag{2.7}$$

Retention is a measure of the membrane's ability to separate components of the feed solution. How the membrane selects which component diffuses varies on the scale of separation (i.e. bulk properties down to molecular properties). The most fundamental membrane property for separation is the pore size as determined by the structure of the membrane material. Other properties include material surface chemistry and charge. These will be discussed in more detail later.

Another way to express the Rf is given by Cheryan (1998):

$$\frac{c_{i,retentate}}{c_{i,feed}} = \left(\frac{V_f}{V_r}\right)^{Rf} \text{ with } \frac{c_{i,retentate}}{c_{i,feed}} = \mathrm{CF} \tag{2.8}$$

$$\log(\mathrm{CF}) = Rf \log(\mathrm{VCR}) \tag{2.9}$$

This equation enables to calculate retention using only retentate data, as opposed to equation (2.7), which is based on both permeate and retentate data.

A disadvantage of Equations (2.5) to (2.7) is that the behaviour of dissolved components is not correctly reflected as their concentration changes with the water content of the system. A better way is given by

$$Rf_{dissolved} = \left(1 - \frac{c_{i,permeate}}{100 - TS_p} \cdot \frac{100 - TS_r}{c_{i,retentate}}\right) \cdot 100 \tag{2.10}$$

where TS is the total solid content and the subscripts p and r refer to the permeate and retentate. This equation correctly describes the extreme cases of totally dissolved particles ($Rf_{dissolved} = 0$) and total rejection ($Rf_{dissolved} = 1$).

Diafiltration

A technique to improve separation between compounds with different retention factors is diafiltration (DF). It increases separation quality through the addition of a solvent (often water) to the feed in order to dilute the concentration of permeating species. For substances that are freely permeating through the membrane and not retained, the concentration in the final permeate equals its concentration in the retentate (the normal process is shown in Fig. 2.6). During diafiltration, the initial feed stream is diluted, so that more dissolved particles small enough to pass the membrane are 'washed out' through the membrane. This process results in a lower concentration of permeating compounds in the retentate while the concentration of fully retained particles remains unaffected, as schematically shown on the right-hand side of Fig. 2.6. Therefore, a more distinct separation can be achieved; however, dilution of the feed means a bigger volume needs to be processed in order to achieve the same degree of concentration, which results in higher energy demands. The DF is a measure of the extent of washing that has been performed. It is based on the volume of diafiltration medium V_d fed to the system relative to the volume of retentate produced V_r:

$$\mathrm{DF} = \frac{V_d}{V_r} \tag{2.11}$$

A common application is the dealcoholisation of wine, where diafiltration is often employed as alcohol is a freely permeating solvent and can be reduced by such solvent dilution and removal.

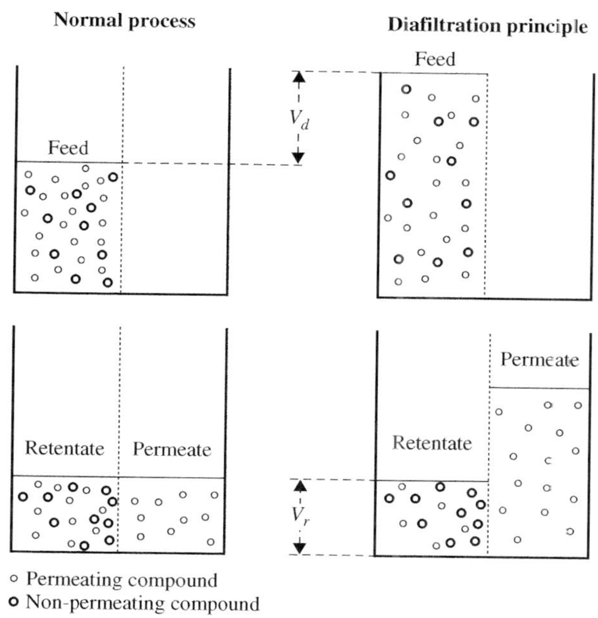

Fig. 2.6 Principle of a diafiltration: normal filtration process (left) and feed stream diluted (right).

Table 2.1 Relation between the diafiltration factor and resulting protein purity and lactose level during ultrafiltration of whey.

Component (g 100 g^{-1})	Diafiltration (DF) factor							
	0	1	2	3	4	5	7	10
Lactose	3.64	1.81	1.2	0.9	0.72	0.6	0.45	0.33
Protein	73.33	84.67	89.29	91.74	93.28	94.34	95.69	96.81

An example from the dairy industry is the production of whey protein isolates, where lactose and milk minerals are diluted to obtain whey protein concentrates of high purity using UF. Assuming 100% protein retention ($Rf = 1$) and 100% lactose permeation ($Rf = 0$), and a final retentate concentration of 10 g 100 g^{-1} protein, then without diafiltration, the total solids of this retentate will consist 3.64 g 100 g^{-1} lactose and 73 g 100 g^{-1} of the total solids are protein. If the feed is mixed 1:1 with water (DF = 1) and then concentrated to 10 g 100 g^{-1} protein, the composition changes; now around 85 g 100 g^{-1} of the total solids are protein. Table 2.1 shows the influence of DF on protein purity. Clearly, the purification effect diminishes with increasing DF. In this example, the protein concentration in the retentate was kept constant, not the volume concentration ratio, the effective VCR increases from 10 in the first column to 110 in the last (Table 2.1). This demonstrates the increased energy demand, and also the washing fluid might need to be cleaned from the lactose in a subsequent processing step.

2.2 Membrane properties based on materials

The variety of commercially available membranes has been largely extended in the last two decades, and overall thermal, chemical and mechanical strength has been improved. Such variety results from application-dependent interactions between feed components and membrane materials, so that membrane materials have been specifically optimised for each application (Melin & Rautenbach, 2007). This also means that laboratory and pilot plant testing is necessary for every new application. In the following, membrane properties will be outlined with considerations regarding membrane-feed systems.

2.2.1 *Membrane structure*

From a structural point of view membranes can broadly be divided into symmetric (or isotropic) and asymmetric (or anisotropic) membranes. As illustrated by Fig. 2.7, symmetric membranes have a similar structural composition and morphology at all positions within, and can be porous or dense. Asymmetric membranes, as shown in Fig. 2.7, are composed of two or more structural planes of non-identical morphologies (Koros *et al.*, 1996). They can consist of one single material or two different materials combined together resulting in thin-film composite membranes. Asymmetric membranes combine the complimentary needs for thin, permeable membranes (high flux) and thick, self-supporting membranes (high mechanical stability). These membranes are made of a very thin active layer on top

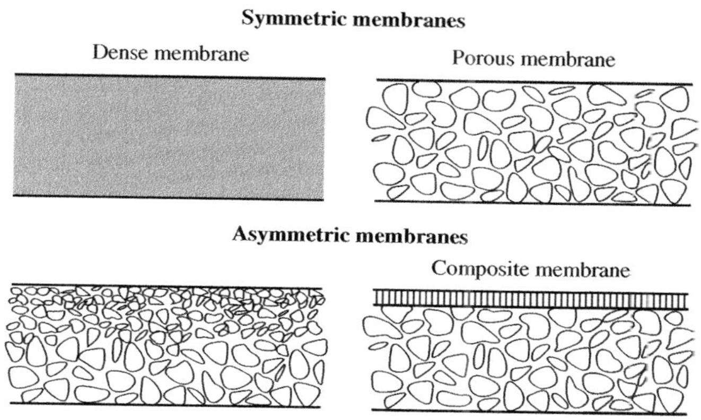

Fig. 2.7 Classification of commercial membranes based on morphology, structure, showing cross-sections.

of a highly porous support layer. The active membrane layer determines its selectivity, but is not stable enough for reliable operation. Also, certain membrane materials would lead to impractically low fluxes if they were thick enough to be self-supporting. The highly porous support layer therefore provides stability and ideally should not influence flux or separation. Fine materials entering and irreversibly fouling the membrane emphasise the advantage of asymmetric membranes. In-pore fouling is widely reduced or even avoided with this type of membrane.

Dense membranes are used for RO, whereas membranes for the other three pressure-based filtration processes are porous, meaning they have microscopically detectable pores (Koros *et al*., 1996). According to the International Union of Pure and Applied Chemistry (IUPAC) classification of membrane pores (Table 2.2), MF membranes and upper-limit UF membranes are macroporous, lower size UF membranes are mesoporous membranes and NF membranes are microporous membranes (Rouquérol *et al*., 1994; Ayral, 2008).

Membrane manufacturers often describe the membrane's ability to retain molecules in terms of its molecular weight cut-off (MWCO), and it refers to the molecular weight of a solute that is 90% retained by a given membrane (Koros *et al*., 1996). This is a highly case-dependent definition since it does not specify operating conditions, and membrane feed interactions narrowing effective pore size are not accounted for. Standardisation of test methods is necessary to achieve reproducible results, and problems arise when MWCO values are used to compare membranes from different suppliers.

Table 2.2 International Union of Pure and Applied Chemistry classification of the pores as a function of their size.

Micropores	Mesopores	Macropores
<2 nm	2–50 nm	>50 nm
Nanofiltration	Ultrafiltration	Microfiltration

Source: After Rouquérol *et al*. (1994).

2.2.2 *Material properties*

The performance of a membrane process depends to a large extent on the properties of the membrane material. This includes properties concerning membrane lifetime and stability, such as

- mechanical strength, e.g. bursting pressure
- chemical resistance, e.g. acceptable pH range, solvent compatibility
- thermal stability.

Synthetic, solid membranes can be made from organic or inorganic materials. Organic membranes are made from an almost unlimited number of polymers or polymer blends. Most common polymers include cellulose, cellulose acetate (CA), polysulphone (PS), polyethersulphone (PES), polyamides (PA), polyvinylidine fluoride (PVDF) and polypropylene (PP). Examples for inorganic membrane materials are those made from ceramics, glass, pyrolysed carbon and metals. Interest in comparatively expensive ceramic membranes is related to the intrinsic characteristics of the used materials like high mechanical strength allowing high pressures to be applied; chemical resistance permitting application in corrosive aqueous media as well as strong chemical cleaning regimes and its high thermal stability allowing for thermal sterilisation. Longer term benefits lie in the prolonged lifetime of ceramic membranes. For example, in the dairy industry ceramic membranes can be used for more than 8 years as opposed to polymeric membranes needing to be replaced every 2 years (Schier & Paar, 2007). Ceramic membranes are also food contact approved, whereas most polymers and blends cannot be in contact with foods.

Membrane polarity and charge

Important considerations for the choice of membrane material depend on the intended application as interactions between the membrane and the feed stream determine performance to a high degree. The membrane polarity is responsible for its hydrophilic/hydrophobic character and therefore determines wettability/permeability by different feed components. With aqueous feeds, the ideal membrane should be hydrophilic (Zheng *et al*., 2008). Also, most proteins are hydrophobic or at least amphoteric and, therefore, adsorb to hydrophobic materials (Bottino *et al*., 2000; Howe & Clark, 2002).

Solute retention can be influenced by the membrane charge due to electrostatic interactions. Generally, membranes should be neutral or of the same charge as the particles being processed to prevent attraction, which can result in fouling. For process design, the pH of the feed stream needs to be considered as the membrane charge is influenced by the pH of the fluid in touch with the membrane surface (Rice *et al*., 2011). Some membrane materials, for example polyamide, can exhibit amphoteric behaviour; here the pH of the cleaning agent should condition the surface with a charge appropriate for the feed being processed in the following production period (Tang *et al*., 2006). To enable operation in a wide pH range, membranes with two active membrane layers of different materials and different isoelectric points (IEP) are being designed. These bi-layered systems are found to perform better over a large pH range than a membrane with only one type of selective layer (Elmarraki *et al*., 2001; Samuel de Lint *et al*., 2006).

For membranes used in electrodialysis, surface charge is crucial for the process to work. Electrodialysis membranes can be cationic or anionic and usually consist of swollen charged gels that have varying amounts of water embedded (Kariduraganavar *et al.*, 2006). A membrane with positive charges is referred to as an anion exchange membrane because it binds and selectively transports anions in the electrical field but excludes cations because of its positive charge. Similarly, a membrane of negative charge is termed a cation exchange membrane (Strathmann, 2010). Another process that requires charged membranes is piezodialysis. This process removes salt from water by applying high pressures on the feed side and using membranes with a charge mosaic. This means the membranes have a close arrangement of positively and negatively charged sites within the same membrane. This multifunctional charge allows all ions independent of their charge to pass through the membrane (Bolto *et al.*, 2010).

Pore statistics

The pore statistics include average pore sizes (as discussed above), pore density (number of pores per unit membrane surface area), pore size distribution, pore shape and the membrane porosity, which is the fraction of the membrane volume occupied by pores (Scharnagl & Schossig, 2005). Standard test methods for determining these properties have not been established and membrane manufacturers can use differing methods for determination, which means that the identical numerical designation of pore size by different manufacturers does not necessarily mean that the membrane pores are identically sized. Additionally, the distribution of pore sizes on every membrane surface has a gradual effect on separation performance, as illustrated in Fig. 2.8. The more precise the pore size distribution, the more distinct the separation of molecules of different molecular weight and, consequently, the sharper the slope of the graph in Fig. 2.8 (Porter, 1990).

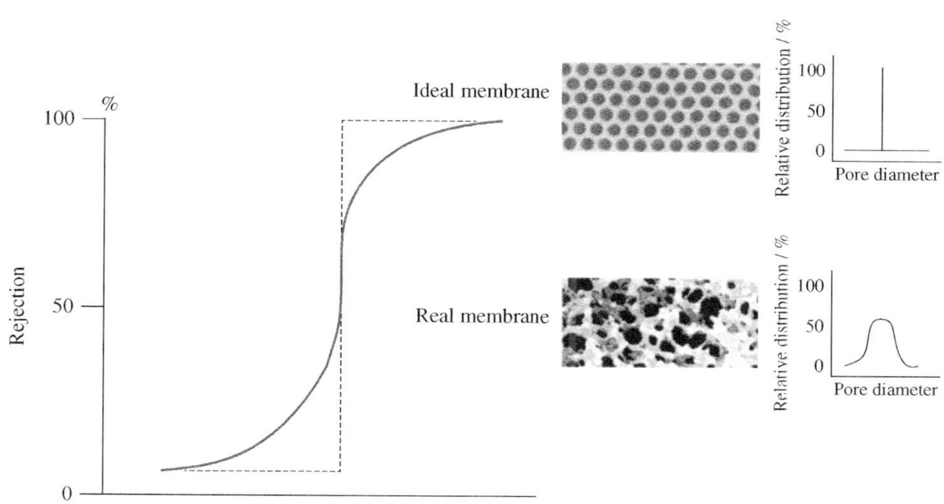

Fig. 2.8 Relationship between particle size of solutes and rejection of a membrane process depending on the membrane pore size distribution; ideal membrane (dashed line) with equally sized pores (top) and real membrane with less narrow distribution of pores (bottom, solid line).

If size distribution of the particles to be separated is of the same order of magnitude as the range of pore sizes on the membrane, there is a high tendency for physical blockage of pores (Kallioinen *et al.*, 2007). This fouling phenomenon will cause a rapid drop in flux in the first few minutes of operation, which is understandable since much of the initial permeation occurs in the larger pores. In contrast, if the pores are considerably smaller than the particles to be separated, the particles will not get caught within the pore entrance, but will roll off the surface under the shear forces. Ideal solutions for membrane separation have large size differences between the retained and permeating substances. For such streams, membranes with pore sizes that are around one-tenth the retained particle size can be selected as a starting point for testing (Cheryan, 1998). In particular for symmetric membranes with long narrow pore channels, this is favourable as particles get stuck more easily in a long pore channel. Nevertheless, often the size range of the components to be separated does not allow for such proportionally small pore sizes. Additionally, a building cake layer changes retention characteristics by reducing pore diameter, which will be discussed in the next section. In fact, pores are often much larger than the particles they retain.

Another vital aspect of membrane performance determined by the membrane pores is the porosity of the membrane. The membrane porosity (ε) is a measure for the fraction of membrane volume (V) filled with pores:

$$\varepsilon = \frac{V_{pores}}{V_{total}} \tag{2.12}$$

Although high porosity is desirable for stable membrane materials, high membrane porosity increases its susceptibility to deformation under pressure, and such compaction decreases the rate of flux (Tarnawski & Jelen, 1986; Kallioinen *et al.*, 2007). The membrane porosity is not only determined by the membrane area covered by pores, but also its tortuosity (τ) defined as (Melin & Rautenbach, 2007):

$$\tau = \frac{L_{pore}}{\delta} \tag{2.13}$$

where L_{pore} is the actual pore channel length and δ the membrane thickness. The concept of tortuosity is illustrated in Fig. 2.9.

In theory, pore shape is often assumed to be ideal (left in Fig. 2.9); however, there are many various pore shapes that are far from being identical. The influence of pore shape (Fig. 2.10) has been studied using micropatterned evenly shaped circular and slotted membrane pores (Bromley *et al.*, 2002; Chandler & Zydney, 2006). Circular pores have

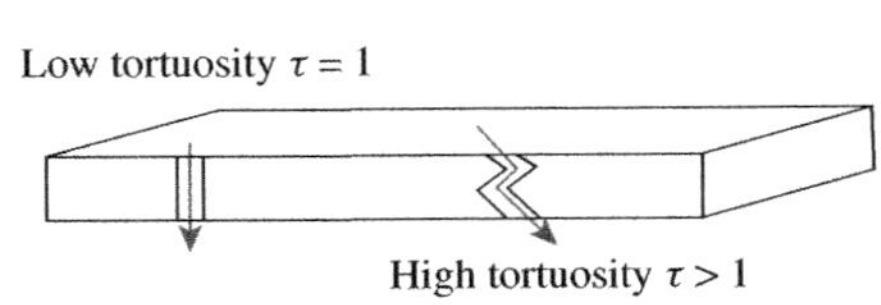

Fig. 2.9 An illustration of membrane tortuosity.

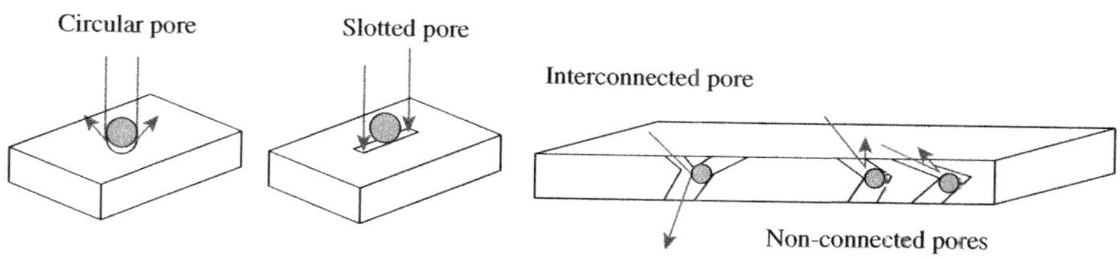

Fig. 2.10 Pore shape and pore channel morphology.

a smaller size at the pore entrance and, thus, get plugged more easily by feed compounds leading to greater surface fouling under similar shear, and other filtration conditions than for a comparable slotted pore membrane. Despite the geometry of the pore entrance, the pore channel morphology can also have a significant effect on the rate of fouling during filtration. With membranes having interconnected pores, less severe fouling occurs due to the filtrate being able to flow under and around any pore blockage within interconnected pores (Chandler & Zydney, 2006).

A more recent development in membrane manufacturing has led to the production of membranes with extremely narrow pore size distributions, high porosity and evenly sized and shaped pores. These membranes are called microsieves and their production is based on photolithographic techniques from the semiconductor industry or phase separation micromoulding (Brito-de la Fuente *et al*., 2010; Bikel, *et al*., 2010). Pore sizes for these membranes can go down to 0.35 μm and microsieves with slotted pore shapes have been tested for cold sterilization of whole milk and clarification of beer, where they showed superior performance to membranes conventionally used for these applications (Brito-de la Fuente *et al*., 2010; Kuiper *et al*., 2002).

2.3 Flux behaviour in pressure-driven membrane operations

During filtration, solids in the feed are brought to the membrane surface by convective transport, and permeating components are removed by passing through the membrane. As a result, local concentration of rejected solutes is higher at the membrane surface than in the bulk. This results in the formation of a concentration gradient from the bulk feed to the membrane surface. This concentration gradient is generally known as concentration polarisation and can lead to a cake layer formation as shown schematically in Fig. 2.11. The nature of this cake layer depends on the type of solid accumulating on the membrane surface and/or within the pores; it can be fairly gelatinous when hydrocolloids such as proteins are rejected by the membrane.

Membrane-solute and solute–solute interactions in the concentrated cake layer can eventually lead to membrane fouling. Membrane fouling manifests itself as a flux decline over time at constant operating conditions (Bottino *et al*., 2000; Tolkach & Kulczik, 2006; Rajniak *et al*., 2008). The decline in performance due to fouling is irreversible, which marks the transition from concentration polarization and requires cleaning to restore membrane performance (Chen *et al*., 1997). If performance does not recover even after cleaning, this is considered irreversible fouling.

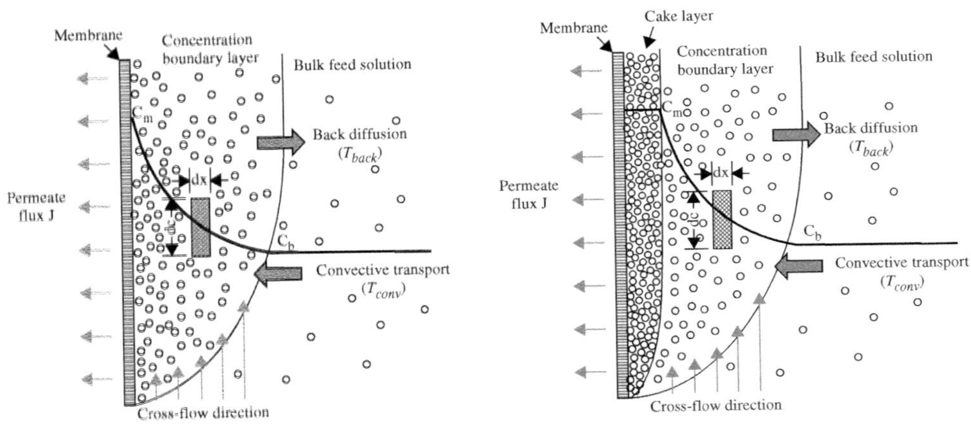

Fig. 2.11 Schematic of concentration polarisation (left) with cake layer formation (right).

2.3.1 *Modelling flux behaviour*

For pressure-driven membrane processes, the applied TMP is the driving force. However, the relation between input (TMP) and output (flux) is not always linear and the effects of pressure on flux will be looked at more closely in the following. As shown in Fig. 2.12, flux behaviour with increasing TMP can roughly be divided into two regions: (a) linear flux behaviour at low TMPs, and (b) pressure-independent flux behaviour at high pressures (Cheryan, 1998).

For an ideal process where concentration polarisation is minimal and no fouling occurs, flux increases linearly with TMP. These conditions are given at low pressures, high shear and low solute concentrations. When pure water/solvent is processed that does not contain any solutes retained by the membrane, flux remains linear with increasing TMP as long as no changes to membrane morphology (i.e. membrane compaction) occur. However, when processing feed streams with a specific solute content, the flux–TMP curve deviates from linearity because of concentration polarisation. The polarisation cake layer is assumed to

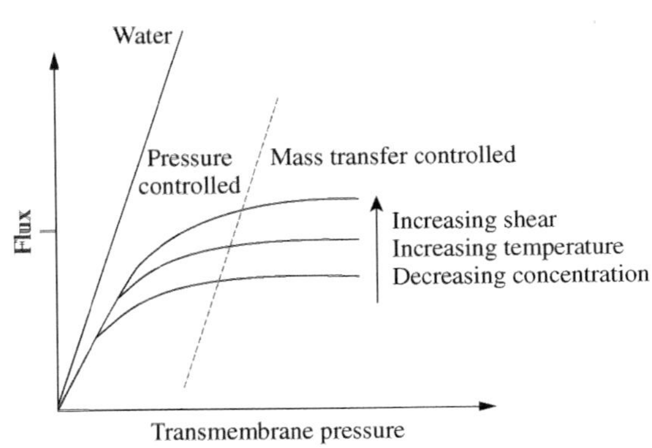

Fig. 2.12 General effect of transmembrane pressure on flux assuming no membrane compaction.

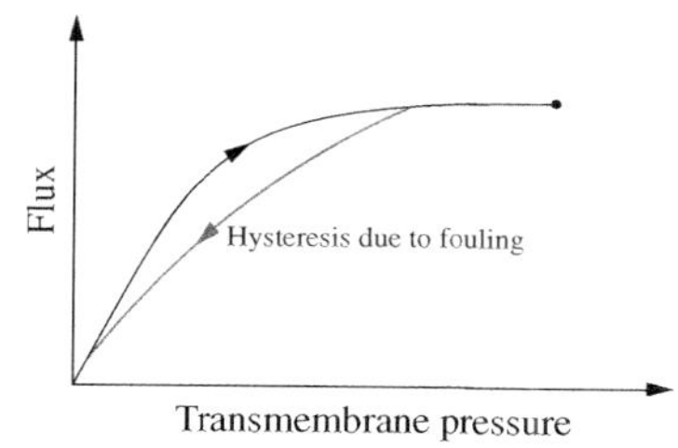

Fig. 2.13 Hysteresis of flux with decreasing pressure.

be dynamic, and changing the operating conditions, such as lowering pressure or feed concentration or increasing cross-flow velocity, which results in higher shear or lifting forces, can bring the system back to the pressure-controlled operating regime (Cheryan, 1998). If fouling has occurred, lowering pressure will result in an irreversible hysteresis of flux as shown in Fig. 2.13. Hysteresis of flux can also occur in the absence of fouling due to flux recovery being time dependent, only if flux does not increase with time when reducing pressure and, therefore, hysteresis is irreversible, this is due to fouling (Bacchin *et al*., 2006).

At higher TMPs, flux plateaus off marking the transition to the pressure independent region due to cake formation, at higher pressure consolidation of the cake layer formed on the membrane surface often occurs (Keskinler *et al*., 2004). The maximum stationary permeation flux which can be reached through increasing TMP is referred to as the limiting flux (Porter, 1972; Chen *et al*., 1997). It depends on the feed processed and the given set of hydrodynamic conditions, and varies over the range of wall shear stress, temperature and concentration.

Flux behaviour under ideal conditions

The Hagen–Poiseuille law describes the volumetric flow (q) through a cylindrical channel under laminar flow conditions. This law relates the dynamic viscosity of the fluid (η), the pressure drop from channel inlet to its outlet (ΔP) and channel dimensions to the flow rate through that channel:

$$q = \frac{\pi \cdot d^4}{128 \cdot \eta \cdot \Delta x} \cdot \Delta P \tag{2.14}$$

where d is the diameter, and Δx the length of the channel.

Assuming an ideal membrane with uniformly distributed evenly sized pores and no feed components accumulating on the membrane surface, this equation can be used to describe the fluid flow through membranes with μm sized pores (Porter, 1990). The pure solvent flux ($J_{solvent}$) without any resistance from accumulating feed components equals the sum of all flows through the individual pores and is given by:

$$J_{solvent} = N \cdot \frac{\pi \cdot d^4}{128 \cdot \eta \cdot L_{pore}} \cdot \mathrm{TMP} \tag{2.15}$$

where N is the number of pores per unit membrane area, L_{pore} is the average pore channel length depending on membrane thickness or the thickness of the active membrane layer for thin film composite membranes and pore tortuosity as discussed before.

Under the simplifying assumption that the value for the membrane porosity (ε) is equal to the total surface area covered by cylindrical pores ($\varepsilon = N\pi r^2$), which is proportional to the inverse square of the pore diameter (Porter, 1990). That is

$$N = \varepsilon \cdot \frac{4}{\pi \cdot d^2} \tag{2.16}$$

where d is the mean pore diameter.

Substitution in the pure solvent flux Equation (2.15) results in the overall relation (Porter, 1990):

$$J_{solvent} = \frac{\text{TMP} \cdot d^2 \cdot \varepsilon}{32 \cdot \eta \cdot L_{pore}} \tag{2.17}$$

According to this model, pure solvent flux is directly proportional to the applied pressure and inversely proportional to the fluid viscosity. As illustrated in Fig. 2.14, the flux–TMP curve increases linearly with increasing pressure. For solvents of different viscosities it is still linear but the slope is lower for high viscosities. Factors affecting viscosity will be discussed later.

However, this model is only adequate for the region of linear flux–TMP relation where no feed components add resistance to the process and concentration polarisation is negligible. When a cake layer has formed on the membrane surface, the actual pore size is not dictated by the membrane pore size any more, but the cake layer which has much smaller pore sizes. Therefore, effective pore size is no longer constant and, although viscous flow still applies, the multiple functional domains need to be accounted for.

Mass-transfer controlled flux behaviour

When a cake layer forms on the membrane surface, performance is determined by the efficiency of minimising boundary layer thickness and enhancing the rate of back-transfer

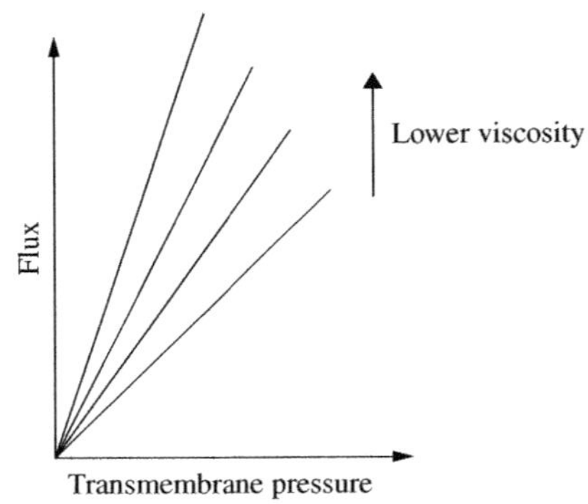

Fig. 2.14 Flux–transmembrane pressure relation for an ideal membrane process, assuming constant viscosity for each graph.

of solute from the membrane surface into the bulk stream. As shown in Fig. 2.11, layer thickness is a result of convective transport from the bulk to the membrane surface and back diffusion of solutes into the bulk stream. The rate at which solutes are brought to the membrane surface (T_{conv}) depends on the permeate flux (J) and solute content brought towards the membrane, which equals the difference of solute concentration in the bulk (c_b) and in the permeate (c_p) (Cheryan, 1998).

$$T_{conv} = J(c_b - c_p) \tag{2.18}$$

Owing to the higher concentration in the cake layer than the bulk stream, diffusion towards the bulk stream occurs. The rate of back transport depends on the diffusion coefficient (D) and the mass balance through the concentration boundary gradient as schematically shown in Fig. 2.11:

$$T_{back} = -D \cdot \frac{dc}{dx} \tag{2.19}$$

where dx is the distance from the membrane or cake layer, and dc is the incremental change in concentration.

At steady state, back diffusion of solute from the membrane surface to the bulk equals the rate of convective transport of solids towards the membrane, so that a stable concentration gradient exists near the membrane surface. Consequently, the two equations can be written as:

$$J \cdot dx = -D \cdot \frac{dc}{c_b - c_p} \tag{2.20}$$

Now this equation can be integrated from the membrane or cake layer surface to the bulk flow, which results in (Ripperger & Altmann, 2002):

$$J = k \cdot \ln \frac{c_m - c_p}{c_b - c_p} \text{ with } k = \frac{D}{\delta_{boundary}} \tag{2.21}$$

where k is the mass transfer coefficient (same units as the flux J) and $\delta_{boundary}$ the thickness of the concentration boundary layer.

In the literature, a 100% retention of solutes is often assumed, so that c_p drops out of the equation, and equation (2.21) results in the commonly known equation of the film theory (Vladisavljević & Rajković, 1999):

$$J = k \cdot \ln \frac{c_m}{c_b} \tag{2.22}$$

The concentration at the membrane surface c_m is the maximum concentration of a flowable particle system. c_m, c_p and c_b are dictated by physico-chemical properties of the feed and thus flux can only be improved by enhancing k, which, in turn, needs to be done in a way that reduces thickness of the boundary layer (Wang & Song, 1999). There is a range of correlations for the mass transfer coefficient k, that relate channel dimensions, flow conditions and different means of mass transfer to the cake layer thickness. Often it is given as a dimensionless number, the Sherwood number, which describes the ratio of convective to diffusive mass transfer (Ripperger & Altmann, 2002). Furthermore, there is no accurate method of measuring the membrane surface solute concentration c_m and therefore this model does not allow prediction of flux.

Retention and, therefore, the membranes selectivity will also change when a deposit layer forms on the membrane surface. Even when the layer does not affect effective filtration pore size, it will still cause local concentration of the solute at the membrane surface to be higher than the bulk solute concentration where samples are taken to analyse retention. Thus, the retentate concentration in the equation for the retention factor needs to be substituted by the concentration at the membrane surface (c_m) to calculate true retention (R_{tr}) (Cheryan, 1998):

$$Rf = 1 - \frac{c_p}{c_r} \rightarrow R_{tr} = 1 - \frac{c_p}{c_m} \tag{2.23}$$

Using the film theory, the true retention can be derived from the determined retention factor (Rf) by the following relationship (Cheryan, 1998):

$$\ln\left[(1 - Rf)/Rf\right] = \ln\left[(1 - R_{tr})/R_{tr}\right] + \frac{J}{k} \tag{2.24}$$

The resistance in series model

The resistance in series concept is an approach that accounts for all cases of flux behaviour. Here, the concentration polarisation and fouling layers are considered as resistances adding to the overall resistance of the process.

Darcy's law was originally formulated to describe fluid flow across a porous medium like micron-porous membranes (Mulder, 1991). The following Darcy-derived form is often used for flux behaviour of membrane processes (Field *et al*., 1995):

$$J = \frac{\text{TMP}}{\eta \cdot R} \tag{2.25}$$

where η is the viscosity of the fluid permeating the membrane at a given operating temperature, and R is the overall hydraulic resistance. This resistance can be broken down in various ways, the following has become common (Field *et al*., 1995; Zondervan & Roffel, 2007):

$$R = R_m + R_c \text{ with } R_c = R_f + R_p \tag{2.26}$$

where R_m is the resistance of the clean membrane, including the resistance of a support layer in composite membranes although this should be negligible due to its high porosity. R_c is the overall cake layer resistance, which is split into R_f the irreversible fouling resistance unaffected by operating parameters and R_p the resistance due to concentration polarization, which includes the concentration boundary layer resistance.

For ideal conditions, without any components accumulating at the membrane surface, the overall resistance equals the clean membrane resistance. In this case the equation is very similar to the Hagen–Poiseuille law as described above, and the pure membrane resistance measured with pure water can be used to monitor long-term performance of membranes and cleaning efficiency. Indeed, in the food industry a short and standardised pure water flux test is often employed after cleaning before the next filtration period is started to ascertain cleaning effectiveness before any sensitive liquid food product gets in touch with the membrane. The membrane resistance is not an appropriate tool to choose between different membrane types for a certain application, because in actual filtration operations the membrane resistance may only be a small portion of the overall resistance and membrane feed interactions need to be taken into account.

Not only looking at the resistance of the membrane alone, but including cake/polarisation layer resistances, this model is used to estimate the influence of operation conditions on fouling phenomena (Tansel *et al*., 2000; Li *et al*., 2007). For this the cake resistance can be determined from the rate of flux decline due to cake build-up, and the contribution of fouling to the overall cake layer resistance can be estimated by measuring pure water flux after fouling has established (Mikulásek *et al*., 2004; Rabiller-Baudry *et al*., 2008).

As will be discussed below, for small retained molecules the osmotic pressure difference across the membrane needs to be accounted for and the pressure term of the resistance in series model can be modified to (Grandison & Lewis, 1996):

$$J = \frac{\mathrm{TMP} - \Delta\Pi}{\eta \cdot R} \tag{2.27}$$

where $\Delta\Pi$ is the osmotic pressure difference across the membrane.

2.3.2 *Influence of chemical potential on the reverse osmosis process*

For pore sizes below 10 nm (mostly NF and RO), permeation is no longer a purely pressure-driven flow through tiny pores but a diffusive process, and the osmotic pressure difference between the two membrane sides needs to be overcome. Fig. 2.15 illustrates the effect of osmosis and the principle of RO, working against the naturally occurring osmotic pressure. Naturally, osmosis occurs when the concentration difference of two solutions that are separated by a semi-permeable membrane results in a gradient of chemical potential across such a membrane. Solvent will flow through the membrane from the lower solute concentration to the compartment of higher solute concentration attempting to outbalance the difference in chemical potential. As a consequence of flow towards the high

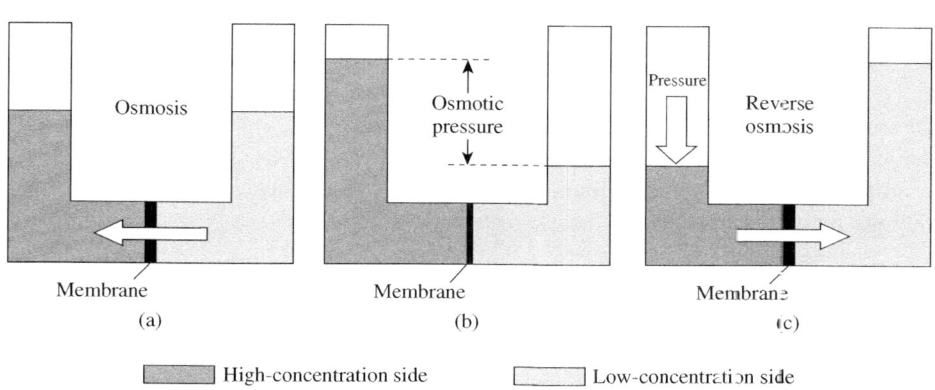

Fig. 2.15 Principle of reverse osmosis filtration.

concentration compartment, the liquid level on the concentrate side increases and results in a hydrostatic pressure difference. When the hydrostatic pressure difference equals the difference in chemical potential, net mass transport will be zero (Fig. 2.15b), or in other words the hydrostatic pressure raises the chemical potential of the concentrate side until it balances. In RO, a hydraulic pressure is induced on the concentrate side forcing the solvent back through the semi-permeable membrane. While solvent passes through the semi-permeable membrane, the concentration on the compressed side increases resulting in an incremental increase in osmotic pressure. This means that the pressure and, consequently, energy required to produce pure water increases with increasing concentration on the compressed side.

For filtration processes, the TMP must exceed the osmotic pressure for flow to occur. The osmotic pressure Π is described by the van't Hoff equation (Mulder, 1991):

$$\Pi = \frac{R \cdot T}{M_i} \cdot c_i \tag{2.28}$$

where R is the universal gas constant, T the absolute temperature, M_i the molar mass and c_i the concentration of component i. The higher the concentration of dissolved solids at the membrane surface and the lower the molecular weight of these, the higher the resulting osmotic pressure (Jiao *et al*., 2004). The van't Hoff equation is valid for low concentrations and low molecular weight molecules. The equation for macromolecular or high concentration solutions is enhanced by a virial expansion where the terms B and C are constants depending on the molecule species (Mulder, 1991):

$$\Pi = \frac{R \cdot T}{M_i} \cdot c_i + B \cdot c_i^2 + C \cdot c_i^3 + \ldots \tag{2.29}$$

If solute dissociation or association occurs, the changes in molecular weight and concentration need to be taken into account (Mulder, 1991). Osmotic pressure does not only influence RO, but also has a bearing on NF to a lesser extent as reflected in the TMP ranges (Fig. 2.1). It is generally assumed to be negligible in the cases of MF, and also for most retained solutes during UF which is due to the high molecular weights of retained solutes.

Apart from osmotic pressure, another factor adding to the chemical potential that needs to be overcome for RO flux to occur is the vapour pressure. The vapour pressure of a liquid is not only determined by its temperature but also the solvent concentration. At equal temperature, as found during RO, the vapour pressure of a solution is less than that of the pure solvent and consequently the vapour pressure on the RO permeate side is higher than on the feed side. Vapour pressure differences have been made useful in rather novel membrane processes that will be discussed later.

Solution diffusion model

A theory explaining transport mechanisms in RO membranes is the solution diffusion model. In this model, the membrane is treated as a non-porous, homogeneous wall, and components in the solution diffuse into and through the membrane. Diffusion occurs due to the existing chemical potential gradient across the membrane, for example increasing total applied pressure leads to an increase in chemical potential and therefore flux. The flow of solvent and solute through the membrane are independent of each other, and water transport through the membrane occurs more rapidly than solute transport. For RO, flux occurs according to (Porter, 1990):

$$J_{water} = B_{water}(\text{TMP} - \Delta\Pi) \tag{2.30}$$

where B_{water} is the water transport coefficient and $\Delta\Pi$ the osmotic pressure difference across the membrane. As equation (2.30) reveals, the osmotic pressure difference across the membrane reduces the driving force and, therefore, flux. The osmotic pressure increases with the concentration of solids at the membrane surface (c_m), which explains why increasing (TMP) does not necessarily increase flux because the resulting concentration and osmotic pressure at the membrane surface increase.

The flow of solute or dissolved solids across the membrane occurs according to (Porter, 1990):

$$J_{solid} = B_{solid}(c_f - c_p) = B_{solid}\Delta c \tag{2.31}$$

where B_{solid} is the solute transport coefficient, Δc the solute concentration gradient across the membrane, c_f and c_p the solute concentrations in the feed and permeate. According to this equation, solute flux is not affected by the total pressure difference, but primarily depends on the concentration of dissolved solids on each side of the membrane. This is not fully true, solute transport is still influenced by the pressure applied which is made use of during piezodialysis, but it only influences RO to a very small extent and can thus be ignored (Bolto *et al.*, 2010).

2.4 Effects of feed characteristics and operating parameter on separation efficiency

There are two ways to define separation efficiency, the actual membrane throughput measured as the flux and the retentivity/selectivity. General characteristics of feed components and operating parameter influencing flux and retentivity will be looked at below.

2.4.1 *Effects of feed components*

Each component of a feed-stream reacts differently with the membrane: (a) conformation; (b) solute shape and size; (c) hydrophobic interactions; (d) solute solubility; and (e) other factors will influence these membrane–solute interactions (Cheryan, 1998). The higher these interactions, the more severe membrane fouling and, consequently, the more significant permeate flux decline (Bottino *et al.*, 2000). Factors leading to membrane fouling will be discussed in Chapter 4, and such influencing performance and retention in general will be discussed here.

Steric effects on retention

Solute shape and size are the most important parameters influencing process retentivity. An illustrative example for the impact of solute shape on retention is the comparison of two molecules of the same molecular weight but different diameters. A smaller diameter can be related to a more linear shape and a bigger diameter occurs with spherical molecules. The smaller diameter of the linear molecule makes it more likely to snake through pores whereas the bigger diameter of the spherical molecule causes it to be retained. This is schematically shown in Fig. 2.16. Molecules with branched compounds show even lower permeability, even at lower molecular weight than those with a linear configuration, which has been related to steric hindrance of the branched groups (Zheng, *et al*., 2008). This shows that the molecular shape rather than the molecular weight determines retention, and the molecular weight cut-off is not a reliable tool to predict retentivity as discussed above (Van der Bruggen *et al*., 1999).

The relation between particle size and membrane pore size is also crucial, as particles which are approximately of same size as the pores may penetrate partially into the pores and block them. Even solutes that are much smaller than the smallest pore on the membrane and, therefore, should be freely permeable, may interact with or bind to the impermeable compounds in the feed (Howe & Clark, 2002). Shape and conformation of macromolecules can be affected by pH, ionic strength, temperature and interactions with other components. Differences in shape can also be due to high shear prevalent at the membrane surface (Cheryan, 1998). The pH of an aqueous solution not only has a bearing on particle size and electrostatic potential of many components in the solution itself, but can also change the surface charge on the membrane as discussed above (Bouzid *et al*., 2008; Rice *et al*., 2011).

Charge and polarity

Charge effects of solutes in the feed also depend on the membrane material; in general, high permeability is obtained with uncharged molecules and uncharged membranes (Van der Bruggen *et al*., 1999). The addition of charged ions to a feed solution of uncharged solutes can result in reduction of retention of the neutral components. This effect is higher

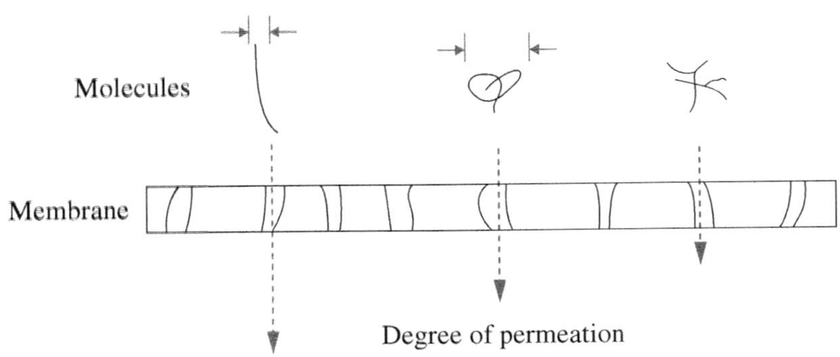

Fig. 2.16 Solute shape and resulting degree of permeation, from left to right decreasing permeability: linear, spherical, branched.

for ions for which the membranes show low retention (e.g. molecules that are much smaller than the pores), and can either be explained through an increase in repulsive forces due to a formed layer or by an increased effective average pore size as a consequence of blockage of smaller pores (Bargeman *et al.*, 2005). The small charged molecules enter membrane pores and block smaller pores rather than bigger ones, and as a result retention of neutral solutes is determined to a larger extent by the bigger pores (Bargeman *et al.*, 2005). This effect has practical implications for nanofiltration of a glucose (natural component) and salt-containing feed solution like, for example, milk/whey UF permeates. These permeates are often concentrated by NF to obtain a lactose-enriched stream.

The entry into the membrane structure is facilitated for dipole molecules. The side of the dipole with opposite charge of the membrane is directed towards the pore which is not a static orientation and, thus, the dipole enters more easily into the membrane (Van der Bruggen *et al.*, 1999). This polarity effect is the same for positively and negatively charged membranes, only the direction of the dipole will be opposite, but the effect of a facilitated entry remains.

Viscosity

The Hagen–Poiseuille law has shown that even under ideal conditions (no concentration polarisation or fouling) membrane process performance is influenced by viscosity. Viscosity of the process stream influences mass transfer and pressure drop from module inlet to outlet: at high viscosity the mass transfer decreases and pressure drop increases. Specifically for membranes that follow viscous flow theory like MF and UF membranes, flux is directly influenced by solvent viscosity, for NF and RO viscosity does not influence flux other than via the bulk flow dynamics. The two main factors influencing viscosity are concentration and temperature. In general, increasing temperature should result in increasing fluxes because of lower viscosity of the permeating streams for Newtonian fluids. Nevertheless, temperature can affect particle–particle interactions, depending on the feed processed, and thus can potentially lead to fouling phenomena, e.g. due to protein denaturation. Shear (induced by feed flow rate) also affects the viscosity of non-Newtonian liquids. Systems like milk, tomato juice and potato starch water are non-Newtonian systems, which means that their fluidity changes non-linearly with shear.

Major food components influencing filtration processes

Although the influence of food components on membrane performance is highly application specific, some general observations include the following.

- Proteins are a major foulant in membrane processing because of their multiplicity of functional groups, charge density, varying degrees of hydrophobicity, and the complex secondary and tertiary structure that allows a protein to interact with other feed components and the membrane itself. Complicating the phenomenon is that all these properties are affected by pH, ionic strength, shear, heat treatment and other environmental factors (Marshall *et al.*, 1993).

- Salts can precipitate on the membrane because of poor solubility or bind to the membrane directly by charge interactions. Interactions with proteins increase protein deposition. Calcium has been identified as a major cause of fouling, not only due to precipitation, but also because its ionic form can act as a salt bridge between membrane and proteins (Cheryan, 1998).
- Lipids can cause problems with slightly hydrophobic membranes, but are not considered a major foulant.
- Sugars can lead to high osmotic pressures, and solutions with high sugar contents often exhibit high viscosities (Hinkova *et al.*, 2002).

2.4.2 *Effects of operating parameters*

Reducing the negative effects of flux decline by optimising process operation is essential, and needs to be done in a way that minimises energy input and simultaneously provides high operational safety. Fig. 2.17 looks at the forces exerted on particles within the feed. These forces influence the thickness of forming cake layer, its porosity and, consequently, in cake-layer controlled systems flux and retention characteristics. There are four major operating parameters affecting the forces on a particle and hence flux as well as retentivity: (a) feed concentration, (b) temperature, (c) TMP and (d) flow rate or turbulence in the feed channel. Concentration in the feed channel is usually dictated by each application, and adjusting concentration to optimise operation is not possible due to production purposes. The effect of temperature, as mentioned before, results in changes of solvent viscosity, but can worsen fouling with heat-sensitive components like proteins. The two main operating parameters governing membrane performance are, therefore, TMP and flow rate as discussed in the following section.

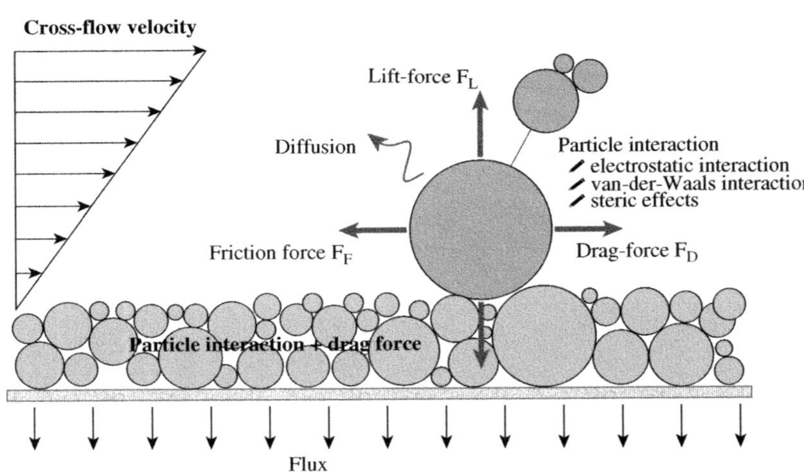

Fig. 2.17 Forces exhibited on a particle during cross-flow filtration (Reproduced from 'Modelling of the Cross-flow Filtration with Membranes' by Professor S. Ripperger of Kaiserslautern Technical University with permission from GeoDict by Fraunhofer ITWM).

Feed flow along the membrane surface induces a wall shear stress (τ_w). The hydrodynamic lift force F_L and the drag force F_D, as shown in Fig. 2.17, are functions of this wall shear stress and responsible for particles being removed from the surface (Altmann & Ripperger, 1997; Piry *et al*., 2008). For submicron particle sizes, diffusion becomes more and more significant and, at sizes below 100 nm, the filtration process is determined mainly by diffusion, whereas at particle sizes bigger than 1 μm filtration is mainly controlled by hydrodynamics (Ripperger & Altmann, 2002). Here, the ratio J/τ_w defines the balance of particle deposition rate (relating to J) and particle removal rate (relating to τ_w) and, therefore, determines particle transport towards the deposit layer (Altmann & Ripperger, 1997). If a critical value of J/τ_w is exceeded, membrane processes become unsteady, thus operating conditions are recommended in which J/τ_w is below that critical value (Le Berre & Daufin, 1996; Grandison *et al*., 2000). This value varies with pore size, membrane material and feed composition, and needs to be determined experimentally for every membrane feed combination (Grandison *et al*., 2000).

Increasing cross-flow is the principal way to reduce deposition layer thickness, but increasing wall shear stress leads to a relatively high pressure drop from module inlet to outlet and increased pumping energy requirements. Also, this pressure drop causes TMP to decrease over the membrane length resulting in a flux profile, whereas the lift force as a function of the wall shear stress length is constant over the membrane (Piry *et al*., 2008). Thus, high deposit layer build-up occurs at the front section of the membrane, whereas the rear part is less affected by particle deposition (Wang & Song, 1999; Piry *et al*., 2008). This in turn results in faster membrane/module ageing at more heavily used areas and, therefore, increases membrane replacement costs. Different means of achieving an evenly distributed TMP have been proposed, all based on establishing an equivalent pressure drop on the permeate side along the module length (Huisman *et al*., 1997; Vadi & Rizvi, 2001; Atkinson, 2005).

Excessively increasing wall shear stress to reduce deposition layer thickness can also have undesired consequences on layer formation and, therefore, performance. Both, the lift force and the filtrate drag force increase with particle size; however, lift force increases with the third power of the particle size and drag force with the first (Altmann & Ripperger, 1997). Consequently, at high values of wall shear stress, larger particles are preferentially removed from the surface compared with smaller ones and the filter cake becomes enriched with small particles, leading to a lower porosity of the cake layer and decreasing effective pore size (Le Berre & Daufin, 1996; Ripperger & Altmann, 2002). Flux of polydisperse suspensions is often similar to that of a monodisperse suspension of the smallest particle (Kromkamp *et al*., 2002). As shown, wall shear stress has both a quantitative and a qualitative effect. An example of its effect on retention is the MF of skim milk to separate caseins from whey protein. It was found that at a low value for J/τ_w (<0.8 L h^{-1} m^{-1} Pa^{-1}), casein retention is low and at high values (>1.0 L h^{-1} m^{-1} Pa^{-1}) whey protein permeation is reduced (Le Berre & Daufin, 1996). Therefore, an ideal value of J/τ_w for the size separation of two species exists. Ripperger & Altmann (2002) found that it is possible to generate specific structures of a cake layer by varying operating parameters and flux rate during filtration start-up. It was also suggested that for filtration of fine particles, the addition of flocculants can aid performance due to the formation of a more porous filtration cake and preventing the deposition of fine particles on the membrane, which would result in a dense cake layer (Nguyen & Ripperger, 2002).

The friction and particle interaction forces are sometimes higher than the hydrodynamic forces which may result in membrane fouling. As stated before, once fouling has occurred, it is irreversible and cannot be combated by adjusting operating parameters. It is the main obstacle for efficient membrane operation and fouling management techniques at different stages of the process have been proposed. A common method to remove cakes formed due to reversible fouling and to restore performance quickly during the process with only short process interruptions is periodic back washing by intermittently reversing the direction of the filtrate flow from the permeate to the feed channel side (Gao *et al.*, 2011; Qaisrani & Samhaber, 2011). However, backwashing is problematic with polymeric composite membranes, as the active membrane layer delaminates from the support layer at higher pressures from the permeate (support layer side) to the feed side. Other strategies to improve flux rates include (a) the introduction of air bubbles into the feed stream to create a highly turbulent two-phase flow creating a scouring effect (Mercier-Bonin *et al.*, 2003); (b) pre-treatment of the feed (remove fouling compounds); and (c) the development of rotating or vibrating membrane modules (Espina *et al.*, 2008).

Critical and sustainable flux operation

In pressure-driven membrane operation, flux decline below the theoretical capacity of the membrane is often experienced (Field *et al.*, 1995; Wang & Song, 1999). Such low flux is commonly related to a high permeation rate during start-up of the membrane process when cake layer formation takes place. The increased filtrate drag force resulting from this high start-up flux leads to a more severe cake layer on the membrane surface and consequently lower operation flux. According to Field *et al.* (1995), who introduced the critical flux concept, the critical flux is a certain flux value below which a decline of flux with time does not occur and above it fouling is observed. More generally, the critical flux can be defined as the permeate flux at which fouling becomes apparent. It is a criterion for the transition from concentration polarization to fouling. The TMP corresponding to this critical value is then called critical TMP (Youravong *et al.*, 2002). Since the introduction of this concept differing definitions of the term 'critical flux' have arisen. Bacchin *et al.* (2006) have proposed to differentiate two different versions, the strong form of the critical flux definition according to which the critical flux is reached when the flux–TMP curve starts to deviate from linearity and the weak form relating it to the first permeate flux for which irreversible fouling appears.

The critical flux is determined by hydrodynamic conditions, pore size and composition of the feed, and is sometimes equal to the limiting flux (Fig. 2.12), but in other cases operating membrane filtration at the limiting flux causes more fouling and, therefore, reduces performance (Grandison *et al.*, 2000). This critical flux concept has lead to a subcritical membrane operation where the flux is initially kept below its critical value during operation start-up and only increased to the critical flux after cake layer establishment has taken place. The critical flux needs to be experimentally tested for each filtration process (Cho & Fane, 2002). This is mostly done by either stepping flux or TMP. The critical flux can then be determined by plotting flux against TMP. Another way to establish whether the balance between particle convection and removal has been shifted towards deposition is to look at the stability of flux when increasing TMP or the stability of TMP when a certain flux has been imposed (Defrance & Jaffrin, 1999). The transition from a dynamic, labile polarisation layer to an irreversibly fouled cake is reached when the other parameter does not remain constant over time.

When dealing with complex feeds, the criteria for a subcritical flux operation is often too strong and a low rather than zero rate of fouling is considered to be indicative of subcritical operation. The term sustainable flux has been introduced to refer to the operational and economic sustainability of a membrane process that minimises fouling to avoid frequent cleaning (Bacchin *et al.*, 2006; Rabiller-Baudry *et al.*, 2008).

2.5 Cross-flow systems

2.5.1 *Background*

A schematic illustration of flow arrangements during cross-flow filtration is given in Fig. 2.18. The cross-current mode refers to a set-up of free permeate discharge which is typical for spiral wound modules. The permeate flows orthogonally to the membrane reducing a pressure gradient along the module on the permeate side. A stirred filtration cell is used for fermentations where high shear of pumping could destroy sensitive bio cells; here the stirring induces shear flow across the filter medium. Generally, the applied flow mode during cross-flow filtration is mostly dictated by the membrane module design in use. It should be noted that these stream configurations are idealised extremes, and module geometry often leads to a combination of these flow regimes.

2.5.2 *Single-pass versus feed-and-bleed operation*

Cross-flow membrane processes can be operated as single-pass filtration or feed and bleed operations, schematic illustrated in Fig. 2.19. In a single-pass operation, the feed passes along the membrane surface once. In feed and bleed operations, a part of the exiting stream is recirculated and mixed with the incoming feed creating in fact a multipass

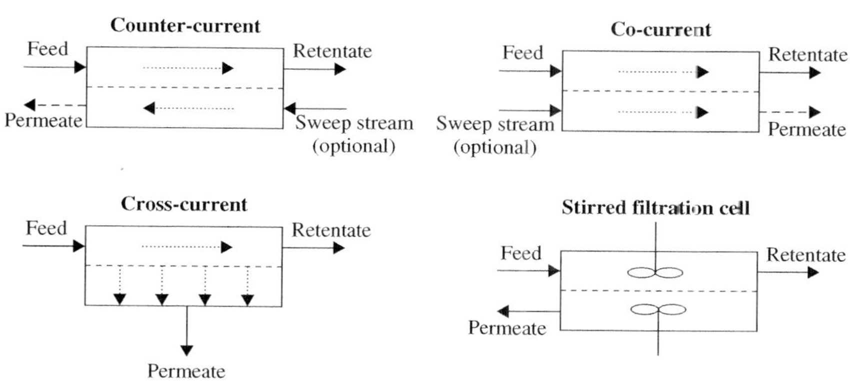

Fig. 2.18 Different operating schemes for cross-flow filtration (Adapted from Koros *et al.*, 1996. Reproduced with permission from Elsevier).

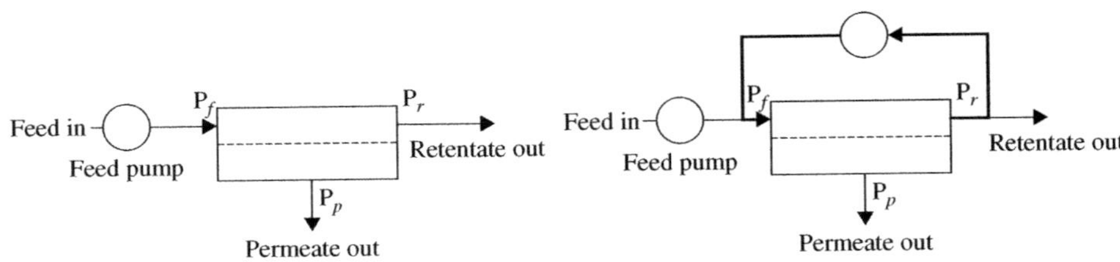

Fig. 2.19 Simplified flow chart of a single-pass operation (left), and feed and bleed (right) cross-flow membrane process.

filtration technique. The start-up of feed and bleed processes is similar to that in dead-end batch systems in that the retentate is initially not withdrawn but completely recycled. After the required solute concentration is reached within the loop, a fraction of the retentate in the loop is continuously bled off. The inflowing feed stream is a fraction of the flow within the recirculation loop, assuring a steady concentration during operation. Volume flow of the retentate bleed stream is controlled by the volume concentration factor and the permeate plus feed flow rates. Disadvantages of feed and bleed configurations are the need for a recirculation pump adding more energy to the plant and the loop continuously operating at a concentration equivalent to the final concentration in the system. The resulting lower average flux than single-pass filtration leads to a higher membrane area required (Melin & Rautenbach, 2007). Advantages include the hydraulic conditions inside the module being independent of the incoming feed flow, so that these can be adjusted in order to minimize fouling (high wall shear stress). Therefore, the feed-and-bleed operation is mostly preferred for high solute applications.

2.6 Recent membrane processes following different operating principles

This section looks at membrane processes that use driving forces other than hydraulic pressure. Despite the presented and most common four pressure-driven membrane processes (MF, UF, NF and RO), other membrane processes are being developed. Most of these have only become viable due to the advances made in material science, new ideas to bring these processes up to an industrial viable stage are being realized and variations of membrane processes as well as combinations thereof are being tested for specific applications.

2.6.1 *Forward osmosis*

In forward osmosis (FO), diffusion-based transport is utilised through a concentration difference of the transported species across a semipermeable membrane. The resulting difference in osmotic pressure is the driving force of this process. To transfer water to the permeate side a stripping (or draw) solution of higher osmotic pressure is required on that side. The main advantages of forward osmosis are the reduced amount of electrical energy needed to establish high hydraulic pressures and less cake layer compaction due to the absence of pressure-forcing solutes against the membrane surface. FO has been tested for fruit juices and water purification (Jiao *et al*., 2004; Mi & Elimelech, 2010). However, limitations are given to acceptable concentrations as very high concentrations can osmotically push (or suck) the water out of the membrane. This dehydration results in the membrane being compacted irreversibly and therefore compromises performance. The ideal FO membrane needs to withstand the mechanical stresses generated during operation without compromising retention and permeability. Thin film composite membranes have been proposed for this purpose (Tiraferri *et al*., 2011). Further, despite the energy-saving feature of the FO unit, energy is needed to recover permeate removed by the draw solution. Also, if the product is contained in the draw solution (e.g. desalinated water), then the draw solution must be non-toxic and/or not contaminate its intended use. For this reason, FO can be more attractive for applications requiring the reject (i.e. concentration), this does not need to be pressurised (unlike RO for example). However, back diffusion of the draw solution must be monitored.

2.6.2 *Osmotic distillation*

The principle of OD is shown in Fig. 2.20. As with FO, OD also requires a stripping solution to maintain an osmotic pressure but, in this case, the membrane is hydrophobic and prevents water from entering the pores. The OD process is driven by a vapour pressure difference that is due to the difference in chemical potential of both sides, and leads to a difference in water activity (Jiao *et al*., 2004). This pressure difference at the vapour–liquid interface causes a vapour transfer from the feed to the stripping solution. The operation pressure needs to be kept below the liquid entry pressure of the membrane in order to prevent a liquid–liquid contact, which can be several hundred kPa for an MF membrane (Curcio *et al*., 2010). New membrane materials with a thin hydrophilic coating are being developed to allow processing of constituents that cause membrane wetting, like fats or tensides (Chanachai *et al*., 2010). OD has also been tested for various juices (Jiao *et al*., 2004).

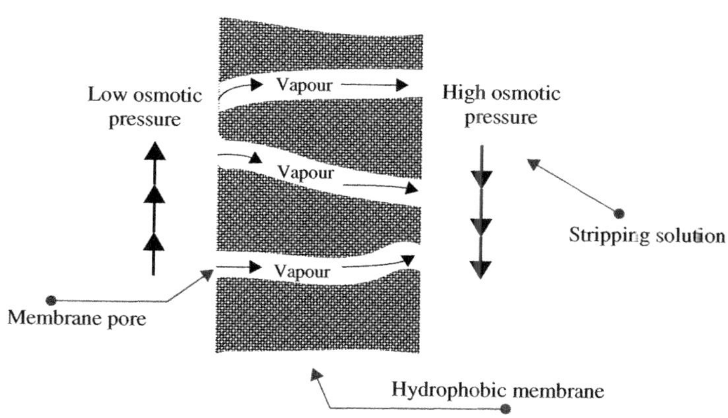

Fig. 2.20 Principle of osmotic distillation.

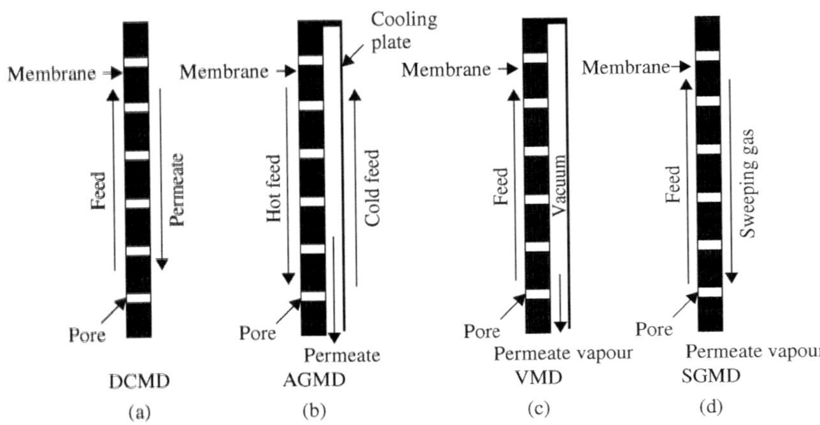

Fig. 2.21 Principle of the four major membrane distillation configurations. (a) DCMD, direct contact membrane distillation; (b) AGMD, air gap membrane distillation; (c) VMD, vacuum membrane distillation; (d) SGMD, sweep gas membrane distillation (Reproduced courtesy of Zhang *et al.* (2009), reprinted from Water Science and Technology, and with permission form the copyright holders, IWA).

2.6.3 *Membrane distillation*

The schematic of membrane distillation (MD) and major configurations of the process are outlined in Fig. 2.21. Similar to OD, MD also works by allowing only water vapour to pass through the membrane. For this the membrane either needs to be hydrophobic or dense, only letting water pass through. In MD, the vapour pressure difference from feed to permeate side results from a temperature difference or by imposing a lower vapour pressure than the feed solution by either vacuum or sweeping with dry gas. Other than FO and OD, MD does not necessarily require a stripping stream and, if stripping water is used, it can be of high purity. As a result, MD can be used to produce ultra-pure water. The temperatures used to drive MD can be of low grade (i.e. 40–80°C) and, thus, allows waste heat and solar energy to be used to drive the process. The origin of MD lies in desalination, and it has also been tested for concentration purposes in the food industry, like fruit juices and different dairy streams (Gunko *et al.*, 2006; Kozák *et al.*, 2009; Hausmann *et al.*, 2011).

There are different ways to condense the permeating vapour on the cold side of the process (Fig. 2.21): a stripping water is used for direct contact membrane distillation (DCMD), a sweep gas for sweep gas membrane distillation (SGMD), a cold surface serves as a condenser during air gap membrane distillation (AGMD) and a vacuum draws the vapour during vacuum membrane distillation (VMD).

In cases where the production of pure water is not the process requirement, the vapour pressure difference across the membrane and therefore the driving force can be increased by increasing the osmotic pressure on the cold side. This essentially creates a coupled operation of membrane distillation and osmotic distillation (Bélafi-Bakó & Koroknai, 2006).

Integration of forward osmosis and membrane distillation

The advantage of low-fouling and low-energy requirements in FO can be combined with the ability of MD to produce high-purity water in one hybrid system (Wang *et al.*, 2011).

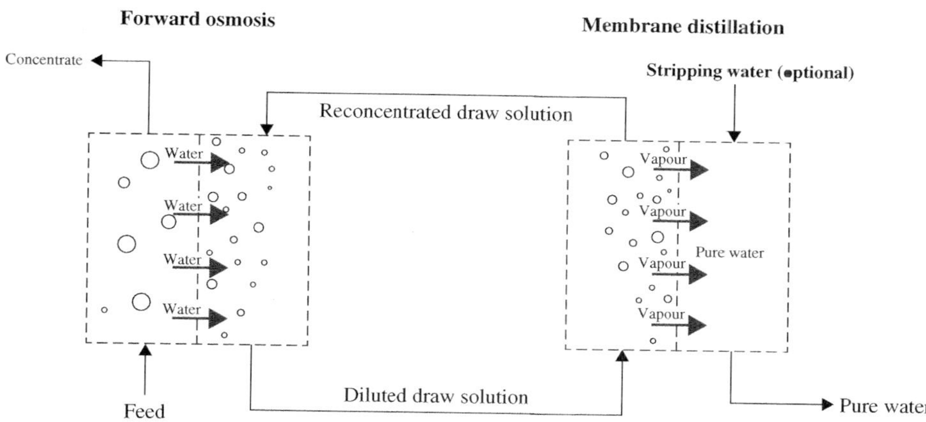

Fig. 2.22 Integrated forward osmosis and membrane distillation hybrid system.

Concentration of the feed solution is achieved in a first stage of FO and the draw solution that has been diluted due to water permeating from the feed is reconcentrated in a subsequent membrane distillation step before being recycled back to the FO unit (Fig. 2.22).

This linked system enables the production of pure water and maintains a high osmotic pressure of the draw solution so that a continuous high performance of the FO unit can be achieved. Also, pressure and temperature requirements are low, which allows for gentle processing of sensitive constituents, in particular proteins.

2.7 Conclusions

The four pressure-driven membrane processes have already become an integral part of food production, and technology improvements are now focused on application-specific performance and membrane durability. The emergence of new membrane technologies that follow different process principles creates opportunities for even more applications. However, new membrane processes require the development of different membrane materials with special characteristics before the full potential of each process becomes apparent. Continued development of improved membrane process designs and membrane materials enables membrane systems to achieve the overall aims of sustainable production and high product quality at reduced costs. Despite that, industrial success depends on the advantages over existing technologies. It can be anticipated that the development of membrane technology will further increase the use of membrane technology in the food industry.

References

Altmann, J. & Ripperger, S. (1997) Particle deposition and layer formation at the cross-flow microfiltration. *Journal of Membrane Science*, **124**, 119–128.

Atkinson, S. (2005) Isoflux membrane technology is efficient and economical. *Membrane Technology*, **8**, 9–10.

Ayral, A. (2008) Introduction to baromembrane techniques. Available from http://water-environment.vin.bg.ac.yu/proceedings/2%20Stana%2015–29%20B5%20texte%20Serbie-%20A-Ayral.doc.

Bacchin, P., Aimar, P. & Field, R.W. (2006) Critical and sustainable fluxes: Theory, experiments and applications. *Journal of Membrane Science*, **281**, 42–69.

Bargeman, G., Vollenbroek, J.M., Straatsma, J., Schroën, C.G.P.H. & Boom, R.M. (2005) Nanofiltration of multi-component feeds: Interactions between neutral and charged components and their effect on retention. *Journal of Membrane Science*, **247**, 11–20.

Bélafi-Bakó, K. & Koroknai, B. (2006) Enhanced water flux in fruit juice concentration: Coupled operation of osmotic evaporation and membrane distillation. *Journal of Membrane Science*, **269**, 187–193.

Bikel, M., Çulfaz, P.Z., Bolhuis-Versteeg, L.A.M., Pérez, J.G., Lammertink, R.G.H. & Wessling, M. (2010) Polymeric microsieves via phase separation microfabrication: Process and design optimization. *Journal of Membrane Science*, **347**, 93–100.

Bolto, B., Hoang, M. & Tran, T. (2010) Review of piezodialysis – salt removal with charge mosaic membranes. *Desalination*, **254**, 1–5.

Bottino, A., Capannelli, G., Monticelli, O. & Piaggio, P. (2000) Poly (vinylidene fluoride) with improved functionalization for membrane production. *Journal of Membrane Science*, **166**, 23–29.

Bouzid, H., Rabiller-Baudry, M., Paugam, L., Rousseau, F., Derriche, Z. & Bettahar, N.E. (2008) Impact of zeta potential and size of caseins as precursors of fouling deposit on limiting and critical fluxes in spiral ultrafiltration of modified skim milks. *Journal of Membrane Science*, **314**, 67–75.

Brito-de la Fuente, E., Torrestiana-Sánchez, B., Martínez-González, E. & Mainou-Sierra, J.M. (2010) Microfiltration of whole milk with silicon microsieves: Effect of process variables. *Chemical Engineering Research and Design*, **88**, 653–660.

Bromley, A.J., Holdich, R.G. & Cumming, I.W. (2002) Particulate fouling of surface microfilters with slotted and circular pore geometry. *Journal of Membrane Science*, **196**, 27–37.

Chanachai, A., Meksup, K. & Jiraratananon, R. (2010) Coating of hydrophobic hollow fiber PVDF membrane with chitosan for protection against wetting and flavor loss in osmotic distillation process. *Separation and Purification Technology*, **72**, 217–224.

Chandler, M. & Zydney, A. (2006) Effects of membrane pore geometry on fouling behavior during yeast cell microfiltration. *Journal of Membrane Science*, **285**, 334–342.

Chen, V., Fane, A.G., Madaeni, S. & Wenten, I.G. (1997) Particle deposition during membrane filtration of colloids: transition between concentration polarization and cake formation. *Journal of Membrane Science*, **125**, 109–122.

Cheryan, M. (1998) *Ultrafiltration and Microfiltration Handbook*, Technomic Publication, Lancaster.

Cho, B.D. & Fane, A.G. (2002) Fouling transients in nominally sub-critical flux operation of a membrane bioreactor. *Journal of Membrane Science*, **209**, 391–403.

Curcio, E., Di Profio, G. & Drioli, E. (2010) Membrane distillation and osmotic distillation. *Comprehensive Membrane Science and Engineering* (eds. D. Enrico and G. Lidietta), pp. 1–20, Elsevier, Oxford.

Defrance, L. & Jaffrin, M.Y. (1999) Comparison between filtrations at fixed transmembrane pressure and fixed permeate flux: application to a membrane bioreactor used for wastewater treatment. *Journal of Membrane Science*, **152**, 203–210.

Elmarraki, Y., Persin, M., Sarrazin, J., Cretin, A. & Larbot, M. (2001) Filtration of electrolyte solutions with new TiO2-ZnAl2O4 ultrafiltration membranes in relation with the electric surface properties. *Separation and Purification Technology*, **25**, 493–499.

Espina, V.S., Jaffrin, M.Y., Frappart, M. & Ding, L.-H. (2008) Separation of casein micelles from whey proteins by high shear microfiltration of skim milk using rotating ceramic membranes and organic membranes in a rotating disk module. *Journal of Membrane Science*, **325**, 872–879.

Field, R.W., Wu, D., Howell, J.A. & Gupta, B.B. (1995) Critical flux concept for microfiltration fouling. *Journal of Membrane Science*, **100**, 259–272.

Gao, W., Liang, H., Ma, J., Han, M., Chen, Z.-l., Han, Z.-S. & Li, G.-b. (2011) Membrane fouling control in ultrafiltration technology for drinking water production: A review. *Desalination*, **272**, 1–8.

Grandison, A.S. & Lewis, M.J. (1996) *Separation Processes in the Food and Biotechnolgy Industries*, Woodhead Publishing, Cambridge.

Grandison, A.S., Youravong, W. & Lewis, M.J. (2000) Hydrodynamic factors affecting flux and fouling during ultrafiltration of skimmed milk. *Le Lait*, **80**, 165–174.

Gunko, S., Verbych, S., Bryk, M. & Hilal, N. (2006) Concentration of apple juice using direct contact membrane distillation. *Desalination*, **190**, 117–124.

Hausmann, A., Sanciolo, P., Vasiljevic, T., Ponnampalam, E., Quispe-Chavez, N., Weeks, M. & Duke, M. (2011) Direct contact Membrane distillation of dairy process streams. *Membranes*, **1**, 48–58.

Hinkova, A., Bubník, Z., Kadlec, P. & Pridal, J. (2002) Potentials of separation membranes in the sugar industry. *Separation and Purification Technology*, **26**, 101–110.

Howe, K.J. & Clark, M.M. (2002) Fouling of microfiltration and ultrafiltration membranes by natural waters. *Environmental Science & Technology*, **36**, 3571–3576.

Huisman, I.H., Johansson, D., Trägårdh, G. & Trägårdh, C. (1997) Design of a cross-flow microfiltration unit for studies of flux and particle transport. *Chemical Engineering Research and Design*, **75**, 508–512.

James, B.J., Jing, Y. & Dong Chen, X. (2003) Membrane fouling during filtration of milk – a microstructural study. *Journal of Food Engineering*, **60**, 431–437.

Jiao, B., Cassano, A. & Drioli, E. (2004) Recent advances on membrane processes for the concentration of fruit juices: a review. *Journal of Food Engineering*, **63**, 303–324.

Kallioinen, M., Pekkarinen, M., Mänttäri, M., Nuortila-Jokinen, J. & Nyström, M. (2007) Comparison of the performance of two different regenerated cellulose ultrafiltration membranes at high filtration pressure. *Journal of Membrane Science*, **294**, 93–102.

Kariduraganavar, M.Y., Nagarale, R.K., Kittur, A.A. & Kulkarni, S.S. (2006) Ion-exchange membranes: preparative methods for electrodialysis and fuel cell applications. *Desalination*, **197**, 225–246.

Keskinler, B., Yildiz, E., Erhan, E., Dogru, M., Bayhan, Y.K. & Akay, G. (2004) Cross-flow microfiltration of low concentration-nonliving yeast suspensions. *Journal of Membrane Science*, **233**, 59–69.

Koros, W.J., Ma, Y.H. & Shimidzu, T. (1996). 'Terminology for membranes and membrane processes (IUPAC recommendation 1996). *Journal of Membrane Science*, **120**, 149–159.

Kozák, Á., Békássy-Molnár, E. & Vatai, G. (2009) Production of black-currant juice concentrate by using membrane distillation. *Desalination*, **241**, 309–314.

Kromkamp, J., van Domselaar, M., Schroën, K., van der Sman, R. & Boom, R. (2002) Shear-induced diffusion model for microfiltration of polydisperse suspensions. *Desalination*, **146**, 63–68.

Kuiper, S., van Rijn, C., Nijdam, W., Raspe, O., van Wolferen, H., Krijnen, G. & Elwenspoek, M. (2002) Filtration of lager beer with microsieves: flux, permeate haze and in-line microscope observations. *Journal of Membrane Science*, **196**, 159–170.

Lawrence, N.D., Kentish, S.E., O'Connor, A.J., Barber, A.R. & Stevens, G.W. (2008) Microfiltration of skim milk using polymeric membranes for casein concentrate manufacture. *Separation and Purification Technology*, **60**, 237–244.

Le Berre, O. & Daufin, G. (1996). 'Skimmilk cross-flow microfiltration performance versus permeation flux to wall shear stress ratio. *Journal of Membrane Science*, **117**, 261–270.

Li, M., Zhao, Y., Zhou, S., Xing, W. & Wong, F.-S. (2007). 'Resistance analysis for ceramic membrane microfiltration of raw soy sauce. *Journal of Membrane Science*, **299**, 122–129.

Marshall, A.D., Munro, P.A. & Trägårdh, G. (1993) The effect of protein fouling in microfiltration and ultrafiltration on permeate flux, protein retention and selectivity: a literature review. *Desalination*, **91**, 65–108.

Melin, T. & Rautenbach, R. (2007) *Membranverfahren – Grundlagen der Modul- und Anlagenauslegung*, Springer, Berlin.

Mercier-Bonin, M., Gésan-Guiziou, G. & Fonade, C. (2003) Application of gas/liquid two-phase flows during cross-flow microfiltration of skimmed milk under constant transmembrane pressure conditions. *Journal of Membrane Science*, **218**, 93–105.

Mi, B. & Elimelech, M. (2010) Organic fouling of forward osmosis membranes: Fouling reversibility and cleaning without chemical reagents. *Journal of Membrane Science*, **348**, 337–345.

Mikulášek, P., Dolecek, P., Smídová, D. & Pospísil, P. (2004) Cross-flow microfiltration of mineral dispersions using ceramic membranes. *Desalination*, **163**, 333–343.

Mulder, M. (1991) *Basic Principles of Membrane Technology*, Kluwer Academic, Dordrecht.

Nguyen, M.-T. & Ripperger, S. (2002) Investigation on the effect of flocculants on the filtration behavior in microfiltration of fine particles. *Desalination*, **147**, 37–42.

Piry, A., Kühnl, W., Grein, T., Tolkach, A., Ripperger, S. & Kulozik, U. (2008) Length dependency of flux and protein permeation in cross-flow microfiltration of skimmed milk. *Journal of Membrane Science*, **325**, 887–894.

Porter, M.C. (1972) Concentration polarization with membrane ultrafiltration. *Product R&D*, **11**, 234–248.

Porter, M.C. (1990) *Handbook of Industrial Membrane Technology*, Noyes Publication, Park Ridge.

Pouliot, Y. (2008) Membrane processes in dairy technology-from a simple idea to worldwide panacea. *International Dairy Journal*, **18**, 735–740.

Qaisrani, T.M. & Samhaber, W.M. (2011) Impact of gas bubbling and backflushing on fouling control and membrane cleaning. *Desalination*, **266**, 154–161.

Rabiller-Baudry, M., Bégoin, L., Delaunay, D., Paugam, L. & Chaufer, B. (2008) A dual approach of membrane cleaning based on physico-chemistry and hydrodynamics: application to PES membrane of dairy industry. *Chemical Engineering and Processing: Process Intensification*, **47**, 267–275.

Rajniak, P., Tsinontides, S.C., Pham, D., Hunke, W.A., Reynolds, S.D. & Chern, R.T. (2008) Sterilizing filtration: Principles and practice for successful scale-up to manufacturing. *Journal of Membrane Science*, **325**, 223–237.

Rice, G., Barber, A.R., O'Connor, A.J., Pihlajamaki, A., Nystrom, M., Stevens, G.W. & Kentish, S.E. (2011) The influence of dairy salts on nanofiltration membrane charge. *Journal of Food Engineering*, **107**, 164–172.

Ripperger, S. & Altmann, J. (2002) Cross-flow microfiltration – state of the art. *Separation and Purification Technology*, **26**, 19–31.

Rouquerol, J., Avnir, D., Everett, D.H., Fairbridge, C., Haynes, M., Pernicone, N., Ramsay, J.D.F., Sing, K.S.W. & Unger, K.K. (1994) Guidelines for the characterization of porous solids. *Studies in Surface Science and Catalysis*, **87**, 1–9.

Samuel de Lint, W.B., Zivkovic, T., Benes, N.E., Bouwmeester, H.J.M. & Blank, D.H.A. (2006) Electrolyte retention of supported bi-layered nanofiltration membranes. *Journal of Membrane Science*, **277**, 18–27.

Scharnagl, N. & Schossig, M. (2005) Charakterisierung von Membranen. *Chemie Ingenieur Technik*, **77**, 517–526.

Schier, G. & Paar, S. (2007) Kaseinstandardisierung mit verschiedenen Membranen. *Technik und Wissenschaft*, **128**, 22–25.

Strathmann, H. (2010) Electrodialysis, a mature technology with a multitude of new applications. *Desalination*, **264**, 268–288.

Tang, C.Y., Kwon, Y.-N. & Leckie, J.O. (2006) Characterization of humic acid fouled reverse osmosis and nanofiltration membranes by transmission electron microscopy and streaming potential measurements. *Environmental Science & Technology*, **41**, 942–949.

Tansel, B., Bao, W.Y. & Tansel, I.N. (2000) Characterization of fouling kinetics in ultrafiltration systems by resistances in series model. *Desalination*, **129**, 7–14.

Tarnawski, V.R. & Jelen, P. (1986) Estimation of compaction and fouling effects during membrane processing of cottage cheese whey. *Journal of Food Engineering*, **5**, 75–90.

Tiraferri, A., Yip, N.Y., Phillip, W.A., Schiffman, J.D. & Elimelech, M. (2011) Relating performance of thin-film composite forward osmosis membranes to support layer formation and structure. *Journal of Membrane Science*, **367**, 340–352.

Tolkach, A. & Kulozik, U. (2006) Transport of whey proteins through 0.1 mm ceramic membrane: phenomena, modelling and consequences for concentration or diafiltration processes. *Desalination*, **199**, 340–341.

Vadi, P.K. & Rizvi, S.S.H. (2001) Experimental evaluation of a uniform transmembrane pressure cross-flow microfiltration unit for the concentration of micellar casein from skim milk. *Journal of Membrane Science*, **189**, 69–82.

Van der Bruggen, B., Schaep, J., Wilms, D. & Vandecasteele, C. (1999) Influence of molecular size, polarity and charge on the retention of organic molecules by nanofiltration. *Journal of Membrane Science*, **156**, 29–41.

Vladisavljević, G.T. & Rajković, M.B. (1999) The effect of concentration dependent viscosity on permeate flux limitation in ultrafiltration. *Facta Universitatis*, **2**, 9–19.

Wang, K.Y., Teoh, M.M., Nugroho, A. & Chung, T.-S. (2011) Integrated forward osmosis-membrane distillation (FO-MD) hybrid system for the concentration of protein solutions. *Chemical Engineering Science*, **66**, 2421–2430.

Wang, L. & Song, L. (1999) Flux decline in cross-flow microfiltration and ultrafiltration: experimental verification of fouling dynamics. *Journal of Membrane Science*, **160**, 41–50.

Youravong, W., Grandison, A.S. & Lewis, M.J. (2002) Effect of hydrodynamic and physicochemical changes on critical flux of milk protein suspensions. *Journal of Dairy Research*, **69**, 443–455.

Zheng, F., Li, C., Yuan, Q. & Vriesekoop, F. (2008) Influence of molecular shape on the retention of small molecules by solvent resistant nanofiltration (SRNF) membranes: A suitable molecular size parameter. *Journal of Membrane Science*, **318**, 114–122.

Zondervan, E. & Roffel, B. (2007) Evaluation of different cleaning agents used for cleaning ultra filtration membranes fouled by surface water. *Journal of Membrane Science*, **304**, 40–49.

3 Commercial Membrane Technology

K. Smith

3.1 Introduction: polymers used in membrane manufacture

All materials used for separations share the property of restricting passage of certain materials in a specific manner. Although a wide variety of polymers can be manufactured into membranes for separation processes, very few of these polymers satisfy enough of the required characteristics for membrane applications to be commercially feasible. Some polymers, such as cellulose acetate, are extremely popular on the basis of high performance in several categories even though they possess severe limitations in other areas.

3.1.1 *Cellulose acetate*

The initial successes in developing semi-permeable membranes by Loeb and Sourirajan involved modifying the cellulose acetate polymer to produce anisotropic membranes possessing high water fluxes along with high salt rejections. As such, cellulose acetate has been the most studied of the membrane polymers. Cellulose acetate polymers (Fig. 3.1) are made of cellulose units with various degrees of acetylation, and are composed of β-1,4 linked glucose units. Each glucose molecule has three free hydroxyl groups that can be replaced with acetyl groups. A completely substituted cellulose molecule has three acetate moieties for a degree of substitution equal to three; however, this compound is insoluble in acetone, thereby causing problems during membrane manufacture. Generally, cellulose acetate polymers used for membrane separation processes have an average of 2.4–2.5 of the hydroxyl groups replaced, and these acetate molecules form rigid, linear polymers of varying length. Optimum chain length for membranes is estimated at 100–300 units with a molecular weight of 25 000–800 000. Cellulose acetate is a hydrophilic polymer.

The chief advantages of cellulose acetate are high water permeability, high salt rejection, good film-forming ability and high mechanical strength. Also, the basic raw material, cotton, is inexpensive. The use of cellulose acetate is limited by its disadvantages, which include instability to extremes of pH and high temperatures, a tendency to be hydrolysed by some organic compounds and micro-organisms, incompatibility with chlorine and a tendency towards compaction.

The intolerance of cellulose acetate to pH extremes is a severe limitation. A pH range of 4–7, generally, is considered acceptable and a pH of 2.5–8 possible for short exposures. Membrane life is shortened by lengthy exposure to either pH extreme, and the effect of

Membrane Processing: Dairy and Beverage Applications, First Edition. A. Y. Tamime.

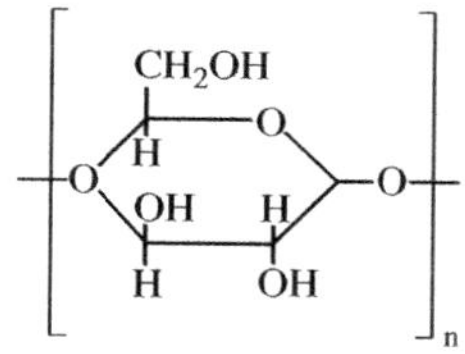

Fig. 3.1 Cellulose acetate polymer.

pH on membrane life also is dependent on temperature. Acidic or alkaline conditions cause hydrolysis of the polymer at the ester linkages leading to changes in membrane performance, such as increased flux and decreased salt rejection for reverse osmosis (RO) membranes. Higher temperatures also increase hydrolysis. The limited alkaline pH range restricts the products that can be handled by cellulose acetate membranes and the types of cleaning solutions used.

Hydrolysis of the cellulose acetate units by micro-organisms through depolymerisation or deacetylation also is possible. The effect, as with extremes of pH, is to alter membrane performance and, consequently, shorten membrane life. Chlorine used as a sanitiser will control micro-organism growth; however, chlorine can reduce membrane life. Cellulose acetate membranes typically can tolerate no more than 1 mg L^{-1} available chlorine.

Difficulties with shortened membrane life from compaction and processing at higher temperatures are less clear. Higher temperatures will accelerate damage caused by certain conditions, such as pH extremes, and use of chlorine. Cellulose acetate membranes typically have an upper temperature limit of 40°C. The low glass transition temperature of cellulose acetate is responsible for the low limits on operating temperature. This restriction limits the temperature of cleaning solutions and requires cooling of some feed materials that could be processed more efficiently at higher temperatures.

Compaction or creep is a thickening of the thin upper membrane surface of an anisotropic membrane caused by the separating surface compressing into the support material beneath. Compaction causes decreased salt rejection and water flux especially for higher flux membranes and membranes used at higher temperatures. Factors, such as type of polymer, manufacturing method, and operating pressure and temperature, are thought to influence compaction rate. Because flux and salt rejection decline logarithmically, change is greatest initially and less rapid with continued use. Use of cross-linking agents and fillers reduces compaction issues with cellulose acetate membranes.

A similar phenomenon results when cellulose acetate membranes are allowed to dry. Drying causes irreversible compaction or collapse of the fine spongy structure, resulting in changes in the degree of matrix crystallinity and loss of permeability. Organic solvents also can cause a similar collapse of the membrane.

Despite these difficulties, cellulose acetate has proven very useful, especially in the area of RO. The combination of high water flux and salt rejection provided by cellulose acetate has been difficult to match with other polymers.

3.1.2 *Polysulphone/polyethersulphone*

Diphenylene sulphone units $[(C_6H_5)_2SO_2]_n$ make up the polysulphone polymer (Fig. 3.2), and result in a rigid polymer with good strength and compaction resistance. Also,

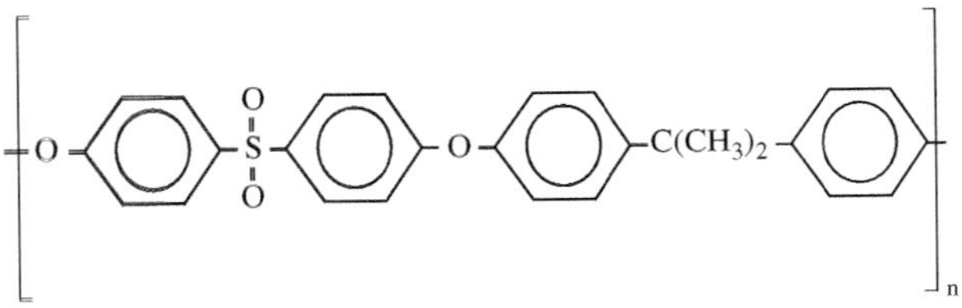

Fig. 3.2 Polysulphone polymer.

diphenylene sulphone is very stable at high temperatures and oxidative conditions. The significant advantages of this polymer compared with other membrane polymers account for the widespread gains in popularity of polysulphone/polyethersulphone (PS/PES). Advantages include a wide pH and temperature tolerance, increased resistance to chlorine compared with cellulose acetate and ease of manufacture into various configurations with a wide range of pore sizes. Although many pore sizes are feasible, it has not been possible to commercially produce PS membranes in the RO range. Another drawback is the lower mechanical strength of PS, limiting it to lower operating pressures.

PS/PSE can withstand temperatures at 120°C and a pH range from 1 to 14 (room temperature). PS membranes can be stored in 50 mg L^{-1} available chlorine, and exposed to 200 mg L^{-1} available chlorine for very short time periods. The temperature, pH and chemical tolerances of PS permit the use of a wider range of cleaning and sanitiser products. PS membranes must be stored wet or manufactured with a wetting agent like Triton-X, sodium dodecyl sulphate or glycerine. The structure of membranes made with PS does not collapse upon drying as occurs with cellulose acetate membranes, but rather a large amount of pressure is required to initiate permeate production or 'break through' after the membrane has dried.

The structure of PS membranes directly below the separating skin is very different from cellulose acetate membranes. In cellulose acetate membranes, the layer beneath the separating skin tends to have a sponge-like structure, whereas PS membranes have finger-like projections of voids. If these voids project into the upper skin, the separating layer can be disrupted or weakened. Pressure also can cause voids to collapse thereby decreasing flux. However, because fingers or voids are less resistant to the flow of permeable constituents, PS membranes will have a greater flux than cellulose acetate membranes with similar porosity but no voids.

3.1.3 *Polyamide*

The term polyamide (PA) applies to a group of polymer-forming compounds each having an amide bond in their structure (Fig. 3.3). PA compounds used for membranes include polybenzimidazole, polybenzamide, polybenzhydrazide, polyurethane and nylon. PA membranes generally have a lower tolerance for chlorine than cellulose acetate. A limit of <0.25 mg L^{-1} available chlorine at pH >8 or 0.1 mg L^{-1} at pH <8 is given for PA versus 1 mg L^{-1} available chlorine for cellulose acetate. A pH range from 4 to 10 (continuous operation) is typical for PA membranes versus 4 to 7 for cellulose acetate. Although

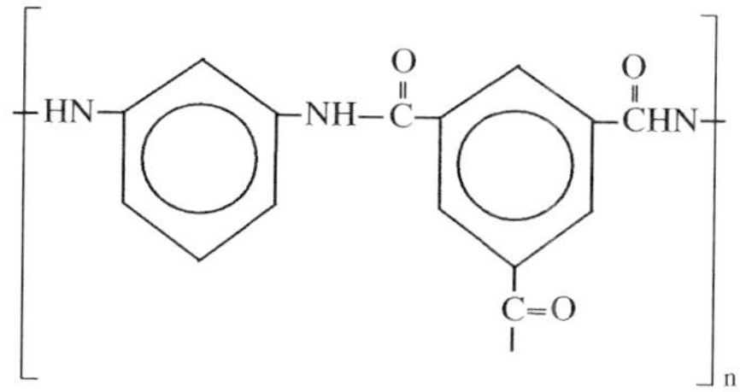

Fig. 3.3 Polyamide polymer.

hydrolysis of amide linkages is possible at extremes of pH, operation and cleaning in a more alkaline range is possible for PA compared with cellulose acetate. Furthermore, PA membranes withstand higher temperatures (~50°C), experience less compaction than cellulose acetate and, unlike cellulose acetate, polyamide RO membranes may be dried and rehydrated without affecting rejection properties. This last property is thought to result from the PA polymer having larger pores under the thin skin, which do not easily collapse when dried and rehydrated. These polymers also have good mechanical strength and resistance to oxidants and, in general, PA membranes are used for RO and nanofiltration (NF).

3.1.4 *Polyvinylidene fluoride*

Polyvinylidene fluoride (PVDF) is in the fluoropolymer family (Fig. 3.4), and the polymer has excellent resistance to hydrocarbons and oxidising compounds, such as chlorine (200 mg L^{-1} available chlorine at pH 10). PVDF membranes can tolerate a pH range 3–10 continuously or pH 2–11.5 during cleaning. They have a maximum operating temperature of ~50°C, and some manufacturers consider PVDF to be a more difficult compound to use for membrane manufacture. Membranes of PVDF are available in the ultrafiltration (UF) and microfiltration (MF) pore ranges.

3.1.5 *Thin-film composites*

A thin-film composite (TFC) membrane consists of a thin layer of one polymer deposited over a thicker layer of another polymer. The resulting membrane is anisotropic, with the thin upper layer separating molecules by size and the thicker bottom layer serving as a support. An intermediate layer may be used to prevent the porous backing material, or lower layer, from penetrating the thin separating layer during casting or manufacture.

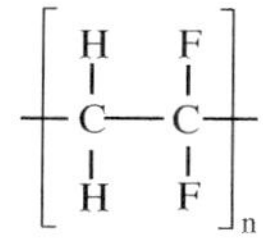

Fig. 3.4 Polyvinylidene fluoride polymer.

For maximum flux, a very thin separating layer over a very porous support layer is desired; however, if in an effort to increase flux, the pores in the bottom layer become too large the separating layer may collapse.

A TFC membrane takes advantage of the desired properties of a polymer while minimising its shortcomings. PA and PVDF membranes can be used for the thin separating layer, with PS/PES membranes generally used for the support material. The combination of PA for separating with PS/PES for support results in a membrane more tolerant to extremes of temperature and pH and more resistant to compaction than cellulose acetate. When polyamide is used, however, the TFC has less chlorine tolerance than cellulose acetate.

Advantages of a TFC membrane composed of PA and PS/PES include: (a) resistance to compaction; (b) ability to dry the membrane; (c) good chemical and thermal stability; (d) excellent salt rejection (for RO); (e) high flux and (f) tolerance to a pH range from 2 to 11. TFC membranes typically can tolerate temperatures to 45°C. Problems include adverse affects with pH values greater than 12 for polyester support material used for some TFCs and intolerance to chlorine. Additionally, the membrane can lack mechanical strength when cast in layers compared with a membrane cast as one unit, and casting in layers is more expensive than casting as a single unit.

3.2 Other materials used for membranes

The previously discussed materials are considered organic in nature. It also is possible to use inorganic materials to produce membranes. Ceramics, metals (stainless steel) and carbon currently are used. The physical form of inorganic materials generally determines separating ability rather than chemical composition as with polymers, and some examples follow.

3.2.1 *Ceramic membranes*

Ceramic membranes, also known as mineral or inorganic membranes, are formed by combining metals, such as aluminium, titanium or zirconium, that are in the form of oxides, nitrides or carbides with support materials. The resulting membranes are typically in the MF and UF range, although some manufacturers have elements that are considered to be in the NF range. Ceramic membranes can be manufactured using sintered spheres made of aluminium, titanium or silica oxides. Asymmetric ceramic membranes are produced by layering spheres of differing diameters onto the inner surfaces of support materials. Support materials in the form of elements or blocks often are composed of aluminium or titanium oxides; however, many other materials including stainless steel, graphite, sintered glass, silicon carbide and porous carbon have been used. The inner surfaces are coated with one or more thin layers of microporous ceramic, and the layers are bonded by sintering to form a monolithic structure between layers and block.

In general, membranes for commercial use are composite in design and the membrane layer that is capable of separating molecules is ~2–1 μm thick. As the thickness of the separating layer increases, the membrane has more mechanical strength; however, flux decreases. Middle layers of the membrane generally are 10–20 μm thick. The porous

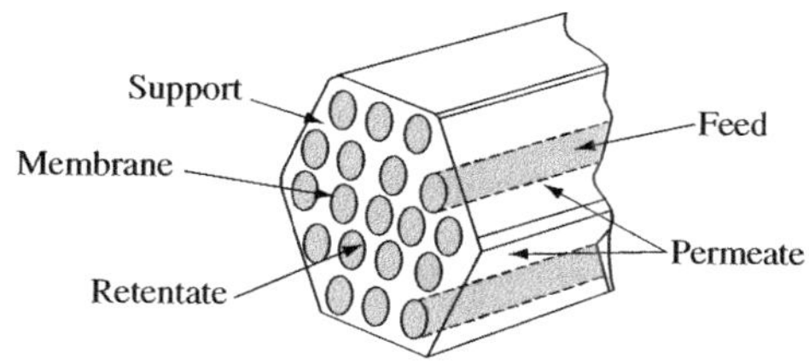

Fig. 3.5 Multichannel ceramic element.

support is the final layer, and has a thickness of ~2 mm with pore diameters of 10 μm. These membranes often have a porosity of 30–60% by volume, which is a compromise between flux and strength. Porosity is the volume of the pores divided by the total volume of the membrane.

The precise design of ceramic elements depends on the manufacturer. Elements often are hexagonal or cylindrical tubes with up to 19 channels going the length of the tube. The diameter of the channels ranges from 2 to 6 mm depending on the manufacturer, and the stainless steel housing commonly may contain 1–19 elements. Disc and plate configurations are available, but they are limited to certain medical and waste water applications. Elements containing several channels are referred to as multichannel (Fig. 3.5). The multichannel design provides increased packing density and mechanical strength in addition to lower maintenance and installation costs. Narrow channels have a lower circulation volume, and require lower recirculation flow and less energy to operate than wider channels.

The support materials used for inorganic membranes can tolerate a wide range of temperatures and pH values, and have good resistance to strong oxidising compounds. This ability, along with the high tolerance of the membrane layer, allows ceramic membranes to be used under more severe processing and cleaning conditions than generally are possible with traditional membrane polymers. Many ceramic membranes can tolerate pH values ranging from 0 to 14, pressures of approximately 3.1 MPa and transmembrane pressure >1.7 MPa. The gasket material generally is the limiting factor rather than the element. Autoclaving and cleaning with steam also may be possible. Additional advantages for ceramic membranes include abrasion resistance, high water flux and ability to back-flush and store the element dry.

Several problems, including high capital costs, have limited the widespread acceptance of ceramic membranes and, in general, they cost three to six times more than traditional polymer membranes. Ceramic membranes also are brittle, relatively sensitive to temperature changes, and are intolerant to phosphoric acid.

3.2.2 *Metallic membranes*

Metallic membranes may be produced using non-woven fibres or sintered stainless steel spheres. The non-woven fibre method has problems maintaining a very narrow pore size distribution and obtaining pore sizes in the micron range; therefore, metallic membranes generally are produced using stainless steel spheres. A sintering process using high pressure forces powdered type 316 stainless steel into a porous tube. A food grade metallic

Fig. 3.6 Metallic membrane.

oxide, such as titanium dioxide, is then circulated through the tube to form a permanently bonded membrane on the inside surface. A small section of a metallic membrane is shown in Fig. 3.6.

Several stainless steel membranes typically are welded into a housing to form a module. Because the elements are welded into place, gaskets are not required. The single piece construction, therefore, has excellent mechanical stability, can tolerate temperatures up to 177–371°C and operating pressures of 1.0–27 MPa, depending on the manufacturer. The stainless steel can tolerate chemical sanitisers and steam in addition to a pH range from 0 to 14.

Metallic membranes can be manufactured in the UF and MF range (0.1–1.0 μm). Membrane tube diameters range from 6 to 25 mm (inner diameter) and modules may be up to 6 m in length. Because metallic membranes are welded into a single unit that cannot be disassembled for visual inspection, they often are not considered suitable for food applications by regulatory groups.

3.3 Membrane configuration

The four general categories of membrane configuration or element design are spiral-wound, tubular, plate and frame and hollow fibre. As with membrane composition, each of these configurations has advantages and disadvantages, and, therefore, some systems are more suitable for some applications than others. Desired design features generally include maximised surface area with a minimised volume, adequate support for the membrane, sanitary design, ability to clean-in-place (CIP) and ease of accessibility for membrane replacement.

3.3.1 *Spiral-wound*

Spiral-wound membranes consist of sheets of materials assembled into layers and rolled around a central permeate tube to form a membrane element (Fig. 3.7). The retentate spacer is a plastic material woven in a diamond or rectangular pattern, which provides the channel for the feed or retentate stream (Fig. 3.8). On either side of the retentate spacer, the membrane surface is followed by backing or support material. Between the backing material from the two sheets of membrane is the permeate mesh, which provides the path for permeate flow into the permeate tube in the centre of the coiled membrane assembly. A cross-section of a spiral-wound membrane showing the path of permeate within a membrane is shown in Fig. 3.9. The feed stream enters the feed channel, and

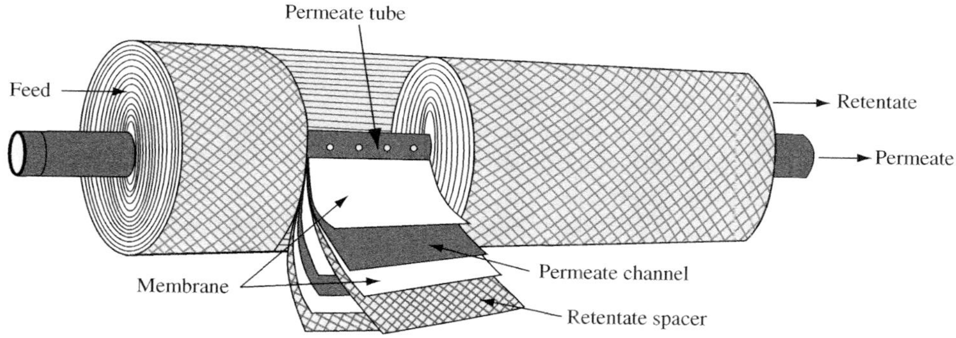

Fig. 3.7 Spiral-wound membrane construction.

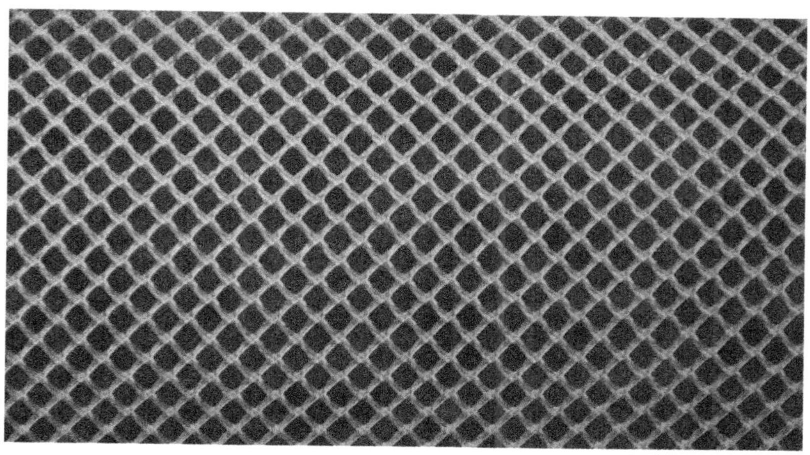

Fig. 3.8 Retentate spacer material.

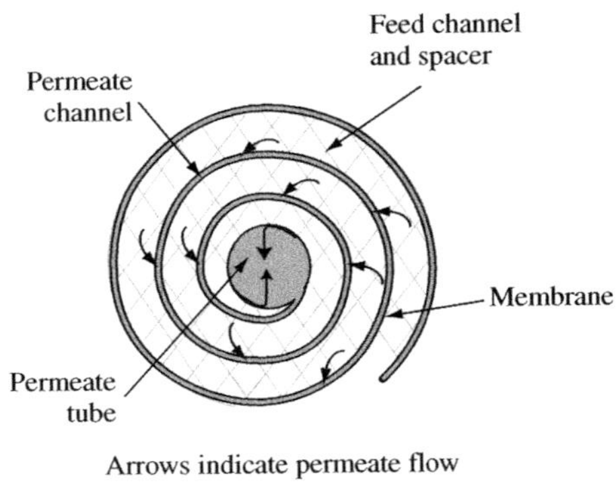

Fig. 3.9 Cross-section of a spiral-wound membrane.

some of the water and solutes that are able to cross into the permeate stream will cross the membrane and enter the permeate channel. The permeate channel then spirals into the permeate tube that collects all of the permeate from a given membrane.

The sheets of membrane vary in size depending on element length and diameter and the thickness of the retentate spacer used. A membrane is sealed on both front and back edges with the permeate mesh in between. The membrane is then wrapped and sealed around the permeate tube, and the two ends glued together to prevent product from flowing into the filtered permeate. Gluing the edges seals the permeate section from the retentate stream; however, this reduces the actual membrane area, and needs to be considered in any calculation of effective membrane area.

The thickness, shape and size of the openings of the retentate spacer vary depending on the application. Spacers typically range from 0.75 to 1.55 mm in thickness and, because the diameter of the membrane is limited by the housing, the use of a thicker retentate mesh effectively reduces the amount of membrane that can be included in the spiral-wound element. The thicker spacers are used for feed streams containing larger particles or having greater viscosity where the disadvantage of reduced membrane area is outweighed by the greater turbulence the spacer creates, which results in increased permeate flux.

Spacers typically have either a diamond or rectangular pattern in various sizes of pattern. A diamond pattern is designed to increase turbulence, which helps to sweep deposits from the membrane surface during operation and, thereby, improve flux. The rectangular pattern is more typically used for products with suspended particles, and provides less turbulence to the feed stream. The size of the diamond or rectangular pattern typically varies from 30 to 80 mm and is referred to as mils.

A spiral-wound element has several additional layers of retentate spacer wound around the outside of the assembled element. Upon insertion into the housing unit, the element unwinds slightly to fill the housing, thus providing rigid support for the membrane. The amount of additional retentate spacer on the outside of the element is important for ensuring proper feed stream/retentate flow.

Several elements may be placed in series in a housing with the permeate tubes joined together by interconnectors. Anti-telescoping devices made of plastic or stainless steel are used to keep the face of elements on the discharge side from telescoping from applied pressure (Fig. 3.10).

Fig. 3.10 Anti-telescoping devices for a spiral-wound membrane.

The advantages of spiral-wound elements include high membrane packing density and low retention volumes. Also, large amounts of membrane can be replaced easily, and parasitic losses from pressure drops are low and spiral-wound membranes can tolerate high pressure. Capital costs for spiral-wound systems are relatively low as is energy consumption for operations. However, a disadvantage of the spiral-wound configuration is the requirement for a relatively clean feed material since large particles can clog the retentate spacer. The ability to control flow within the element also can be limited, and large pressure differences across the membrane are not tolerated. Spiral-wound membranes can be difficult to clean when badly fouled.

3.3.2 *Tubular*

Tubular elements have a tube in shell design with each shell or housing containing from 1 to 37 tubes depending on system design (Fig. 3.11). Individual tubes may range in length from 1.2 to 5 m with a diameter of 5–25 mm. In general, the membrane surface is on the inside surface of the tube (Fig. 3.12), and the feed and retentate streams flow through the inside of the tubes with the permeate collected in the outer shell. End caps can connect the tubes in parallel or series, and assembled units typically are 3 m in length with a diameter of 76 mm.

The membrane may be cast directly onto a porous support tube or manufactured separately and manually placed inside the porous support. The supports are made of several types of material, including perforated stainless steel, and may be self-supporting or require additional external supports.

Ceramic elements are considered to be a tubular design. The channels typically have a smaller diameter than tubular designs made of polymeric membranes and, although not

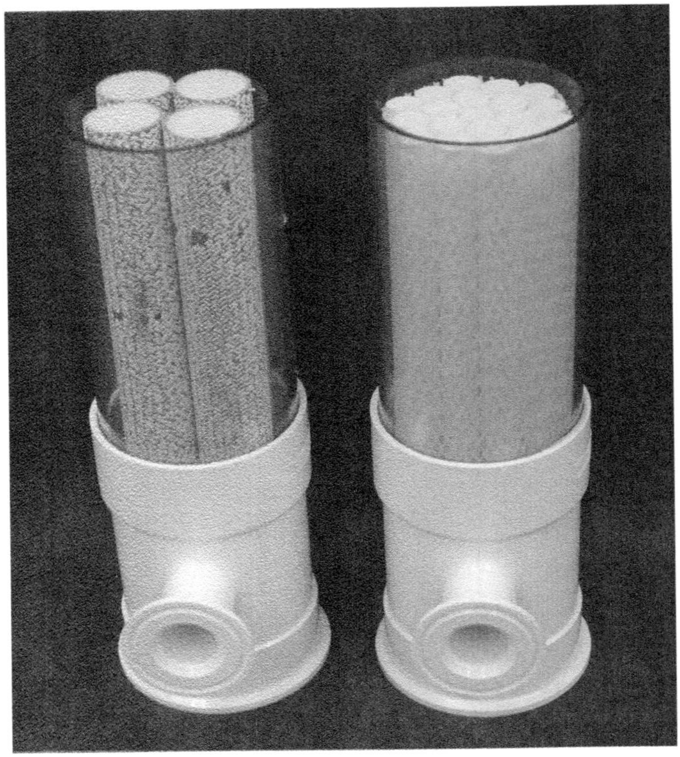

Fig. 3.11 Tubular elements.

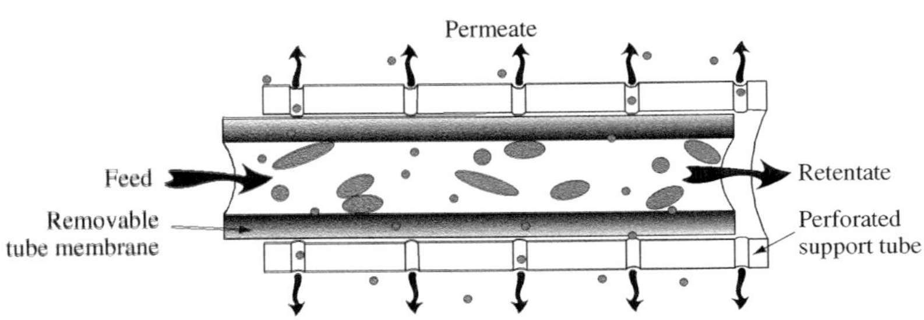

Fig. 3.12 Flow in a tubular element.

permitted for food processing, carbon and stainless steel elements also have a tubular design.

The tubular design has several advantages. The large tube diameter permits processing of fluids containing larger particle sizes and having higher viscosities than other systems allow. Good control of feed stream flow also is possible, and it is claimed that these systems are readily cleaned because of the more open design of the tubes. Nevertheless, several important disadvantages have limited the use of tubular systems. The design has the lowest membrane area to volume ratio of the four systems and, therefore, tubular designs require the most space for a given membrane surface area by comparison. The low packing density also results in a high retention volume, which makes processing small volumes of product impractical. The recommendation for turbulent flow along with substantial pressure drops through the system makes tubular designs highest in energy use.

3.3.3 *Hollow fibre*

The difference between hollow-fibre and tubular designs is the diameter of the membrane element. Generally, hollow fibres have inside diameters of 0.04–3 mm compared with tubular elements with inside diameters of 5–25 mm. Hollow fibres with smaller diameters are used in applications requiring higher operating pressures, whereas larger diameter fibres are preferred for lower pressure situations, and they may sometimes be referred to as capillary fibres rather than hollow fibres. The separating surface often is on the inside wall of the hollow fibre, and the feed stream is circulated through the inside of the hollow fibre with the permeate collected in the outer shroud (Fig. 3.13). The flow pattern may be reversed with feed circulated in the outer housing, and the permeate is collected inside the hollow fibres for some applications.

Because many fibres are packed into a housing, hollow fibres are self-supporting, and no additional support material other than the housing is used. Depending on the manufacturer, each housing may contain 50 to thousands of hollow fibres, which are aligned with the ends of the fibre bundle sealed in resin to separate permeate and retentate streams (Fig. 3.14). The length of fibre bundles depends on the application, where the longer bundles are used for dilute solutions and the shorter bundles are used for more viscous and higher solids materials. Generally, each bundle or cartridge is placed in parallel since the pressure drops are too severe when used in series, and commercial cartridges typically are 137 cm in length with a diameter of 15–30 cm.

The main advantages of hollow fibres are an extremely high packing density and low retention volume. Because the hollow fibres are self-supporting, capital costs are reduced, and it is possible to back-flush the system. Hollow-fibre systems also have a relatively low power consumption compared with other membrane designs and, therefore, are very economical to operate.

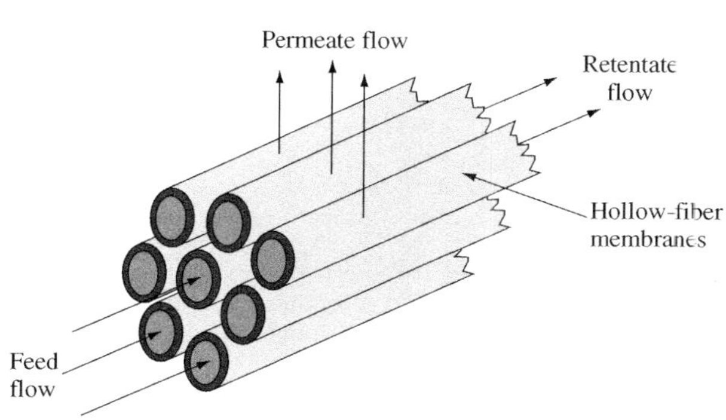

Fig. 3.13 Flow in a hollow-fibre element.

Fig. 3.14 Hollow-fibre element.

Applications for hollow-fibre configurations are limited by the strength of the fibres. Hollow fibres cannot tolerate large pressure changes within the system as can occur when long cartridges are used for processing very viscous materials. Too high a pressure or feed rate also can cause fibres to burst, requiring replacement of the entire cartridge, although kits are available to seal individual fibres. Replacement costs are high since a large amount of membrane is in each cartridge. Because of the small diameter of the hollow fibres, feed streams must be filtered to prevent plugging of fibres and cleaning problems have been reported by some researchers.

3.3.4 *Plate and frame*

In a plate and frame system, membranes are manufactured into sheets and placed between plates which form channels for permeate and retentate streams (Fig. 3.15). There are many ways to arrange system hardware and flow through the system. Generally, a support plate is sandwiched between two sheets of membrane with the separating surface of the membrane facing away from the plate. The support plate provides the channel for the permeate

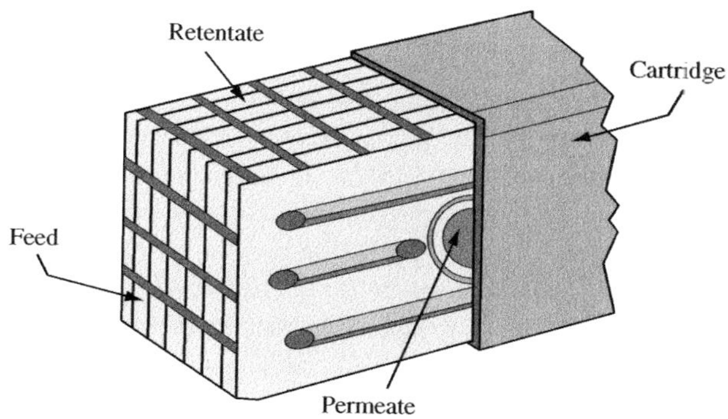

Fig. 3.15 Flow in a plate and frame element.

stream, and on either side of the packet are spacers followed by another membrane-support plate-membrane packet. The permeate stream from each support plate may be drawn off separately, or permeate from several plates may be drawn off together. The depth of the retentate feed stream/retentate channel depth can range from 0.5 to 1.5 mm depending on the product processed. Membrane sheets are rectangular, round or elliptical depending on the manufacturer. Originally, units were designed to operate vertically with feed flowing from bottom to top. The horizontal design currently is used in the food industry.

An advantage of the plate and frame system is the ease of replacing or removing individual membranes. Clear permeate tubes coming from individual support plates allow determination of which membrane has failed, and each membrane can be removed for hand cleaning. Products with higher viscosity also may be processed by plate and frame systems.

The plate and frame design is intermediate in several operating characteristics. Although packing density is similar to tubular designs, plate and frame systems have a lower hold-up volume by comparison, and energy consumption for plate and frame operation is between spiral-wound and tubular systems. However, a large drawback to the plate and frame design is the amount of labour and hardware required for operation. Although individual membranes may be replaced or cleaned, doing so is labour intensive since a large number of seals and connections are involved in addition to extensive plumbing. Some operators feel these systems are not easily cleaned as other membrane designs.

3.4 Modes of operation

Several operating modes for separation systems are possible depending on the desired end product and system considerations, such as amount of membrane available, feed volume, concentration factor required, desired product residence time and ancillary equipment available. Diafiltration (DF) is a special operating method used to achieve higher percentage concentrations of a retained component than is possible with standard UF or MF. Separation systems can operate in either a batch or continuous mode, with the continuous mode divided further into single-pass, recirculation or multistage designs.

3.4.1 *Diafiltration*

DF is used to obtain a higher percentage concentration of a retained component, often protein, than standard UF or MF can achieve. Discontinuous and continuous modes of DF are possible and, in both cases, water is added to the retentate, so with further UF/MF additional permeate species are removed from the retentate. Ultimately, the final retentate has a higher concentration of the desired retained component than would be possible with conventional UF/MF. The limitation of standard UF/MF in achieving higher concentrations of a retained component are due to the increase in viscosity of the retentate stream as material is removed. At some point the feed stream is too viscous to properly flow through the feed channel. The addition of water reduces the viscosity and osmotic pressure of the feed stream thereby permitting more of the permeability material to cross the membrane into the permeate. Depending on the degree of purity desired, the amount of diafiltered water can be very significant compared with the original amount of feed material.

In continuous DF (Fig. 3.16) water is added at the same rate permeate is withdrawn so original feed volume is maintained. Water addition continues until the desired component purity is achieved. In contrast, during discontinuous DF (Fig. 3.17), the feed stream or retentate is ultrafiltered/microfiltered to a given volume. Water then is added to restore the retentate to the original feed volume, and the resulting solution is ultrafiltered/microfiltered again. The cycle of UF/MF, water addition and UF/MF is repeated until the desired component purity is achieved.

Discontinuous and continuous DF can achieve the same degree of purification. The continuous method has a higher average flux overall, but requires more water for the same degree of purification than discontinuous DF. When processing time and efficiency are important, concentrating the initial feed 3:1 by UF/MF followed by DF and additional

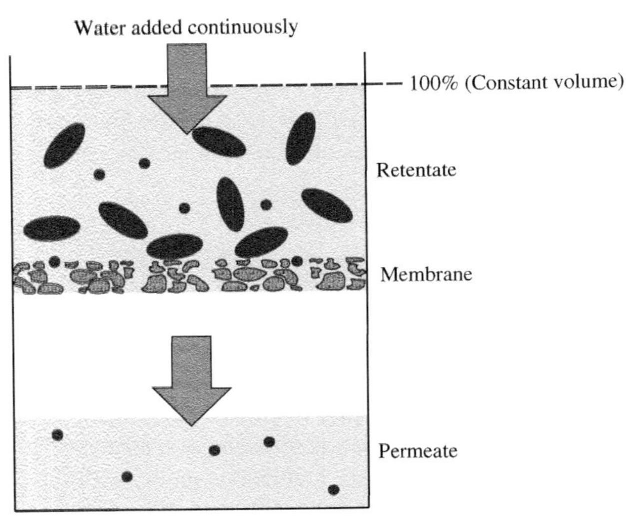

Fig. 3.16 Continuous diafiltration.

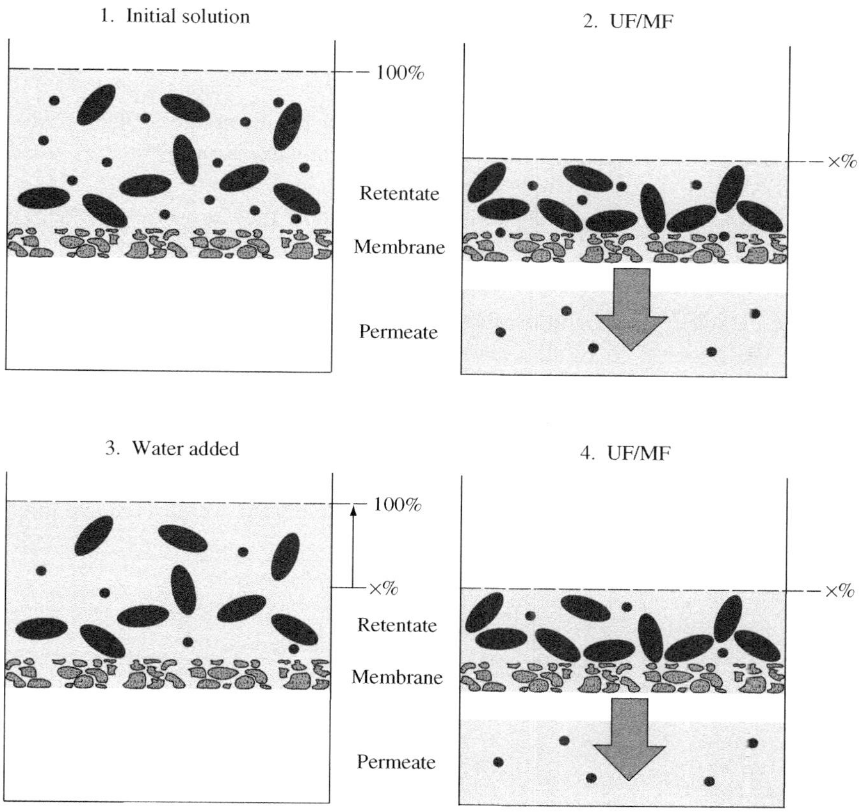

Fig. 3.17 Discontinuous diafiltration. UF/MF, ultrafiltration/microfiltration.

UF/MF to achieve the desired final product concentration is recommended. Flux during DF, generally, is higher than during the preceding UF/MF.

3.4.2 *Batch design*

Batch systems are characterised by the feed stream remaining within the system until all of the retentate is at the desired concentration or purity. Only the permeate stream is removed from the process. Thus, there are several types of batch operation designs. The differences in the batch operation designs involve the method of retentate recycling. Retentate may be completely recycled, partially recycled or the system may be operated in a dead-end design. In all cases the retentate stream is recycled until the desired final product concentration is achieved.

Advantages of the batch design compared with continuous systems include

- lower capital investment
- less complex valve system
- increased yields

- higher average flux rates
- Simplicity of design and flexibility are additional advantages

Processing material has a longer residence time in a batch operation than continuous systems, and, therefore, increased growth of micro-organisms is possible with the batch designs.

There are distinct batch design operating systems, and they are briefly described as follows.

- *Completely recycled* – The entire retentate stream is diverted back to the feed tank containing the material to be processed in a completely recycled design (Fig. 3.18) and, off the three batch methods, the completely recycled design requires the least membrane area and processing time for a given total solids concentration.
- *Partially recycled* – The partially recycled design (Fig. 3.19) diverts a portion of the retentate stream back to the feed tank with the remainder fed directly to the membranes. The portion of the retentate stream returning to the membrane forms a recycle loop. The

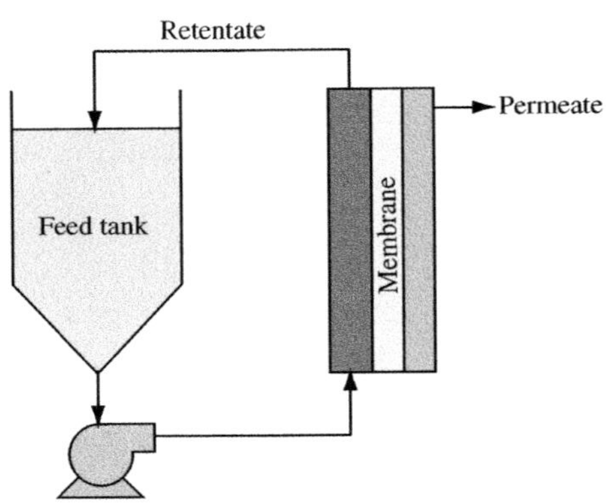

Fig. 3.18 Completely recycled batch design.

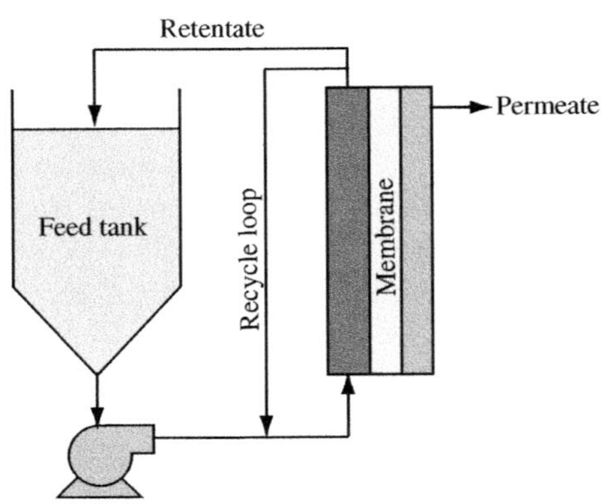

Fig. 3.19 Partially recycled batch design.

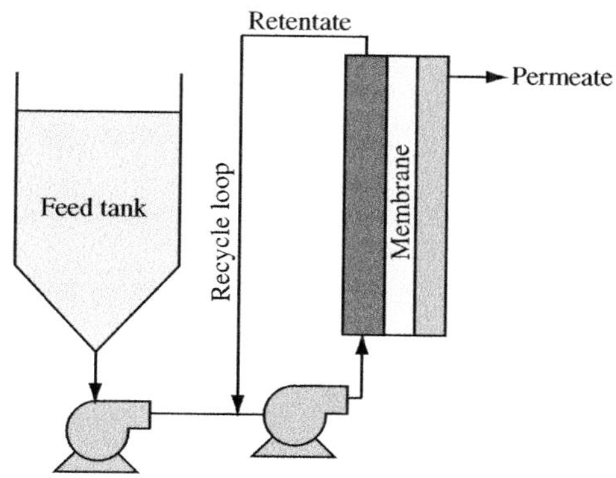

Fig. 3.20 Dead-end batch design.

partially recycled design can continuously process a feed stream, and requires smaller feed tanks than the completely recycled process.

- *Dead-end* – The batch design (Fig. 3.20) diverts the entire retentate stream to the membranes, i.e. there is no return loop to the feed tank for the retentate stream. The dead-end method is used for start-up of a continuous recirculation design or to deliberately induce fouling.

3.4.3 *Continuous design*

Continuous designs have a portion of the retentate continually removed from the process compared with batch designs where all of the retentate remains in the system until completion of processing. As with batch processes, permeate is withdrawn continually in continuous designs. There are several general categories of continuous designs including single pass, recirculated (feed and bleed) and multistage. Continuous designs are more suited for commercial operations where large volumes of material are processed. An important microbiological safety consideration is the decreased residence time for material in a continuous operation compared with a batch design. The designs are briefly described as follows.

- *Single-pass design* – In a single-pass continuous operation (Fig. 3.21), the feed stream passes through the system once with the resulting retentate having the desired component concentration when exiting the system. Because the retentate stream is not recycled, the retained particles have the shortest residence time of any processing mode discussed. The short residence time is advantageous for handling labile products or products where microbiological growth is a problem. Low energy consumption is an additional advantage for this design, whereas the disadvantages include: (a) the large amount of membrane area required to achieve the desired product concentration in one pass; and (b) the inability to handle products with high total solids or viscosity.
- *Recirculation (feed and bleed)* – A recirculation design or feed and bleed system (Fig. 3.22) is a continuous separation process resembling both dead-end and partially recycled batch operations. The start-up procedures for the recirculation method require operating the system in a dead-end batch mode until the retentate reaches the desired

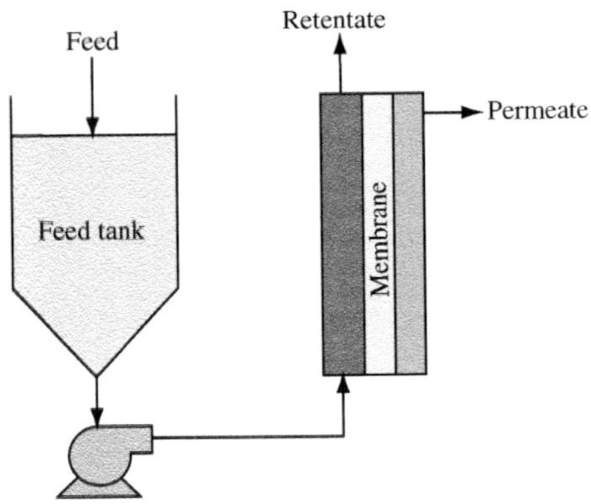

Fig. 3.21 Single-pass continuous design.

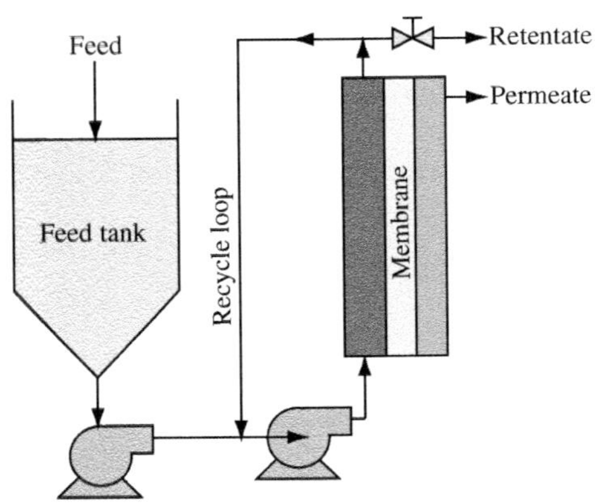

Fig. 3.22 Recirculation (feed and bleed) continuous design.

concentration level. The retentate is then withdrawn from the system at a rate equal to the rate that fresh feed is introduced. The feed and bleed method provides a continuous supply of retentate containing the desired level of total solids; however, because membranes are continually handling feed streams near the desired levels of total solids, flux for this design is lower than a comparable batch system and, therefore, additional square meters of membrane are required.

- *Multistage* – Because of limitations with batch and continuous recirculation methods, large membrane filtration operations generally are multistage systems. By varying the number of modules, stages and patterns of flow through the system a large variety of multistage designs are possible. In all cases, flux is similar to batch systems, while increasing the number of stages decreases product residence time. A design using several stages and continuous recirculation (Fig. 3.23) is similar to the previously discussed recirculation configuration; however, in this system, the recycled retentate is

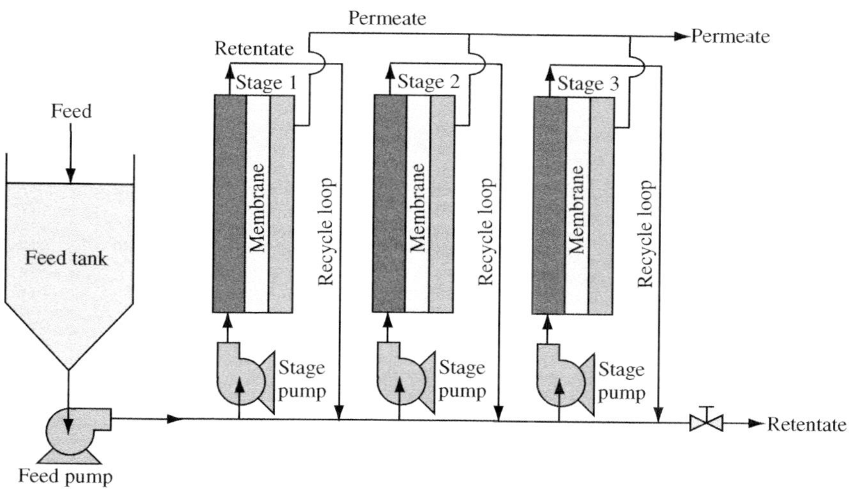

Fig. 3.23 Multistage continuous design.

at a concentration less than desired in all but the final stage. The multistage design has a higher operating flux overall and requires less membrane surface area than continuous recirculation with one stage. Also, the residence time and hold-up volume are less than with batch processing. Three stages are considered a minimum for feasible operation with 7–10 stages common.

3.5 Conclusion and future developments

The polymers used for membranes have remained largely unchanged in recent years as have membrane designs. The areas of greatest innovation have been in the understanding of the operation of MF systems. The ability to maintain the MF separating capability rather than having the membranes revert to a UF-type separation with fouling is a key to future applications of MF. In addition, progress will continue in modification of membrane polymers to improve resistance to pH, chemicals and temperature.

Suggested literature

Baker, R.W. (2000) *Membrane Technology and Applications*, McGraw-Hill, New York.

Cabasso, I. (1980) Practical aspects in the development of a polymer matrix for ultrafiltration. *Ultrafiltration Membranes and Applications* (eds. A.R. Cooper), pp. 57–78, Plenum Press, New York.

Cheryan, M. (1998) *Ultrafiltration and Microfiltration Handbook*, Technomic Publishing Co, Lancaster.

Friedlander, H.Z. & Litz, L.M. (1971) Membranes for pressure permeation. *Membrane Processes in Industry and Biomedicine* (ed. M. Bier), pp. 73–99, Plenum Press, New York.

Nielsen, W. (2000) *Membrane Filtration and Related Molecular Separation Technologies*, APV Systems, Silkeborg.

Paulson, D.J., Wilson, R.L. and Spatz, D.D. (1984) Crossflow membrane technology and its applications. *Food Technology*, **38**, 77–87.

Renner, E. & M.H. Abd El-Salam, M.H. (1991) *Application of Ultrafiltration in the Dairy Industry*, Elsevier Applied Science, New York.

Wagner, J. (2001) *Membrane Filtration Handbook – Practical Tips and Hints*, 2nd edn. Osmonics, Inc. Minnetonka.

Zeman, L.J. & A. L. Zydney, A.L. (1996) *Microfiltration and Ultrafiltration – Principles and Applications*, Marcel Dekker, Inc., New York.

4 Membrane Fouling, Cleaning and Disinfection

L.L.A. Koh, M. Ashokkumar and S.E. Kentish

4.1 Introduction

Fouling is a major problem in membrane operations as it directly affects productivity. The main effect of fouling is a sharp reduction in permeate flux that cannot be resolved while the process is running (Eykamp, 1995). Non-productive and costly cleaning cycles are required in order to restore the original flux (Cheryan, 1986). The rate and extent at which fouling occurs thus directly influences the economic feasibility of membrane technology (James *et al*., 2003).

In the past few decades, there has been a rapid rise in research to understand the effects of fouling. Various cleaning methods for cross-flow filtration have been developed in a bid to establish cost-effective cleaning methods for the restoration of membrane separation efficiency after fouling. This chapter reviews the cause and effects of fouling with an emphasis on reducing fouling effects with effective cleaning procedures; detergent recovery systems are also discussed.

4.2 Flux reduction

During membrane filtration, the flux through the membrane decreases over time. It has been found that the flux–time behaviour is governed by both concentration polarisation (CP) and fouling. These two phenomena bring about additional resistance to filtration and occur predominantly on the feed side of the membrane surface (Mulder, 1996). As both concentration polarisation and fouling have flux reduction effects, it is difficult to separate the individual contribution of these two effects. Fig. 4.1 shows the flux–time comparison for both CP and fouling. The rapid flux decline at early stages of filtration is due to CP whereas a gradual decrease in flux in the second stage signifies the build up of deposits on or in the membrane pores. At this stage, a cake layer is gradually formed. The flux then approaches a steady value as the effects of fouling are fully developed. Any further flux decline is due to cake compression and consolidation (Marshall & Daufin, 1995; Rice, 2008).

It is important to note that membrane compaction can also lead to a decrease in permeate flux. However, depending on the type of membrane used and the applied transmembrane pressure (TMP), membrane compaction does not usually show a significant impact. It has

Membrane Processing: Dairy and Beverage Applications, First Edition. A. Y. Tamime.

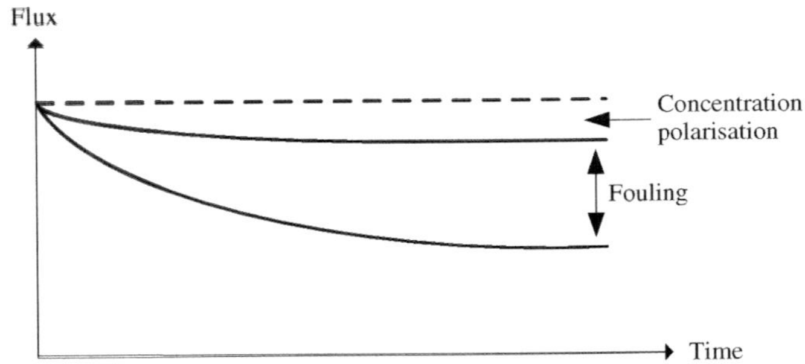

Fig. 4.1 Permeate flux behaviour with respect to time for both concentration polarisation and fouling (Adapted from Mulder, 1996. Reproduced with permission from Springer).

been found that fouling, not membrane compaction, is the main reason for a decrease in flux when a TMP of less than 0.3 MPa is applied on an ultrafiltration (UF) membrane (Renner & Abd El-Salam, 1991).

4.2.1 *Membrane resistance*

The convective flux through the membrane, in the absence of osmotic pressure effects, is given as the following equation:

$$J = \frac{\Delta P}{\mu_P R'_T} \tag{4.1}$$

where J is the permeate flux ($m^3\ m^{-2}\ s^{-1}$), ΔP is the pressure drop across the membrane (Pa), μ_P is the viscosity of the permeate (Pa s^{-1}) and R'_T is the total resistance of flow through the filtration unit (m^{-1}).

In an ideal case, the only resistance towards mass transfer is due to the membrane and this is known as membrane resistance, R'_M. However, as mentioned earlier, CP is prevalent in all filtration systems and the highly concentrated boundary layer at the membrane surface also exerts resistance to mass transfer, R'_{CP}. A gel layer may be formed when the concentration of retained solutes at the membrane surface is very high. This gel layer exerts a resistance R'_G. It is also possible for some solutes to block the pores of a porous membrane. This pore-blocking resistance is denoted as R'_b. In some cases, adsorption of solute to the membrane surface is also observed, providing a resistance R'_A. Therefore, the term R'_T takes into account the resistance due to the membrane, CP, gel layer, pore blockage and adsorption. In this section, both CP and fouling ($R'_G + R'_b + R'_A$) are explained in great detail.

$$R'_T = R'_M + R'_{CP} + R'_G + R'_b + R'_A \tag{4.2}$$

4.2.2 *Concentration polarisation*

When a pressure is applied, flux through a cross-flow membrane is characterised by two distinct flow regions – the pressure-controlled flow region and mass transfer-controlled

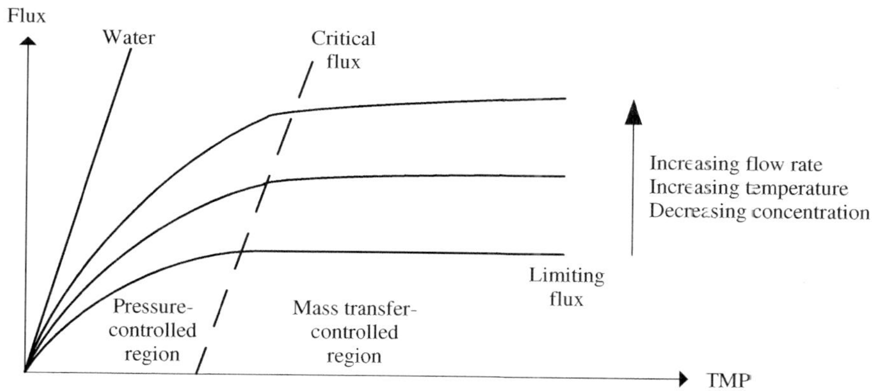

Fig. 4.2 Pressure and mass transfer controlled regions for cross-flow membrane filtration (Adapted from Cheryan, 1998. Reprinted with permission from Taylor and Fancis Group LLC).

region (Fig. 4.2). In both regions, the flux is below that of water due to an increase in viscosity. At low TMP, the flux increases linearly with TMP because the flux is governed by the rate of solute permeating through the membrane. Operating in this pressure-controlled flow region is desirable as there is reduced cleaning demand. However, a large membrane area is required to achieve high permeate flow rates due to low pressures and capacity.

As the pressure increases, concentration polarisation is brought about by a build-up in concentration of retained solutes near the membrane surface due to the rejection of these species. The concentration build-up generates a diffusive back-flow of solute to the bulk solution, in accordance to Fick's Law (Richardson *et al.*, 2002; Askew *et al.*, 2008). This is also known as back-diffusion. The boundary layer consisting of the accumulation of retained solute is called the CP boundary layer and it exerts a resistance towards mass transfer of solute from the bulk solution to the membrane surface (Mulder, 1996). This causes the gradient of the flux versus TMP curve to fall as operation moves into a mass transfer controlled region.

As the pressure increases further, the flux eventually stabilises as seen in the mass transfer controlled region. This is known as the limiting flux, which is independent of pressure and determined entirely by mass transfer considerations. This limiting flux is different from the critical flux, which indicates the onset of fouling and is reached before the limiting flux (Field *et al.*, 1995; Rice, 2008). Critical flux is dependent on particle size, hydrodynamics and membrane-colloid interactions (Howell, 1995). As shown in Fig. 4.2, operating conditions, such as flow rate, temperature and concentration, can be optimised to increase permeate flux in both flow regions.

In dead-end filtration, the solute concentration at the membrane surface and the thickness of the CP boundary layer increases throughout the run resulting in a continuous decline in permeate flux (Zeman & Zydney, 1996). However, in most cross-flow filtration processes, CP begins almost immediately at the start of a filtration run and is stable throughout filtration (Mulder, 1995; Makardij *et al.*, 1999). At steady state, the rate of back-diffusion is equal to the difference between the molar flux of convective transport of the solute towards the membrane and the rate of permeate flux, as shown in Equation 4.3 and Fig. 4.3

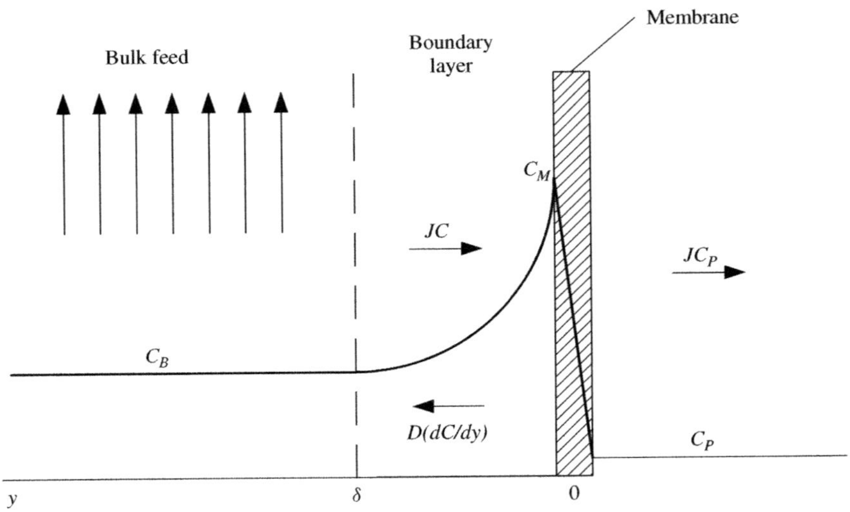

Fig. 4.3 Schematic representation of film model theory for concentration polarisation. *C* is the solute concentration in the bulk (*B*), at the membrane (*M*) and in the permeate (*P*) while *J* is the volumetric flux and *D* is the diffusion coefficient (Adapted from Mulder, 1996. Reproduced with permission from Springer).

(Mulder, 1996). This description of steady state mass transfer in the CP boundary layer is known as the film model theory.

$$JC - JC_P = -\mathcal{D}\,\frac{dC}{dy} \tag{4.3}$$

where J is the membrane flux ($m^3\ m^{-2}\ s^{-1}$), C is the concentration of solute ($kg\ m^{-3}$), C_P is the concentration of solute in the permeate ($kg\ m^{-3}$), $\mathcal{D}$ is the diffusion coefficient of the solute ($m^2\ s^{-1}$) and y is the distance from the membrane (m).

Equation 4.3 assumes that the diffusion coefficient remains constant at all concentrations. In the absence of fouling, integrating the equation from $C = C_B$ = bulk concentration of the solute at $y = \delta$, the thickness of the concentration polarisation boundary layer, to $C = C_M$ = concentration of solute at the membrane surface at $y = 0$ gives Equation 4.4 (Richardson *et al*., 2002).

$$\frac{J\delta}{\mathcal{D}} = \ln\left(\frac{C_M - C_P}{C_B - C_P}\right) \tag{4.4}$$

At this stage, the mass transfer coefficient, k_{CP}, is defined to be the ratio $\mathcal{D}/\delta$. k_{CP} is assumed to be independent of solute concentrations and flux (Zeman & Zydney, 1996). Equation 4.5 is the classic equation used to quantify concentration polarisation during UF. It is only applicable to mass transfer-controlled systems whereby the permeate flux is independent of pressure.

$$J = k_{CP}\ln\left(\frac{C_M - C_P}{C_B - C_P}\right) \tag{4.5}$$

Assuming total rejection of the solute is achieved at the membrane surface, $Rej = 1$ and $C_P = 0$, Equation 4.5 becomes:

$$J = k_{CP} \ln \left(\frac{C_M}{C_B} \right) \tag{4.6}$$

The ratio C_M/C_B is known as the concentration polarisation modulus. Experimentally, k_{CP} is determined by obtaining J at varying C_B; a plot of J against ln C_B gives a gradient of $-k_{CP}$. k_{CP} can also be obtained by using mass transfer correlations. These correlations for various membrane modules and flow regimes take the general form of

$$Sh = aRe^b Sc^c \left(\frac{d_H}{L} \right)^e \tag{4.7}$$

$$Sh = \frac{k_{CP} d_H}{\mathcal{D}}; \ Re = \frac{\rho v d_H}{\mu_F}; \ Sc = \frac{\mu_F}{\rho d_H} \tag{4.8}$$

where Sh is the Sherwood number, Re is the Reynolds number, Sc is the Schmidt number, d_H is the hydraulic diameter (m), L is the length of the tube or duct (m), ρ is the density of the feed fluid (kg m^{-3}), μ_F is the dynamic viscosity of the feed fluid (Pa s^{-1}), v is the cross-flow velocity (m s^{-1}) and, a, b, c and e are empirical constants. The hydraulic diameter is defined as

$$d_H = \frac{4 \times \text{cross-sectional area of flow}}{\text{wetted perimeter of channel}} \tag{4.9}$$

For a tube or hollow-fibre membrane module, d_H is equal to the tube diameter, d_{tube}. The hydraulic diameter for a slit-shaped channel is equal to twice the channel spacing, $2H$. For a membrane module with a feed side spacer in the channel, the hydraulic diameter is a function of spacer porosity, ε, and specific surface area, s (Zeman & Zydney, 1996):

$$d_H = \frac{4\varepsilon}{\frac{2}{H} + (1 - \varepsilon)s} \tag{4.10}$$

For a membrane module with a tube, slit or spacer configuration, the Reynolds number is calculated using Equation 4.8. However, for stirred cells and rotating modules, the Reynolds number is expressed in terms of angular velocity, ω. The Reynolds number for a stirred cell of diameter d is then calculated as (Prudich *et al*., 2008):

$$Re = \frac{\rho \omega d^2}{\mu} \tag{4.11}$$

For a rotating-cylinder with an inner radius of r and gap diameter of d_{gap} (Prudich *et al*., 2008),

$$Re = \frac{2\rho \omega r d_{gap}}{\mu} \tag{4.12}$$

Table 4.1 Typical values of the parameters *a*, *b*, *c* and *e* for use in Equation 4.7 for various membrane modules.

Geometry	Flow	*a*	*b*	*c*	*e*	Reference
Tube	Laminar	1.62	0.33	0.33	0.33	Lévêque (1928)
Slit	Laminar	1.86	0.33	0.33	0.33	Lévêque (1928)
	Turbulent	0.023	0.80	0.33	–	Gekas and Hallström (1987)
	Turbulent	0.023	0.875	0.25	–	Deissler (1961)
Spacer	–	0.664	0.50	0.33	0.50	Da Costa *et al*. (1994)
Stirred cell	Laminar	0.23	0.567	0.33	–	Smith *et al*. (1968)
Rotating	Laminar	0.75	0.50	0.33	0.42	Holeschovsky and Cooney (1991)

Source: Adapted from Prudich *et al*. (2008). Reproduced with permission of McGraw Hill.

Typical values of the empirical constants *a, b, c* and *e* in Equation 4.7 have been tabulated by Prudich *et al*. (2008). This is shown in Table 4.1. For spacer-filled membrane modules, the constant a in Equation 4.7 is dependent on the geometry of the spacer, and the term L is taken as the spacer mesh size (Da Costa *et al*., 1994; Zeman & Zydney, 1996).

As the feed concentration is kept constant in most operations, improving the mass transfer coefficient is the only solution to overcoming concentration polarisation (Cheryan, 1986). From the correlations, it is observed that the mass transfer coefficient is a function of cross-flow velocity, diffusion coefficient of the solute, density of the feed solution and filtration configuration and dimensions. This observation is important as concentration polarisation is a reversible phenomenon whereby its effects can be minimised by varying the operating conditions during filtration, which ultimately increases the rate of mass transfer of the solute. For example, increasing the cross-flow velocity increases the mass transfer coefficient and thus the flux. Operating at higher temperature or lower feed concentration also increases flux through the CP boundary layer (Cheryan, 1998); a review of these aspects follows.

Gel polarisation model

In some cases, the concentration of solute at the membrane surface reaches the upper limit of the solubility of the solute in the solvent. At this point, the solute precipitates out and a gel layer develops. This critical solute concentration is known as the gel concentration or gel point (C_G). The gel concentration is dependent on the type, size, shape and chemical structure of the solute. Once the gel is formed, the solute concentration within this layer remains constant at C_G, as shown in Fig. 4.4. The gel layer formed at the membrane surface exerts an additional resistance to mass transfer of solute permeating through the membrane. The description of mass transfer, in this case, is known as the gel polarisation model. Under such conditions, $C_M = C_G$ and Equation 4.13 becomes (Richardson *et al*., 2002):

$$J = k_{CP} \ln \left(\frac{C_G}{C_B} \right) \tag{4.13}$$

Similar to k_{CP} determination in the film model theory, a plot of J against $\ln C_B$ gives a gradient of $-k_{CP}$. C_G is obtained graphically whereby the intercept of the straight line on the x-axis, which corresponds to zero flux, is $\ln C_G$.

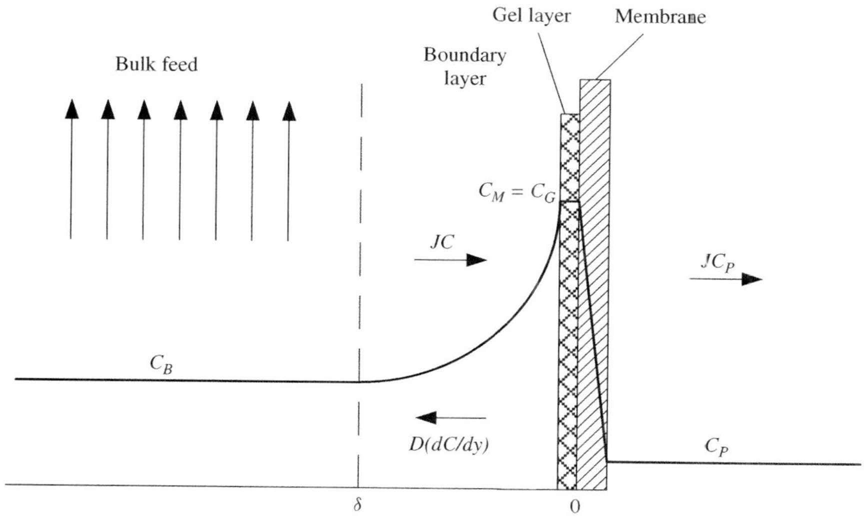

Fig. 4.4 Schematic representation of gel polarisation model (Adapted from Mulder, 1996. Reproduced with permission from Springer).

Although this model is widely used in the analysis of pressure-independent permeate flux during protein UF, the model cannot provide information in the pressure-dependent regime. Different authors have also reported very different values of the gel concentration for a given solute as the gel concentration appears to be dependent on the membrane and membrane configuration used, bulk concentration and cross-flow velocity (Nakao *et al.*, 1979; Mulder, 1996; Zeman & Zydney, 1996). Finally, this model is only applicable to macromolecular solutions and colloidal suspensions. The flux of low molecular weight solutes is primarily controlled by osmotic pressure.

Osmotic pressure model

The osmotic pressure model assumes that the filtrate flux is determined by the difference between the applied TMP and the osmotic pressure difference of the retained solutes (Zeman & Zydney, 1996; Muthukumaran, 2005). The flux equation is written as:

$$J = \frac{\Delta P - \Delta \pi}{\mu_P R'_T} \tag{4.14}$$

where $\Delta\pi$ is the osmotic pressure difference across the membrane $(=\pi(C_M) - \pi(C_P))$. The osmotic pressure of most macromolecular solutions is expressed in the form of a virial expansion as shown in Equation 4.15:

$$\pi = fC_B + gC_B^2 + hC_B^3 \tag{4.15}$$

where f, g and h are osmotic virial coefficients. The coefficient f describes the van't Hoff law, which is a linear relationship that exists in very dilute concentration of low molecular weight solutions. f can be written in terms of the solute molecular weight, $f = RT/MW$, where R is the ideal gas constant (8.314 J mol^{-1} K^{-1}), T is the absolute temperature (K) and MW is the molecular weight of the solute (kg mol^{-1}).

4.2.3 *Fouling*

Fouling is defined as the deposition of particles on the membrane surface or in the pores of the membrane. Examples of fouling include adsorption, pore blockage, precipitation and cake formation. The main effects of fouling are the decline in permeate flux and membrane selectivity. These effects are brought about by the irreversible alteration in membrane properties caused by physical and chemical interactions between the membrane and foulants (Zeman & Zydney, 1996). It is observed that porous membranes are highly susceptible to fouling; thus, fouling occurs mainly in microfiltration (MF) and UF (Mulder, 1996).

As mentioned earlier, fouling, similar to CP, has a flux-reducing effect, giving rise to confusion between the two phenomena. However, unlike CP, fouling does not occur within minutes but occurs slowly with time. In addition to flux decline, membrane fouling may also lead to an alteration in the chemical properties of the membrane, such as membrane selectivity. For example, in the separation of casein from whey proteins by MF, the fouling layer can totally control membrane selectivity and is often referred to as a dynamic or secondary membrane. This causes MF membranes to behave like UF membranes, resulting in high rejection of both casein and whey proteins. Cheryan (1998), Cheryan & Alvarez (1995) and Lawrence *et al*. (2008) recommend operating at high cross-flow velocities (4–7 m s^{-1}) and low TMP (less than 40 kPa) to minimise the formation of a dynamic membrane by increasing wall shear stress for casein MF using ceramic membranes.

As fouling is an irreversible phenomenon, the physical and chemical changes that develop during the process continue throughout filtration (Makardij *et al*., 1999). Changes in operation conditions during filtration may increase the flux temporarily but are ineffective in the long run. Cleaning or replacement of the membrane is thus required to alleviate the problem.

A long-term flux decline may not necessarily suggest fouling as an alteration in the feed stream compositions also leads to a decrease in flux. For example, the feed stream becomes more concentrated if the process is operated in a batch mode. This concentrated and viscous feed is harder to process thereby resulting in a lower permeate flux.

Downtime due to membrane cleaning, use of chemical cleaning agents and membrane replacement are costly consequences of membrane fouling. For this reason, membrane fouling mechanisms and optimisation of cleaning cycles are hot topics in membrane research. Many empirical models have been developed to predict fouling phenomena of ideal solutions. However, it is hard to apply these models to real systems due to complex physical and chemical factors.

Pore blockage models

Blockage of pores occurs either internally or on the membrane surface. When the solute is smaller than the pore size, it is able to diffuse through the pores of the membrane and adsorb to the walls of the pores. This phenomenon is known as internal pore blockage, which reduces the effective pore diameter and prevents the passage of other solutes. This standard pore blockage model is given by Equation 4.16 (Jacob *et al*., 1998).

$$J = \frac{J_o}{1 + it^2} \tag{4.16}$$

where J_o is the initial flux ($m^3\ m^{-2}\ s^{-1}$), i is a constant that characterises the fouling process (s^{-2}) and t is time (s).

In the worst-case scenario, complete pore blockage takes place, causing a drastic flux decline. In the complete pore blockage model, it is assumed that each particle blocks the pores with no superposition of particles. The permeate flux is given by Muthukumaran (2005):

$$J = J_o \exp(-\beta t) \text{ and } \beta = \alpha u_o \tag{4.17}$$

where α is the membrane surface area blocked per unit of total volume permeated through the membrane ($m^2\ m^{-3}$) and u_o is the mean initial velocity of the filtrate ($m\ s^{-1}$).

Cake filtration model

A cake layer is formed on the membrane surface when the retained solutes interact or bind with each other, with other components or with the membrane itself. This phenomenon is commonly observed during filtration of a feed that consists of proteins, organics and colloids. In this cake filtration model, the hydraulic resistance provided by the cake of particles (R'_c) is assumed to be directly proportional to the mass of cake (m_c) (Muthukumaran, 2005):

$$R'_c = \frac{R''}{A} m_c \tag{4.18}$$

where R'' is the specific cake layer resistance ($m\ kg^{-1}$) and A is the effective membrane area (m^2). Assuming the feed concentration is kept constant, membrane filtration through a cake layer is modelled using the following equation (Richardson *et al*., 2002):

$$\frac{t}{V} = \frac{\mu R'_m}{A \Delta P} + \frac{\mu C_B R'' V}{2A^2 \Delta P} \tag{4.19}$$

where t is time and V is the volume of filtrate (m^3). A plot of t/V against V yields a straight line.

Combined pore blockage and cake filtration model

Ho & Zydney (2000) developed a model that accounts for both pore blockage and cake layer formation in the same system. They assumed that the initial flux decline arises from pore blockage by large aggregates on the membrane surface. They also assumed that some fluid continues flowing through the blocked pores with increasing resistance over blocked regions as more particles attach to the membrane surface. The flux through the blocked pores (J_b) is:

$$J_b = \frac{\Delta P}{\mu (R'_M + R'_b)} \tag{4.20}$$

where R'_b is the resistance of deposit that forms over the blocked surface (m^{-1}). Over time, a cake layer is gradually formed and the resistance to flow increases. This change in flow resistance is represented as

$$\frac{dR'_b}{dt} = f'R''J_bC_B \tag{4.21}$$

where $f'R''$ represents the rate of increase of the cake layer resistance with respect to time (kg m^{-2} s^{-1}). The filtrate flux through the fouled membrane is equal to the sum of the flow rate through the open and blocked pores and is expressed as Equation 4.22. In this equation, the permeate flux is characterised by the pore blockage parameter (γ), the initial resistance of the deposit (R'_{bo}) and the cake growth factor ($f'R''$).

$$\frac{J}{J_o} = exp\left(-\frac{\gamma\Delta PC_B}{\mu' R_m}t\right) + \frac{R'_m}{R'_m + R'_b}\left[1 - exp\left(\frac{\gamma\Delta PC_B}{\mu R'_m}t\right)\right] \tag{4.22}$$

Ho & Zydney (2000) showed that their model is in agreement with experimental data for the constant pressure filtration of bovine serum albumin (BSA) solutions over a range of protein concentrations and TMP. Therefore, the need to use different models for different fouling mechanisms is not required as a smooth transition from pore blockage to cake filtration regimes can be determined using this model.

A plot of t/V against V shows the transition in fouling mechanisms with time, as seen in Fig. 4.5. The initial convex curve is attributed to pore blockage. This is followed by a linear region brought about by cake filtration. After a long period of time, cake compression occurs, which is indicated by a concave curve. During cake compression, there is an increase in filtration resistance and a decrease in cake porosity. A compact skin layer is formed on the membrane surface.

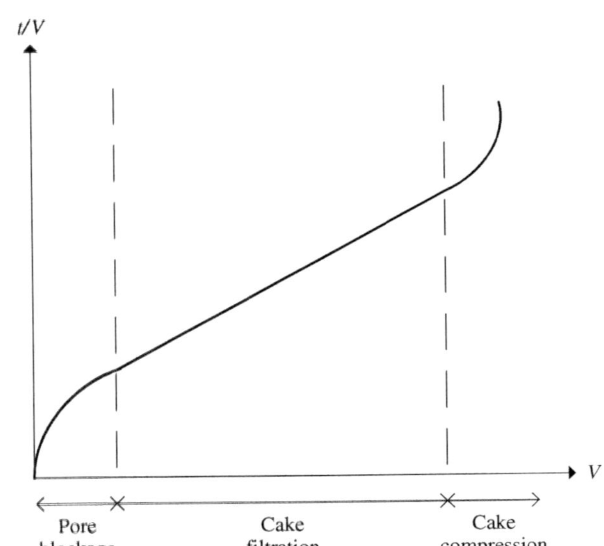

Fig. 4.5 Graphical representation of various fouling mechanisms (Adapted from Muthukumaran, 2005).

Table 4.2 Chemical composition of various components on a deposit of an ultrafiltration (UF) membrane fouled with depectinised passion fruit juice.

Component	Amount (g 100 g^{-1})
Pectin	34.6
Sucrose	21.6
Glucose	20.4
Fructose	10.8
Cellulose	3.5
Hemicellulose	1.5
Citric Acid	1.2

Source: Data taken from Chiang & Yu (1987).

4.2.4 *Fouling in the beverage industry*

The fouling behaviour of a membrane in the fruit juice industry is dependent on the type of fruit, its age and the range of fruits used. In fruit juice processing, the main foulant is pectin and the minor foulants are tannins, proteins, sugars, cellulose and hemicelluloses (Cheryan, 1986; Jönsson & Trägårdh, 1990; Alvarez *et al*., 1996; de Bruijn & Bórquez, 2006; He *et al*., 2007). In apple juice processing, depectinisation, or hydrolysis of pectins by the addition of pectic enzymes, is performed before membrane filtration to improve yield and clarification through reductions in retentate viscosity and osmotic pressure (Cheryan, 1986; Zeman & Zydney, 1996). However, even after depectinisation, pectin is still found to be a major foulant, whereas soluble starch, dextran and laminarin are found to have no impact on filtration (Girard & Fukumoto, 2000). Table 4.2 shows the composition of a sample of the deposit obtained from a UF membrane processing depectinised passionfruit juice.

4.2.5 *Fouling in the dairy industry*

The main foulants in the dairy manufacturing industry are proteins and calcium phosphate. Proteinaceous fouling is caused by the deposition and adsorption of native and denatured dairy proteins onto the membrane surface or within the pores. A gelatinous cake layer of protein is thus formed, creating resistance to bulk flow. Marshall & Daufin (1995) have shown that protein adsorption is influenced by the hydrophobicity of the membrane surface and the electrostatic forces between the charged proteins and membrane. At the initial stages of fouling, protein–membrane interactions dictate the fouling behaviour. However, at the later stages, protein–protein interactions determine this behaviour. Calcium also plays a role in protein cake formation by forming protein–protein and protein–membrane bridges.

Precipitation of salts and scaling of the membrane surface occurs when sparingly soluble salts in the feed stream are concentrated during filtration beyond their solubility. Calcium phosphate is a dominant salt foulant in milk systems due to supersaturation in the aqueous phase. It displays reverse solubility whereby its solubility decreases with increasing temperature (Rice, 2008). The degree of calcium phosphate precipitation also increases with increasing pH as calcium becomes less soluble (Rice, 2008). Thus, low temperatures and pH are utilised in industry to reduce calcium phosphate scaling on the membrane surface and within the pores. Calcium phosphate precipitation during membrane filtration can also be reduced upon addition of calcium-sequestering agents such as ethylenediamine tetraacetic acid (EDTA) or citrate to the feed. This reduces the ionic calcium concentration before filtration.

Nyström *et al*. (1995) have reported that uncharged components are unlikely to cause fouling. This suggests that lactose does not foul membranes, which is also substantiated by van der Horst (1995) who observed no lactose fouling during filtration of whey components. Lipids are also found to have no impact on membrane fouling unless the feed has a high lipid content (Kulozik, 1995; Marshall & Daufin, 1995).

4.3 Membrane cleaning and disinfection

Cleaning and disinfection of filtration units are essential in the food and beverage industries so as to maintain high hygiene standards. Fouling and deposit formation not only reduce flux and productivity of the system, they also promote the growth of bacteria and micro-organisms, which is highly unacceptable for safety and hygiene standards. Cleaning removes any deposits formed or accumulated in the system, thereby preventing the growth of micro-organisms. Therefore, the main aim of cleaning is to prevent contamination of the desired product during manufacture and to remove material that will affect a smooth and efficient operation of the processing and ancillary equipment (Walstra *et al*., 1999).

Excessive fouling is also expensive as the product manufactured may be contaminated and therefore unsalable. High concentrations of cleaning agents and large volumes of water are needed, adding to the expense. A greater load is imposed on the sewage system due to an increase in the volume of wastewater produced (Anonymous, 2003; Walstra *et al*., 2006).

A surface that is visually clean may not be microbiologically clean; hence, disinfection is necessary to destroy micro-organisms that are present so that the quality of the product manufactured is not compromised. It is important to note that absolute destruction of micro-organisms is not possible, but disinfection is able to bring the number of micro-organisms to a low enough level that is deemed microbiologically clean by safety and hygiene regulations (Walstra *et al*., 2006).

Disinfection prior to cleaning is ineffective as it is not designed to remove accumulated deposits (Spreer, 1998; Walstra *et al*., 2006). Both cleaning and disinfection can be performed simultaneously if the choice and concentrations of cleaning agents, pH and temperature of the cleaning solution and time for the procedure are satisfactory.

4.3.1 *Cleaning methods*

There are five main cleaning methods: mechanical, hydraulic, electrical, enzymatic and chemical. The choice of the cleaning method depends on the configuration of the membrane module, nature of the deposit and limitations of the membrane used.

Mechanical cleaning

Mechanical cleaning is a traditional cleaning method that is not extensively used. It involves working with brushes or sponge balls. These items must be properly cleaned and disinfected before and after use and stored in a dry place so as to reduce any microbial growth. In mechanical cleaning, operating conditions, such as temperature, velocity, type of flow regime, pressure and time, must be optimised. It is the cheapest cleaning method and is frequently used when the internal surface of the equipment is rough. Although mechanical cleaning is good for removing soft biological and organic matter, it is only applicable to surface cleaning and not the removal of material within the pores of the membrane (Zeman & Zydney, 1996). Therefore, mechanical cleaning must be performed in conjunction with other methods, such as chemical cleaning, for an effective removal of foulants. Mechanical cleaning can only be applied to tubular filtration systems (Mulder, 1995).

Hydraulic cleaning

Examples of hydraulic cleaning methods are back-flushing and transmembrane pressure pulsing. In most cases, hydraulic cleaning methods can also be used to improve flow hydrodynamics during filtration. Hollow-fibre membranes can be cleaned by back-flushing at regular time intervals. In back-flushing, a negative transmembrane pressure is applied to flush deposits out of the pores of the membrane and off the membrane surface. It can be performed during the filtration process, where permeate is used to back-flush the system or during the cleaning cycle with either water or cleaning agent. Back-flushing during filtration improves permeate flux and delays the need for cleaning. Fig. 4.6 shows a graph of the effects of back-flushing on permeate flux during filtration. A possible setback in hydraulic cleaning is the control of back-pressure in large systems (Caric *et al*., 2009).

Transmembrane pressure pulsing is a modification of back-flushing. It involves the application of back-pressure in a rapid pulse every 10–30 s throughout the process (Zeman & Zydney, 1996). The following texts on hydraulic cleaning are recommended for further reading (Rodgers & Sparks, 1993; Redkar & Davis, 1995; Zeman & Zydney, 1996; Cheryan, 1998).

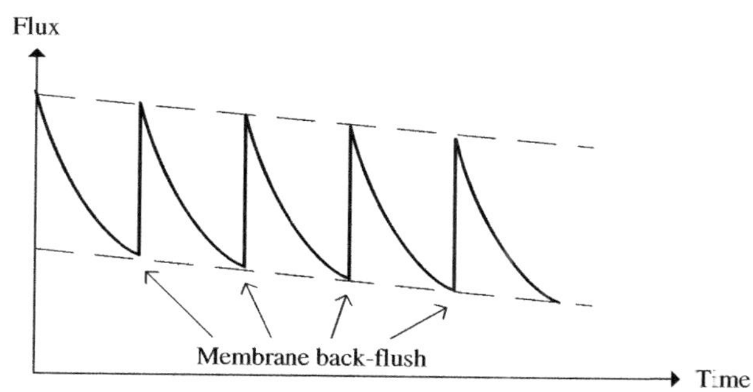

Fig. 4.6 Graphical representation of the change in permeate flux with periodic back-flushing during filtration.

Electrical cleaning

Electrical cleaning makes use of a pulsed electric field. Electric pulses are applied periodically for short durations, resulting in the movement of foulants, as charged particles, away from the membrane. Similar to back-flushing, electrical cleaning can be done without interrupting the filtration process. An added advantage of electric cleaning over back-flushing is that it involves no permeate loss. The requirement of an electrically conducting membrane, however, is a disadvantage to the system and so it is not widely used (Mulder, 1995).

Enzymatic cleaning

Enzymatic cleaners are typically used to remove proteinaceous fouling and can be considered as a type of cleaning agent. Most enzymatic cleaners consist of proteases as protein foulants are hard to remove. Proteases cleave peptide bonds in proteins at specific sites. Upon rinsing, these small polypeptide fragments are easier to remove from the membrane surface due to a decrease in the number of interaction sites between the protein and the membrane (Smith & Bradley Jr., 1987; Zeman & Zydney, 1996). To remove fats, lipases are used: lipases hydrolyse ester bonds, producing free fatty acids that readily solubilise in alkaline solution. In addition, amylases are employed to cleave glucose linkages in starch and, upon cleavage, smaller polysaccharides, which are soluble in aqueous solution, are produced and removed (Zeman & Zydney, 1996).

Enzymatic cleaners usually have only a small operating pH and temperature range. Optimal exposure time varies from 20 min to 1 h (Argüello *et al.*, 2002, 2003; Muñoz-Aguado *et al.*, 1996). This slow processing time makes the use of enzymatic cleaners undesirable due to a decrease in production time. They also tend to be expensive. In addition, industry operators are sceptical about using enzymatic cleaners as product quality can be affected if there is any residual enzyme remaining at the end of the cleaning process (Caric *et al.*, 2009). To reduce the possibility of enzymatic contamination in the product, Askew & te Poele (2008) suggested that acid is added and circulated for 15 min at the end of enzymatic cleaning to inactivate any residual enzymes. A rinsing step is then performed followed by alkaline cleaning and final rinsing.

Enzymatic cleaners can be particularly useful when membranes are severely fouled or are sensitive to pH, temperature and/or chemicals (Smith & Bradley Jr., 1987; Trägårdh, 1989; Coolbear *et al.*, 1992; Rucka *et al.*, 1996; Zeman & Zydney, 1996; D'Souza & Mawson, 2005; Turkiewicz *et al.*, 2006; Caric *et al.*, 2009). As enzymatic cleaners operate at mild temperatures and neutral pH, their use can extend membrane life significantly and reduce the amount of chemicals and energy needed (Argüello *et al.*, 2003). Caric *et al.* (2009) have also noted that the use of enzymatic cleaner reduces salt load in factory wastewater, which is important in many systems. Enzymatic cleaners can be easily tailor-made for cleaning optimisation (Argüello *et al.*, 2003).

An optimum enzyme concentration for cleaning has been reported by Muñoz-Aguado *et al.* (1996), beyond which the enzyme acts more as a foulant. They have also found that for the removal of proteinaceous fouling in dairy processing equipment, cleaning with a protease prior to a surfactant cleaning cycle generates better results than either step alone, particularly if the membrane is rinsed at 40°C between the two cycles. The same authors also observed that majority of the sites for binding of detergents in globular whey proteins are hidden in the structural matrix and upon enzymatic cleaning, these sites are exposed by the scission of protein strands, facilitating later detergent action.

Chemical cleaning

Chemical cleaning is the most popular cleaning method. It involves the use of cleaning agents, or detergents, to dissolve or remove deposits and fouling materials. The main solvent used for dissolution and dilution of cleaning agents is water. It is also used as a rinsing agent in between cycles of cleaning agents. Steam may also be required as an energy source and for sterilisation (Spreer, 1998).

According to Plett (1985) and Trägårdh (1989), chemical cleaning occurs in six stages. The first stage is the reaction of chemical cleaning agents on foulants in the bulk solution. This stage is prevalent if the degree of fouling is high, leading to a large consumption of chemical cleaners in the bulk solution. Hence, it is important to remove as much foulants as possible before the addition of chemical cleaners. In the second stage, chemical cleaning agents are transported from the bulk solution to the fouled surface. This is followed by the penetration of chemical cleaners into the fouled layer, which forms the third stage of chemical cleaning. A cleaning agent with low surface tension is desired for efficient transportation of cleaners into the fouled layer.

The fourth stage involves physico-chemical transformations and chemical reactions between the cleaning agents and the fouled layer. Stabilisation of foulants by dispersion, emulsification or changing their solubility is an example of physico-chemical transformation. Some examples of chemical modification of foulants are the saponification of fats and oils, the hydrolysis of proteins and the sequesteration and chelation of metal ions. In all cases, the foulants are displaced from the membrane by breaking down and dispersing insoluble particles or converting insoluble particles to water-soluble products.

In the last two stages, the products of the cleaning reactions and remnants of chemical cleaners are transported back to the cleaned membrane surface and into the bulk solution. They are then easily removed during circulation of bulk solution. The following sections provide more detail with respect to chemical cleaning.

4.3.2 *Chemical cleaning factors*

Before cleaning of the membrane commences, a number of factors must be considered and optimised as they influence the performance of a cleaning regime and the properties of the membrane. These factors include the nature of the deposit or foulant, the type and concentration of cleaning agent to be used, the duration of cleaning, the temperature, the nature of the equipment surface and the flow regime. The cleaning procedure and its frequency must be optimised not only to ensure a high cleaning efficiency but also to prolong membrane life and to minimise process downtime, cleaning and wastewater treatment costs. It is also desirable that the cleaning operation is not harmful to operating staff and does not pollute the environment (Walstra *et al.*, 2006). This section begins with a detailed discussion on the various factors affecting the performance of a cleaning cycle.

Nature of the deposit

The nature of the foulant deposit determines the type of cleaning agent required. It is important to identify the constituents in the deposit and their relative concentrations so as to develop an effective cleaning procedure that removes loosely bound deposits first and prevents their redeposition onto the surface. Knowledge on the level of fouling is also useful; the level of fouling depends on production time and the fluid being processed. For an unfamiliar filtration process, a membrane analysis, often called a *membrane autopsy*, is usually required to characterise the nature of deposits and the extent of fouling. A typical membrane analysis employs the use of scanning electron microscopy (SEM) whereby surface and cross-sectional views of a fouled membrane are examined.

The two main types of foulant deposits are organic and inorganic. Examples of organic deposits are proteins, fats, fruit cells and sugar. Calcium phosphate and calcium caseinate are examples of inorganic foulants. The degree of heat denaturation and acid coagulation of deposits can change the physical condition of the deposit, hence requiring different cleaning agents. For example, in the dairy industry, it is observed that protein deposition is more prevalent at low pH while greater mineral deposition is observed at high pH (Walstra *et al*., 2006).

Type of cleaning agents

For effective cleaning of fouled membranes, a good chemical cleaning agent is required. The solution should have the following properties (Walton, 2008; Watkinson, 2008).

- Wetting and penetration power – surface tension of the cleaning solution is low so as to allow contact between the foulant and penetration through the fouled layer.
- Organic dissolving power – proteins are broken down into small, soluble molecules.
- Emulsifying power – fats and oils are broken down into small particles, which are dispersed in the cleaning solution.
- Sequestering power – metal cations are deactivated by forming soluble metal complexes.
- Dispersing power – aggregates are broken down into smaller particles, which are easily suspended in the cleaning solution and removed from the system.
- Suspending power – insoluble particles are suspended in the bulk solution and prevented from re-deposition on cleaned surfaces.
- Rinsing power – cleaning solution and suspended particles are flushed from the system easily and completely without leaving any traces of cleaning agents and degraded foulant material.
- Compatibility – the cleaning solution does not degrade the membrane material or corrode surrounding equipment.
- Buffering capacity – the pH of the cleaning solution remains constant during cleaning.

However, there is no single cleaning agent that can provide all the properties mentioned above. It is thus often necessary to incorporate two or more chemicals into a built or formulated cleaning agent. A formulated cleaning agent typically consists of a mixture of alkalis or acids, sequestrants and surfactants. The remainder of this sub-section examines the cleaning properties of these components.

Acids

Acids are used to remove mineral scales and deposits that are hard to remove using alkaline cleaning agents. They are particularly effective at removing calcium scales and metal oxides. These components react with the acid to form a soluble salt. For example, metal carbonates and oxides are converted to soluble metal chlorides when hydrochloric acid is used. Similar reactions occur when nitric and phosphoric acids are used.

Inorganic acids, such as hydrochloric and sulphuric acids, are the cheapest and simplest to use. However, the corrosive properties of hydrochloric and sulphuric acids on membrane modules and stainless steel make them unfavourable for cleaning. The most common inorganic acid used is phosphoric acid. It is less corrosive and a strong chelating agent, facilitating the removal of metal cations. Phosphorus acid is also a good buffer, providing good pH control during cleaning (Zeman & Zydney, 1996). These properties make phosphoric acid a good acid cleaner but it is expensive compared with other inorganic acids. Nitric acid is effective for cleaning of organic and biological foulants as it is a strong oxidant, which causes nitration of organic compounds (Zeman & Zydney, 1996). It is commonly used in dairy cleaning but care must be taken as nitric acid is also corrosive towards module materials.

Organic acids are much weaker than inorganic acids, hence, unlike inorganic acids, they do not damage membrane modules and ancillary equipment. Most organic acids have good buffering capacity. Citric acid is the most common organic acid used as an acid cleaner. As citric acid is a good chelating agent, it is effective at removing calcium precipitates (Zeman & Zydney, 1996).

Alkalis

Alkaline cleaning agents are effective in removing proteinaceous materials, fats and oils by hydrolysis of proteins and esters, neutralisation of acidic materials and dispersion of colloids. Examples of alkaline cleaning agents include hydroxides, carbonates, silicates and phosphates.

Sodium hydroxide or caustic soda is the most commonly used cleaning agent because of its low cost and high alkalinity. Although sodium hydroxide is an effective saponification agent, it is a poor buffer (Zeman & Zydney, 1996). Hence, a highly concentrated solution of sodium hydroxide is required at the start of cleaning, which is undesirable as the membrane module and its ancillary equipment have low pH tolerances. Sodium hydroxide also has poor rinsing power: addition of appropriate surfactants to the cleaning agent is required so as to allow easy and rapid removal of sodium hydroxide from surfaces (Watkinson, 2008).

Sodium carbonate or soda ash is the cheapest alkaline cleaning agent. It has a high alkali reserve – a low concentration of sodium carbonate is required for saponification and neutralisation. Sodium carbonate forms an insoluble salt with calcium, improving soap solubility but increasing the possibility of calcium carbonate deposit formation on the membrane surface (Zeman & Zydney, 1996).

Silicates generate silicic acid when dissolved in water. Silicic acid, which is insoluble in water, exists as colloidal particles that remain suspended in the aqueous solution (Zeman & Zydney, 1996). These particles are able to suspend foulants and prevent their redeposition. Upon rinsing with water, silicic acid and the suspended foulants are easily removed from the system. Silicates also have good buffering capacity and inhibit corrosion of aluminium by strong alkalis, making them an effective alkaline cleaner for use on soft metals (Watkinson, 2008).

There are two main types of phosphates: (a) orthophosphates; and (b) complex phosphates. Orthophosphates, such as trisodium phosphate, have high alkalinity and soften water by precipitation of insoluble metallic salts (Watkinson, 2008). Complex phosphates soften water by sequestering metal ions, forming soluble complexes. They are also able to disperse large colloidal particles, keeping them in suspension and preventing their coagulation. This process is known as peptisation. A disadvantage of complex phosphates is that they hydrolyse to form simpler phosphates, which tend to form insoluble salts with divalent metal ions (Zeman & Zydney, 1996). In addition, phosphates cause eutrophication, making them undesirable for use.

Surfactants

A surfactant molecule has two distinct parts: (a) a hydrophilic portion, which is soluble in water; and (b) a hydrophobic portion, which makes the molecule soluble in hydrophobic liquids, such as oil. This amphiphilic property of the surfactant means that it is able to remove oil-based deposits from the surface – the hydrophobic part of the surfactant binds to the organic foulants and the resultant surfactant-coated deposit becomes soluble in water because of the hydrophilic nature of the surfactant head group.

The properties of a given surfactant are dependent on the relative length of the hydrophilic and hydrophobic portions: surfactants with long hydrophobic groups solubilise hydrophobic foulants easily but have low water solubility and vice versa. For optimal cleaning, a good balance between the hydrophilic and hydrophobic portions is required.

Surfactants are divided into three groups, each with different properties: (a) anionic; (b) cationic; and (c) neutral. Anionic surfactants have negatively charged hydrophilic groups with good wetting and dispersing power (Watkinson, 2008). Common anionic surfactants contain carboxylates, sulphonates, sulphates and phosphates as hydrophilic head groups. Most anionic surfactants have an optimal pH of 10.5 (Zeman & Zydney, 1996). Cationic surfactants have a positively charged hydrophilic group when dissolved in water. Although they have good bactericidal action, cationic surfactants have poor detergent qualities (Watkinson, 2008). Generally, cationic surfactants are used with acid cleaning agents. Non-ionic surfactants do not ionise in solution and are stable in both acidic and alkali solutions. A typical non-ionic surfactant consists of a long-chain hydrophilic polyethylene oxide (PEO) group and an alcohol, alkylphenyl or a polypropylene oxide chain (Cahn & Lynn Jr., 1978; Zeman & Zydney, 1996).

Sequestrants

Sequestrants form coordinate bonds to metal atoms, producing water-soluble complexes, which have cyclic or ring structures. They are effective in removing insoluble metal cations from the solution. The most effective organic sequestrant is EDTA and sodium tripolyphosphate is the most common inorganic sequestrant (Zeman & Zydney, 1996; Watkinson, 2008). Fig. 4.7 shows the chemical structure of an EDTA–metal complex.

Concentration of cleaning agents

Every system requires a specific concentration of cleaning agent for optimum removal of foulants and deposited materials. Exceeding this optimum concentration may result in

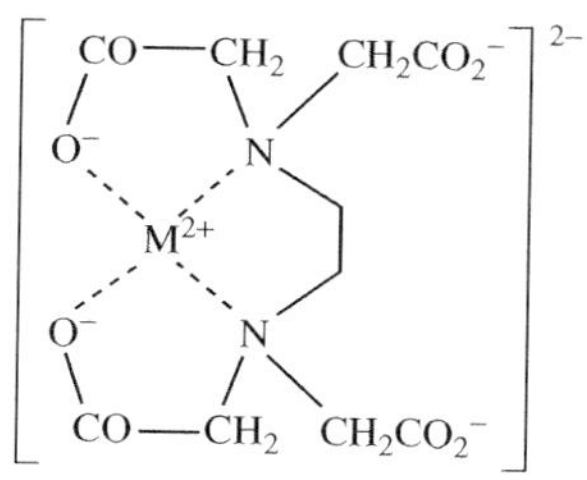

Fig. 4.7 Diagram of an ethylenediamine tetraacetic acid (EDTA)–metal complex where EDTA, which is an aminocarboxylic acid with a chemical formula of $C_{10}H_{16}N_2O_8$, forms multiple bonds with a metal cation, M^{2+} (Adapted from Howard & Wilson, 2000. Reproduced with permission from John Wiley & Sons).

poor cleaning performance, high wastewater treatment costs and membrane deterioration (Bartlett *et al.*, 1995; Bird & Bartlett, 1995; Makardij *et al.*, 1999; Mohammadi *et al.*, 2002). Cleaning agents are expensive; hence, high concentrations also increase cleaning costs. The optimum concentration of cleaning agent required is a function of the type and level of fouling, limitations of the membrane and its ancillary equipment.

Cleaning efficiency generally increases with increasing cleaning solution pH to an optimum value and then decreases. As shown in Fig. 4.8, Muthukumaran *et al.* (2005a) reported that maximum cleaning efficiency occurs at pH 12 or 0.4 g 100 g^{-1} sodium hydroxide, which is consistent with the results published by other researchers (Kim *et al.*, 1993; Bartlett *et al.*, 1995; Madaeni & Mansourpanah, 2004). On contact with sodium hydroxide, the deposit becomes swollen, which increases with increasing concentration or pH (Bartlett *et al.*, 1995). At maximum cleaning efficiency, the deposit has maximum voidage as a result of weak ionic bonds in the deposit. This high permeability of the deposit is attributed to a change in the ionic environment from being positively charged at neutral pH to negatively-charged at pH 12 (Muthukumaran, 2005). Further increase in sodium hydroxide concentration leads to a reduction in efficiency due to a reduction in the deposit voidage.

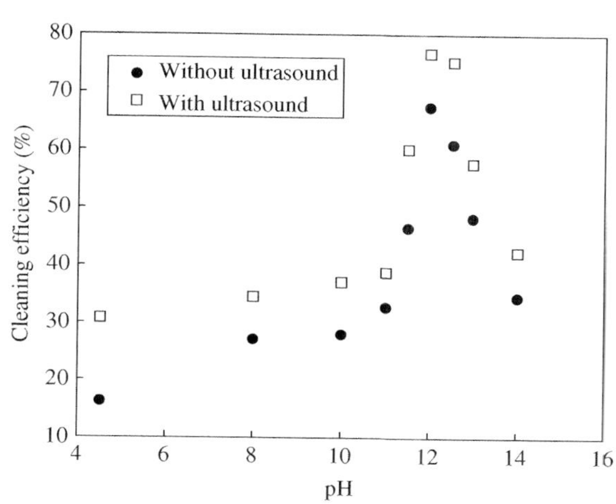

Fig. 4.8 Effect of sodium hydroxide pH on cleaning efficiency in both the absence and presence of ultrasound (Adapted from Muthukumaran *et al.*, 2005a. Reproduced with permission from Elsevier).

During the design of a cleaning procedure, pH is a more important aspect than concentration in the determination of the strength of solutions and their effects on the membrane (D'Souza & Mawson, 2005). pH is checked frequently during cleaning to ensure that satisfactory cleaning is performed and no residual cleaning agents is present.

Duration of cleaning

In order to obtain the optimum cleaning effect and minimise accompanying utilities, the duration of the cleaning cycle must be optimised. Cleaning duration is highly dependent on the level of fouling obtained, temperature of the solution, cleaning agent used and size of the plant.

Dissolution of a deposit may require a long time if a thick layer has accumulated on a surface. This is due to the long time required for the cleaning agent to diffuse through the layer. Most chemical cleaning agents act within 10–30 min whereas enzymatic cleaners take more than 20 min. Prolonged cleaning after the optimum time will cause refouling (Cheryan, 1998). In most cases, cleaning duration is optimised experimentally on laboratory- or pilot-scale membrane units, as longer than required cleaning times reduce subsequent pure water flux (Bird & Bartlett, 1995; Muñoz-Aguado *et al*., 1996; Makardij *et al*., 1999). Operators recommend using several cleaning steps successively with fresh solutions (Cheryan, 1998). Cleaning cycles will be further explained in Section 4.3.4.

Temperature

Chemical reactions and the dissolution of deposits proceed faster at higher temperatures. Diffusion coefficients are also greater at higher temperatures due to lower liquid viscosity and higher Reynolds number. Thus, the rate of chemical dissolution increases. This is particularly true for chemical cleaning, where a higher flux recovery is achieved in a shorter time as the cleaning temperature is increased (D'Souza & Mawson, 2005; Walstra *et al*., 2006). However, at higher temperatures, other factors, such as mineral deposition, may be compromised. For example, in a basic solution, calcium phosphate deposition is greater at high temperatures (Walstra *et al*., 2006). Membrane limitations should also be taken into account as high temperatures can damage the membrane structure.

Membrane limitations

Operating and cleaning conditions must strictly adhere to certain limitations in order to prolong membrane life. The most common limitations are pH, temperature and chlorine concentration. Processing fluids at higher than recommended temperatures will distort and cause irreversible damage to the membranes whereas operating out of pH tolerances will affect membrane-binding properties and destroy the membrane support layer. Generally, organic membranes, which are made of polymeric materials, have lower tolerances than inorganic membranes.

Inorganic membranes have high chemical and temperature resistance, which makes developing a cleaning procedure easy. A ceramic membrane can be operated from pH 1 to pH 13 and temperatures greater than 200°C (Zeman & Zydney, 1996; Askew *et al*., 2008). Although inorganic membranes are able to tolerate strong acids and bases and are unrestricted by chlorine concentration, care must be taken when choosing cleaning agents because of the limited stability of other parts of the membrane modules, such as seals and connectors. Askew *et al*. (2008) have noted that phosphoric acid is not to be used on ceramic membranes as phosphate ions can replace aluminium oxide present in the membrane and make it brittle. The negative impact of phosphoric acid on alumina membranes has also been observed by Trägårdh (1989). Thus, it is suggested that other mineral or organic acids are used during acid cleaning.

Organic membranes are cheaper but have low chemical and temperature resistances. The main limitation of an organic membrane is its susceptibility to chlorine concentration. First-generation membranes that were made of cellulosic materials were hard to clean because of a low temperature tolerance of ~30°C and a narrow pH operating range of 2–8 (Lewis, 1996b; Zeman & Zydney, 1996). The structural integrity of cellulosic membranes is weakened at low pH due to the degradation of β-glucosidic linkages in the backbone of the cellulose material (McCray & Glater, 1985; Zeman & Zydney, 1996). Cellulose acetate membranes also have low hypochlorite tolerance – chlorine readily oxidises cellulose acetate, making the membrane structure weak. To maintain the structural integrity of cellulose acetate membranes, Zeman & Zydney (1996) suggested that the chlorine concentration should be less than 50 mg L^{-1} for short periods and limited to 1 mg L^{-1} for long-term exposure. A weak cellulose membrane structure is also observed by many authors (Cheryan, 1986; Askew *et al*., 2008) when high caustic cleaning agents are used, as cellulose acetate will deacetylate at high pH causing the cellulose layer to be destroyed.

Membranes that are based on polysulphones or polyethersulphones are more chemically and thermally stable than cellulosic membranes. Zeman & Zydney (1996) have suggested that this stability is due to the electronic deactivation of aromatic rings by adjacent sulphone groups. These membranes have high temperature tolerance – the maximum temperature the membrane can withstand varies from 60°C to 100°C. A broader pH operating range of 1–13 is also reported by Zeman & Zydney (1996). Similar to cellulose acetate membranes, active chlorine oxidises sulphone to sulphonate groups, which weaken the membrane structure. Polysulphone-based membranes are able to withstand chlorine concentrations of 200 $\mu g\ g^{-1}$ for short term exposure and 50 $\mu g\ g^{-1}$ for long periods (Lewis, 1996a). The absence of isopropylidene group has made polyethersulphone membranes more chemically stable than polysulphone membranes (Zeman & Zydney, 1996).

Nanofiltration (NF) and reverse osmosis (RO) membranes are typically made from aromatic polyamides. These membranes have a broader pH operating range from 4 to 11 than cellulosic membranes (Fell, 1995). However, they have low chlorine tolerance and are also sensitive to other oxidants which degrade the membrane, resulting in a decrease in salt rejection. Therefore, when in use, the feed and water supply must be dechlorinated prior to filtration.

When designing a cleaning procedure, operators must ensure that the pH, temperature and chlorine concentration are within these membrane tolerances. Membrane specification data sheets provided by the membrane supplier or manufacturer should be consulted.

Nature of the surface

Cleaning agents should not corrode the surface of the equipment. Thus, it is of paramount importance to ensure that the cleaning agent is appropriate for the equipment material.

Metals are highly resistant to physical attack, such as sudden or large temperature drops. They are, however, susceptible to chemical attack, which will lead to corrosion and damage to the material. For example, stainless steel is highly resistant to most chemicals, except for strong acids. Glass and plastics are able to withstand chemical attack, but may be destroyed by physical stress.

For effective cleaning, the surface should also be smooth with no dead spaces, cracks and sharp corners, which are hard to reach by chemical cleaning. Accumulation of deposits in these areas is highly plausible even after vigorous cleaning. These areas promote the growth of micro-organisms. In such cases, mechanical cleaning methods will have to be employed. Therefore, when designing and manufacturing the equipment, a smooth surface is recommended as it ensures easy removal of deposits.

In commercial spiral-wound membrane modules, a feed spacer promotes fluid turbulence at low cross-flow velocities, reducing concentration polarisation and fouling while increasing mass transfer and permeate flux (Schwinge *et al*., 2000; Muthukumaran *et al*., 2005b). However, it has been found that they do not improve cleaning efficiency. Muthukumaran *et al*. (2005b) investigated the effect of spacers on membrane cleaning, as shown in Fig. 4.9, and found that in the presence of spacers, the cleaning efficiency decreases even with increasing cross-flow velocities. This is attributed to the entrapment of particles in the crevices of the spacers.

Flow regime

A good flow regime is required to remove dispersed deposit from the membrane module. Turbulent flow is desired as greater mass transfer is achieved due to a thinner boundary layer, which limits the transport of dispersed deposit. Local pressure fluctuations are also produced in a turbulent flow regime. This pressure change near deposited materials will loosen and dislodge some deposits from the bulk. Deposits are also further broken down in the bulk solution. Jönsson & Trägårdh (1990) and D'Souza & Mawson (2005) recommend a higher circulation flow rate and lower pressure than those used under process operating conditions in order to obtain a good hydraulic cleaning effect.

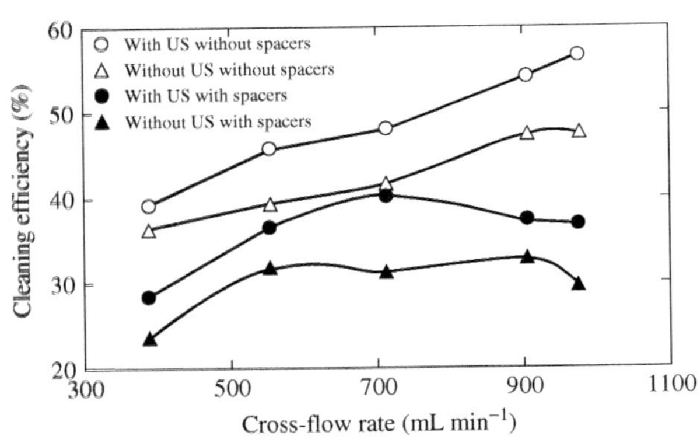

Fig. 4.9 Effect of spacers on cleaning efficiency of a polysulphone ultrafiltration (UF) membrane fouled with 6 g 100 g^{-1} whey solution and cleaned with 0.1 M sodium hydroxide and 15 mM sodiym dodecyl sulphate. US denotes ultrasonic treatment (Adapted from Muthukumaran *et al*., 2005a. Reproduced with permission from Elsevier).

4.3.3 *Disinfection*

In the food and dairy industries, membrane filtration systems are disinfected daily (Zeman & Zydney, 1996). Disinfection destroys all pathogenic micro-organisms, thereby preventing product contamination during manufacture. As mentioned in Section 4.3 Membrane cleaning and disinfection, complete destruction of micro-organisms is impossible. However, disinfection reduces their number to a low level, which is deemed satisfactory. As disinfectants are effective only if the micro-organisms are accessible, cleaning must be performed before disinfection commences. Incorporation of disinfection into the cleaning cycle is highly economical due to decreased utilities cost and chemical consumption. However, this is only applicable to systems with loosely bound deposits that are easily removed during pre-rinsing. Heat or chemicals may be incorporated for an added disinfection effect. High temperatures, supplied using steam, enhance denaturation and destruction of micro-organisms. Drainage and drying of equipment are also improved.

Sodium hypochlorite is extensively used in the disinfection of membrane systems. Sodium hypochlorite dissociates in water to give hypochlorous acid, which is bactericidal, and sodium hydroxide. The best bactericidal effect is achieved at a pH <5 (Walstra *et al.*, 2006). However, at pH 5, hypochlorous acid dissociates further to yield free chlorine in the form of OCl^- or chlorine gas (Zeman & Zydney, 1996; Walstra *et al.*, 2006). Hypochlorous acid also reacts with hypochlorite, therefore, to maintain the stability of the sodium hypochlorite during storage, a high pH ~ 8 is required (Walstra *et al.*, 2006). This is achieved by adding a small amount of sodium hydroxide to prevent consumption of hypochlorite by hypochlorous acid. During application, the disinfectant is diluted with water, which lowers the pH and, hence, increases its effectiveness. As mentioned earlier, most membranes have low chlorine resistance; thus, a very dilute hypochlorite solution is adequate. Sodium hypochlorite is used extensively in the dairy industry.

A number of oxidising agents have also been used as disinfectants. Examples of such oxidising agents are hydrogen peroxide and peracetic acid. Peracetic acid is used in many biological filtration systems as it acts rapidly and decomposes into a non-toxic product (Zeman & Zydney, 1996). Currently, it is used in the disinfection of haemodialysis membranes (Zeman & Zydney, 1996).

Sodium bisulphite is also a common disinfectant. Similar to sodium hypochlorite, the activity of sodium bisulphite is dependent on pH; it is more effective in acidic solutions of pH less than 4 (Smith & Bradley Jr, 1985). Zeman & Zydney (1996) have reported the high compatibility of sodium bisulphite to membrane systems that are incompatible with strong oxidising agents. Rowley (1992) has observed the swelling of cellulose acetate membranes due to sodium bisulphite, leading to an increase in permeate flux and solute passage.

4.3.4 *Cleaning procedures*

Throughout all food and beverage industries, cleaning processes tend to follow similar principles and can be done either manually or cleaning-in-place (CIP). A CIP system is defined as the circulation of cleaning liquid through processing lines, pipes, processing and ancillary equipment without dismantling any part of the system. A typical cleaning procedure consists of the follow steps:

- removal or recovery of product from the system
- pre-rinse
- cleaning with detergent
- intermediate rinse
- second detergent recirculation (optional)
- intermediate rinse
- disinfection (optional)
- final rinse

In the first step, the product is removed from the system. This is mandatory so as to minimise product loss, facilitate cleaning and reduce the load on the sewerage system (Bylund, 1995). This can be performed by draining the product under gravity. Typically, the product is rinsed out of the system using water and collected in a separate tank in both manual and CIP operations. The step is performed until both the retentate and permeate are clear and of neutral pH. This step is often incorporated with the pre-rinse step.

Pre-rinse cycles often utilise water and are controlled by a timer, which is set to a long period to ensure maximum product removal from the system. Walton (2008) observed that pre-rinse cycles often used recovered water from the intermediate rinse step to reduce total water consumption. Upon pre-rinse, a significant increase in pure water flux through the membrane is observed because of the removal of loosely bound deposits.

After product removal and rinsing the system with water, the system is cleaned with cleaning agents, which dislodge, dissolve and remove foulants and deposited materials. Cleaning agents are dissolved in water. It is of paramount importance to add the cleaning agents gradually to avoid chemical shock loading in the system. For the selection of cleaning agents, see Section 4.3.2. This step is operated under recirculation mode for a specific period of time, temperature and pressure. Both time and temperature are determined experimentally on a laboratory scale.

An intermediate rinse with water is necessary to remove traces of cleaning agents and suspended foulants from the system. To ensure complete removal and prevent contamination of the next step, which is either detergent recirculation or disinfection, the system is flushed with water for long periods of time. Detergent solution is also recovered and reused.

Some systems utilise a second detergent cycle, where two different cleaning agents are used in succession, to obtain satisfactory cleaning. This is mandatory for systems that have both organic and inorganic deposits. For example, in dairy processes, where there are both protein and mineral deposits, alkaline cleaning is performed in the first detergent recirculation step to remove protein deposits followed by cleaning with an acid in the second detergent recirculation cycle to remove mineral deposits. A further alkaline step may be used as well.

Identical to the previous intermediate rinse step, the second intermediate rinse step uses water to remove traces of cleaning agents and suspended foulants from the system. Softened water is preferred as it prevents limescale deposition on cleaned surfaces.

Disinfection is optional if the system is properly cleaned with both acidic and basic cleaning agents; upon acidic and basic cleaning, the system is microbiologically clean to a large extent. However, disinfection is definitely necessary for manufacturing of products such as milk. As mentioned in Section 4.3.3, disinfection can be performed thermally, chemically or both.

A final rinse is required to prevent contamination and product spoilage. Similar to all other rinsing steps, this is performed with water

As an example, cleaning of a membrane unit used for the filtration of apple juice is as follows (Girard & Fukumoto, 2000):

- product removal
- pre-rinse with water
- alkali recirculation at 50°C and pH 10.0–11.0
- intermediate rinse with water
- alkali recirculation at 50°C, pH 10.0–11.0 and 220 $\mu g\ g^{-1}$ of free chlorine
- final rinse with water

Alkali recirculation with chlorine is required to remove brown tannins on the membrane (Girard & Fukumoto, 2000).

A typical cleaning cycle in the dairy industry is as follows:

- product removal
- pre-rinse with water
- alkali recirculation at 50°C and pH 10.8–11.0
- intermediate rinse with water
- acid recirculation at 50°C and pH 2.0–2.2
- intermediate rinse with water
- alkali recirculation at 50°C, pH 10.8–11.0 and 180 $\mu g\ g^{-1}$ of free chlorine
- final rinse with water

As dairy processes have both protein and mineral deposits, two detergent recirculation steps are required. The third step, which involves alkali recirculation, is performed to remove fat and protein while acid clean is used to remove minerals. In cases where the bulk of the deposits is proteinaceous, free chlorine will be required to ensure effective removal of protein deposits. Sodium hypochlorite is typically used. As acid tends to tighten the pores of the membrane and give low permeate flux upon start-up, the cleaning cycle generally finishes off with an alkali clean before a final rinse is performed.

Generally, the cleaning efficiency is checked at the end of the cycle. This is conducted by measuring the pure water flux at a defined temperature, pressure and flow rate. This measured flux is compared with the initial pure water flux obtained before a process run, which is expressed as the percentage flux recovery; 100% flux recovery is rarely obtained. A low percentage indicates insufficient cleaning while a high percentage suggests efficient cleaning. However, a high water flux does not guarantee a good operational flux – measuring product permeate flux in the following run is a better indication of membrane cleaning efficiency (Trägårdh, 1989).

4.3.5 *Chemical cleaning agents recovery and reuse*

Cleaning solutions are either discharged from the facility after a single use or reused before being discarded. To reduce cleaning costs, recovery and reuse of chemical cleaning agents is favourable as a smaller amount of cleaning agent is purchased. Other than economic reasons, it is also desirable to reuse cleaning solutions from an ecological standpoint. Currently, the recovery of caustic cleaning solutions is of greater economic significance than that of acidic solutions because sodium hydroxide is used in much larger quantities (Novalic *et al*., 1998a, 1998b).

Reuse of caustic cleaning solutions results in an increase in concentration of suspended solids and chemical oxygen demand (COD), which can decrease cleaning effectiveness. However, cleaning solution reuse also leads to a decrease in surface tension, which facilitates cleaning. In dairy manufacturing facilities, the decrease in surface tension of reused caustic cleaning solution is attributed to the hydrolysis of fat and proteins by sodium hydroxide, which produces surfactant molecules (Alvarez *et al*., 2007). The same researchers also observed that at an identical operating temperature of 50°C, a regenerated caustic cleaning solution with low surface tension and no suspended solids operates as efficiently as a commercial cleaning agent with a faster cleaning rate.

Cleaning solutions are regenerated using MF, UF and NF, where the contaminants are retained and the regenerated cleaning solution permeates the membrane. These membrane processes are operated to control the composition of the permeating cleaning solution but also to reduce caustic rejection and the volume of waste produced (Gésan-Guiziou *et al*., 2007). The same researchers also found that MF is the cheapest approach, and results in a recovered caustic solution with high surfactant concentration and low surface tension. However, MF yields high COD levels compared with UF and NF, which may alter cleaning efficiency and kinetics. In this case, UF, which is more costly, is more appropriate for cleaning solution regeneration. Alternatively, Fleming (2006) suggested that cleaning chemicals can be recovered by a dead-end membrane configuration, which forces 100% of the water through the membrane. Cross-flow may be applied, occasionally, to flush out any contaminants that have deposited on the membrane surface.

4.4 Recent developments

In recent years, the main research focus has been in the field of new membrane materials and fouling minimisation or control; little research has been conducted on membrane cleaning. In general, membrane filtration processes and systems have been modified to enhance membrane performance and reduce fouling. Examples of such modification include feed pre-treatment, insertion of turbulence promoters and ultrasonic enhancements. In this section, these modifications are explained briefly. A summary of methods, principles, advantages and disadvantages for membrane performance enhancements is found in Brans *et al*. (2004).

In feed pre-treatment, the feed solution is treated and its properties are altered before filtration to stabilise or remove foulants. Heat treatment, pH adjustment, addition of complexing agents, precipitation and pre-microfiltration are examples of the methods used in feed pre-treatment. Hickey *et al*. (1980) investigated the effects of heat and pH on UF, and concluded that the increase in feed temperature and pH prior to UF led to an increase in permeate flux as bulk calcium phosphate crystals, which would otherwise precipitate in the pores of the membrane, were formed and readily flushed from the system. However, this method was not used as there was a decrease in protein retention (Brans *et al*., 2004).

Similarly, addition of citrate to the feed reduces the concentration of calcium and phosphate ions, leading to a decrease in the amount of calcium phosphate precipitates formed during filtration (Fox & McSweeney, 1998).

Addition of turbulence promoters in a filtration unit increases the shear rate near the membrane surface due to an enhancement in turbulence and back-transport. Hence, particle deposition is prevented and fouling is reduced. Vibratory shear-enhanced filtration and rotating disc modules are some examples of turbulence promoters. Akoum *et al.* (2005; 2002) have observed an increase in permeate flux by 300–400% in these systems when compared with a standard spiral-wound membrane. However, vibrating equipment is expensive and difficult to scale up (Brans *et al.*, 2004).

In ultrasonic flux enhancement, the filtration unit is installed alongside an ultrasonic unit. The use of an ultrasonic field induces localised flow disturbances through vibrations and cavitations and enhances turbulence near the membrane surface, hence reducing the effects of CP. Muthukumaran *et al.* (2004, 2005a, 2006, 2007) have found that the use of ultrasonics at low power levels can significantly enhance permeate flux during filtration of whey due to bubble collapse and acoustic streaming at the membrane surface, which generates microjets that scour the surface and promotes the back-transport of deposits to the bulk solution. Although ultrasonic enhancement seems promising, there is a discrepancy in the literature regarding the structural integrity of the membrane when exposed to ultrasonic waves. Masselin *et al.* (2001) and Wang *et al.* (2005) observed damage when polymeric membranes were used whereas other studies showed no structural changes to the membrane surface (Chai *et al.*, 1998; Muthukumaran, 2005; Muthukumaran *et al.*, 2006, 2007; Chen *et al.*, 2006). Muthukumaran *et al.* (2005a, 2005b, 2006) have concluded that impacts to the structural integrity of the membrane during filtration with ultrasonic enhancement are a function of the power intensity of the ultrasonic field and the distance between the transducer and membrane. Scale-up of the unit is difficult and large-scale filtration with ultrasonic enhancement may also be expensive due to high energy consumption (Brans *et al.*, 2004).

The use of ultrasound has also been considered as an alternative cleaning method. This cleaning mechanism is similar to that for fouling minimisation: microjets, which are generated by ultrasonic energy, scour the membrane surface, breaking down layers of deposits into smaller particles while preventing their redeposition and keeping them in suspension. Muthukumaran *et al.* (2005a) found that ultrasonic cleaning improves cleaning efficiency of polysulphone UF membranes that were fouled with whey solutions. Most workers observe greater cleaning efficiencies at low frequencies and high acoustic powers (Wakeman & Tarleton, 1991; Li *et al.*, 2002; Kobayashi *et al.*, 2003; Lamminen *et al.*, 2004; Muthukumaran *et al.*, 2005a). In addition, the ultrasonic effect was shown to be more significant in the absence of surfactants and was less dependent on temperature and pressure (Watkinson, 2008). Similar to ultrasonic enhancement, there are inconsistent results regarding the structural integrity of the membrane upon ultrasonic cleaning.

4.5 Conclusions

Even though there are many studies on fouling mechanisms and basic cleaning principles, the development of an effective cleaning procedure for a particular system is often by trial and error. The current emphasis in this area is on fouling minimisation and control, and developing new membrane materials that are more resistant to cleaning and disinfection agents. Recent developments in fouling minimisation, such as the addition of turbulence promoters and acoustic methods, are promising at the laboratory-scale; however, most of these developments have scale-up and economic issues that require further investigation.

4.6 Nomenclature

A	Effective membrane area	m^2
a	Empirical constant	–
b	Empirical constant	–
C	Concentration of solute	$kg\ m^{-3}$ or $g\ L^{-1}$
c	Empirical constant	–
d	Diameter	m
e	Empirical constant	–
f	Osmotic virial coefficient	$m^2\ s^{-2}$
$f'R''$	Cake growth factor	$kg\ m^{-2}\ s^{-1}$
g	Osmotic virial coefficient	$m^5\ kg^{-1}\ s^{-2}$
H	Channel spacing	m
h	Osmotic virial coefficient	$m^8\ kg^{-2}\ s^{-2}$
i	Fouling characterisation constant	s^{-2}
J	Permeate flux	$m^3\ m^{-2}\ s^{-1}$ or $L\ m^{-2}\ h^{-1}$
k	Mass transfer coefficient	$m\ s^{-1}$
L	Length of tube	m
m	Mass	kg
MW	Molecular weight	$kg\ mol^{-1}$
R	Ideal gas constant	$8.314\ J\ mol^{-1}\ K^{-1}$
R'	Resistance	m^{-1}
R''	Specific cake layer resistance	$m\ kg^{-1}$
r	Inner radius	m
Re	Reynolds number	–
Rej	Rejection	–

s	Specific surface area	$m^2\ m^{-3}$
Sc	Schmidt number	–
Sh	Sherwood number	–
T	Absolute temperature	K
t	Time	s
u	Mean velocity of the filtrate	$m\ s^{-1}$
V	Volume of the filtrate	m^3
v	Cross-flow velocity	$m\ s^{-1}$
y	Distance from the membrane	m
α	Membrane surface area blocked per unit of total volume permeated through the membrane	$m^2\ m^{-3}$
$\mathcal{D}$	Diffusion coefficient of the solute	$m^2\ s^{-1}$
δ	Thickness of the concentration polarisation boundary layer	m
ΔP	Transmembrane pressure drop	Pa
$\Delta\pi$	Osmostic pressure difference	Pa
ε	Spacer porosity	–
γ	Pore blockage parameter	$m^2\ kg^{-1}$
μ	Dynamic viscosity	Pa.s
π	Osmotic pressure	Pa
ρ	Density of feed fluid	$kg\ m^{-3}$
ω	Angular velocity	$rad\ s^{-1}$

Subscript

A	Adsorption
B	Bulk
b	Blocked pores
C	Cake of particles
CP	Concentration polarisation
F	Feed

G	Gel layer
gap	Gap
H	Hydraulic
M	Membrane
o	Initial
P	Permeate
T	Total
tube	Tube

References

Akoum, O., Jaffrin, M.Y. & Ding, L.-H. (2005) Concentration of total milk proteins by high shear ultrafiltration in a vibrating membrane module. *Journal of Membrane Science*, **247**, 211–220.

Al-Akoum, O., Ding, L.H. & Jaffrin, M.Y. (2002) Microfiltration and ultrafiltration of UHT skim milk with a vibrating membrane module. *Separation and Purification Technology*, **28**, 219–234.

Alvarez, N., Gésan-Guiziou, G. & Daufin, G. (2007) The role of surface tension of re-used caustic soda on the cleaning efficiency in dairy plants. *International Dairy Journal*, **17**, 403–411.

Alvarez, V., Andres, L.J., Riera, F.A. & Alvarez, R. (1996) Microfiltration of apple juice using inorganic membranes: Process optimization and juice stability. *The Canadian Journal of Chemical Engineering*, **74**, 156–162.

Argüello, M.A., Álvarez, S., Riera, F.A. & Álvarez, R. (2002) Enzymatic cleaning of inorganic ultrafiltration membranes fouled by whey proteins. *Journal of Agricultural and Food Chemistry*, **50**, 1951–1958.

Argüello, M.A., Álvarez, S., Riera, F.A. & Álvarez, R. (2003) Enzymatic cleaning of inorganic ultrafiltration membranes used for whey protein fractionation. *Journal of Membrane Science*, **216**, 121–134.

Askew, C.E., te Poele, S. & Skou, F. (2008) Membrane filtration. *Cleaning-in-Place: Dairy, Food and Beverage Operations* (ed. A.Y. Tamime), 3rd edn, pp. 195–222, Blackwell Publishing, Oxford.

Bartlett, M., Bird, M.R. & Howell, J.A. (1995) An experimental study for the development of a qualitative membrane cleaning model. *Journal of Membrane Science*, **105**, 147–157.

Bird, M.R. & Bartlett, M. (1995) CIP optimisation for the food industry: relationships between detergent concentration, temperature and cleaning time. *Food and Bioproducts Processing*, **73**, 63–70.

Brans, G., Schroën, C.G.P.H., van der Sman, R.G.M. & Boom, R.M. (2004) Membrane fractionation of milk: state of the art and challenges. *Journal of Membrane Science*, **243**, 263–272.

de Bruijn, J. & Bórquez, R. (2006) Analysis of the fouling mechanisms during cross-flow ultrafiltration of apple juice. *LWT – Food Science and Technology*, **39**, 861–871.

Bylund, G. (1995) *Dairy Processing Handbook*, Tetra Pak Processing Systems, Lund, Sweden.

Cahn, A. & Lynn Jr., J.L. (eds.) (1978) *Surfactants and Detersive Systems*, John Wiley and Sons, New York.

Caric, M., Akkerman, J.C., Milanovic, S., Kentish, S. E. & Tamime, A.Y. (2009) Technology of evaporators, membrane processing and dryers. *Dairy Powders and Concentrated Products* (ed. A.Y. Tamime), pp. 99–148, Blackwell publishing, Oxford.

Chai, X., Kobayashi, T. & Fujii, N. (1998) Ultrasound effect on cross-flow filtration of polyacrylonitrile ultrafiltration membranes. *Journal of Membrane Science*, **148**, 129–135.

Chen, D., Weavers, L.K., Walker, H.W. & Lenhart, J.J. (2006) Ultrasonic control of ceramic membrane fouling caused by natural organic matter and silica particles. *Journal of Membrane Science*, **276**, 135–144.

Cheryan, M. (1986) *Ultrafiltration Handbook*, Technomic publishing, Lancaster.

Cheryan, M. (1998) *Ultrafiltration and Microfiltration Handbook*, Technomic Publishing, Lancaster.

Cheryan, M. & Alvarez, J.R. (1995) Food and beverage industry applications. *Membrane Separations Technology: Principles and Applications* (eds. R.D. Noble and S.A. Stern), pp. 415–465, Elsevier, New York.

Chiang, B.H. & Yu, Z.R. (1987) Fouling and Flux Restoration of Ultrafiltration of Passion Fruit Juice. *Journal of Food Science*, **52**, 369–371.

Coolbear, T., Monk, C., Peek, K., Morgan, H.W. & Daniel, R.M. (1992) Laboratory-scale investigations into the use of extremely thermophilic proteinases for cleaning ultrafiltration membranes fouled during whey processing. *Journal of Membrane Science*, **67**, 93–101.

Da Costa, A.R., Fane, A.G. & Wiley, D.E. (1994a) Spacer characterization and pressure drop modelling in spacer-filled channels for ultrafiltration. *Journal of Membrane Science*, **87**, 79–98.

Deissler, R. (1961) Analysis of turbulent heat transfer, mass transfer and friction in smooth tubes at high Prandtl and Schmidt numbers. *Advances in Heat and Mass Transfer* (ed. Hartnett, J. P.), McGraw-Hill, New York.

D'Souza, N.M. & Mawson, A.J. (2005) Membrane cleaning in the dairy industry: a review. *Critical Reviews in Food Science and Nutrition*, **45**, 125–134.

Da Costa, A.R., Fane, A.G. & Wiley, D.E. (1994b) Spacer characterization and pressure drop modelling in spacer-filled channels for ultrafiltration. *Journal of Membrane Science*, **87**, 79–98.

Eykamp, W. (1995) Microfiltration and Ultrafiltration. *Membrane Separations Technology: Principles and Applications* (eds. R.D. Noble and S.A. Stern), pp. 1–43, Elsevier, New York.

Fell, C.J.D. (1995) Reverse Osmosis. *Membrane Separations Technology: Principles and Applications* (eds. R.D. Noble and S.A. Stern), pp. 113–142, Elsevier, New York.

Field, R.W., Wu, D., Howell, J.A. & Gupta, B.B. (1995) Critical flux concept for microfiltration fouling. *Journal of Membrane Science*, **100**, 259–272.

Fleming, S. (2006) *Membrane Filtration of a Product*, Technical Bulletin, Johnson Diversey, Inc., Sturtevant, WI.

Fox, P.F. & McSweeney, P.L.H. (1998) Salts of Milk. *Dairy Chemistry and Biochemistry*, 239–264, Blackie Academic & Professional, London.

Gekas, V. & Hallström, B. (1987) Mass transfer in the membrane concentration polarization layer under turbulent cross flow: I. Critical literature review and adaptation of existing Sherwood correlations to membrane operations. *Journal of Membrane Science*, **30**, 153–170

Gésan-Guiziou, G., Alvarez, N., Jacob, D. & Daufin, G. (2007) Cleaning-in-place coupled with membrane regeneration for re-using caustic soda solutions. *Separation and Purification Technology*, **54**, 329–339.

Girard, B. & Fukumoto, L.R. (2000) Membrane processing of fruit juices and beverages: A review. *Critical Reviews in Food Science and Nutrition*, **40**, 91–157.

He, Y., Ji, Z. & Li, S. (2007) Effective clarification of apple juice using membrane filtration without enzyme and pasteurization pretreatment. *Separation and Purification Technology*, **57**, 366–373.

Hickey, M.W., Hill, R.D. & Smith, B.R. (1980) Investigations into the ultrafiltration and reverse osmosis of wheys 1. The effects of certain pretreatments. *New Zealand Journal of Dairy Science and Technology*, **15**, 109–121.

Ho, C.-C. & Zydney, A.L. (2000) A combined pore blockage and cake filtration model for protein fouling during microfiltration. *Journal of Colloid and Interface Science*, **232**, 389–399.

Holeschovsky, U.B. & Cooney, C.L. (1991) Quantitative description of ultrafiltration in a rotating filtration device. *AIChE Journal*, **37**, 1219–1226.

van der Horst, H.C. (1995) Fouling of organic membranes during processing of dairy liquids. *Fouling and Cleaning in Pressure Driven Membranes*, Special Issue No. 9504, pp. 36–52, International Dairy Federation, Brussels.

Howard, W.L. & Wilson, D. (2000) Chelating agents. *Kirk-Othmer Encyclopedia of Chemical Technology*, Vol. 5, pp. 708–738, John Wiley & Sons, New York.

Howell, J.A. (1995) Sub-critical flux operation of microfiltration. *Journal of Membrane Science*, **107**, 165–171.

Jacob, J., Prádanos, P., Calvo, J.I., Hernández, A. & Jonsson, G. (1998) Fouling kinetics and associated dynamics of structural modifications. *Colloids and Surfaces A: Physicochemical and Engineering Aspects*, **138**, 173–183.

James, B.J., Jing, Y. & Chen, M.D. (2003) Membrane fouling during filtration of milk – a microstructural study. *Journal of Food Engineering*, **60**, 431–437.

Jönsson, A.-S. & Trägårdh, G. (1990) Ultrafiltration applications. *Desalination*, **77**, 135–179.

Kim, K.-J., Sun, P., Chen, V., Wiley, D.E. & Fane, A.G. (1993) The cleaning of ultrafiltration membranes fouled by protein. *Journal of Membrane Science*, **80**, 241–249.

Kobayashi, T., Kobayashi, T., Hosaka, Y. & Fujii, N. (2003) Ultrasound-enhanced membrane-cleaning processes applied water treatments: influence of sonic frequency on filtration treatments. *Ultrasonics*, **41**, 185–190.

Kulozik, U. (1995) Water rinsing in reverse osmosis and ultrafiltration. *Fouling and Cleaning in Pressure Driven Membranes*, Special Issue No. 9504, pp. 136–150, International Dairy Federation, Brussels.

Lamminen, M.O., Walker, H.W. & Weavers, L.K. (2004) Mechanisms and factors influencing the ultrasonic cleaning of particle-fouled ceramic membranes. *Journal of Membrane Science*, **237**, 213–223.

Lawrence, N.D., Kentish, S.E., O'Connor, A.J., Barber, A.R. & Stevens, G.W. (2008) Microfiltration of skim milk using polymeric membranes for casein concentrate manufacture. *Separation and Purification Technology*, **60**, 237–244.

Lévêque, M. D. (1928) Les lois de la transmission de chaleur par convection. *Annales de Mines*, **13**, 201–409.

Lewis, M.J. (1996a) Pressure-Activated Membrane Processes. *Separation Processes in the Food and Biotechnology Industries: Principles and Applications* (eds. A.S. Grandison and M.J. Lewis), pp. 88–90, Woodhead Publishing, Cambridge.

Lewis, M.J. (1996b) Ultrafiltration. *Separation Processes in the Food and Biotechnology Industries: Principles and Applications* (eds. A.S. Grandison and M.J. Lewis), pp. 105–116, Woodhead Publishing, Cambridge.

Li, J., Sanderson, R.D. & Jacobs, E.P. (2002) Ultrasonic cleaning of nylon microfiltration membranes fouled by Kraft paper mill effluent. *Journal of Membrane Science*, **205**, 247–257.

Madaeni, S.S. & Mansourpanah, Y. (2004) Chemical cleaning of reverse osmosis membranes fouled by whey. *Desalination*, **161**, 13–24.

Makardij, A., Chen, X.D. & Farid, M.M. (1999) Microfiltration and ultrafiltration of milk: some aspects of fouling and cleaning. *Food and Bioproducts Processing*, **77**, 107–113.

Marshall, A.D. & Daufin, G. (1995) Physico-chemical aspects of membrane fouling by dairy fluids. *Fouling and Cleaning in Pressure Driven Membranes*, Special Issue No. 9504, pp. 8–35, International Dairy Federation, Brussels.

Masselin, I., Chasseray, X., Durand-Bourlier, L., Lainé, J.-M., Syzaret, P.-Y. & Lemordant, D. (2001) Effect of sonication on polymeric membranes. *Journal of Membrane Science*, **181**, 213–220.

McCray, S.B. & Glater, J. (1985) Effects of hydrolysis on cellulose acetate reverse-osmosis transport coefficients. *Reverse Osmosis and Ultrafiltration* (eds. S. Sourirajan and S. Matsuura), 141–151, American Chemical Society, Washington DC.

Mohammadi, T., Madaeni, S.S. & Moghadam, M.K. (2002) Investigation of membrane fouling. *Desalination*, **153**, 155–160.

Mulder, M.H.V. (1995) Polarization phenomena and membrane fouling. *Membrane Separations Technology: Principles and Applications* (eds. R.D. Noble and S.A. Stern), pp. 45–84, Elsevier, New York.

Mulder, M. (1996) Polarization phenomena and membrane fouling. *Basic Principles of Membrane Technology* (eds. R.D. Noble and S.A. Stern), pp. 416–464, Kluwer Academic, Dordrecht.

Muñoz-Aguado, M.J., Wiley, D.E. & Fane, A.G. (1996) Enzymatic and detergent cleaning of a polysulfone ultrafiltration membrane fouled with BSA and whey. *Journal of Membrane Science*, **117**, 175–187.

Muthukumaran, S. (2005) *Ultrasonic Enhancement of Dairy Ultrafiltration Membrane Processes*, PhD thesis, The University of Melbourne, Melbourne.

Muthukumaran, S., Yang, K., Seuren, A., Kentish, S., Ashokkumar, M., Stevens, G.W. & Grieser, F. (2004) The use of ultrasonic cleaning for ultrafiltration membranes in the dairy industry. *Separation and Purification Technology*, **39**, 99–107.

Muthukumaran, S., Kentish, S., Lalchandani, S., Ashokkumar, M., Mawson, R., Stevens, G.W. & Grieser, F. (2005a) The optimisation of ultrasonic cleaning procedures for dairy fouled ultrafiltration membranes. *Ultrasonics Sonochemistry*, **12**, 29–35.

Muthukumaran, S., Kentish, S.E., Ashokkumar, M. & Stevens, G.W. (2005b) Mechanisms for the ultrasonic enhancement of dairy whey ultrafiltration. *Journal of Membrane Science*, **258**, 106–114.

Muthukumaran, S., Kentish, S.E., Stevens, G.W. & Ashokkumar, M. (2006) Application of ultrasound in membrane separation processes: a review. *Reviews in Chemical Engineering*, **22**, 155–194.

Muthukumaran, S., Kentish, S.E., Stevens, G.W., Ashokkumar, M. & Mawson, R. (2007) The application of ultrasound to dairy ultrafiltration: The influence of operating conditions. *Journal of Food Engineering*, **81**, 364–373.

Nakao, S.-I., Nomura, T. & Kimura, S. (1979) Characteristics of macromolecular gel layer formed on ultrafiltration tubular membrane. *AIChE Journal*, **25**, 615–622.

Novalic, S., Dabrowski, A. & Kulbe, K.D. (1998a) Nanofiltration of caustic and acidic cleaning solutions with high COD: part 1. Recycling of sodium hydroxide. *Journal of Food Engineering*, **38**, 125–132.

Novalic, S., Dabrowski, A. & Kulbe, K.D. (1998b) Nanofiltration of caustic and acidic cleaning solutions with high COD: part 2. Recycling of HNO_3. *Journal of Food Engineering*, **38**, 133–140.

Nyström, M., Kaipia, L. & Luque, S. (1995) Fouling and retention of nanofiltration membranes. *Journal of Membrane Science*, **98**, 249–262.

Plett, E.A. (1985) Cleaning of fouled surfaces. *Second International Conference on Fouling and Cleaning in Food Processing* (eds. D.B. Lund, B. Hallström, E.A. Plett, C. Sandu and G. Trägårdh), pp. 286–311, University of Wisconsin, Madison.

Prudich, M.E., Chen, H., Gu, T., Gupta, R.B., Johnston, K.P., Lutz, H., Ma, G. & Su, Z. (2008) Alternative separation processes. *Perry's Chemical Engineers' Handbook* (eds. D.W. Green and R.H. Perry), 8th edn, pp. 36–71, McGraw-Hill, New York.

Redkar, S.G. & Davis, R.H. (1995) Cross-flow microfiltration with high-frequency reverse filtration. *AIChE Journal*, **41**, 501–508.

Renner, E. & Abd El-Salam, M.H. (1991) Membrane fouling – problem and treatment. *Application of Ultrafiltration in the Dairy Industry*, pp. 79–96, Elsevier, London.

Rice, G. (2008) *Membrane Separation of Calcium Salts from Dairy Ultrafiltration Permeates*, PhD thesis, The University of Melbourne, Melbourne.

Richardson, J.F., Harker, J.H. & Backhurst, J.R. (2002) Membrane separation process. *Coulson and Richardson's Chemical Engineering*, 5th edn, pp. 437–474, Butterworth-Heinemann, Oxford.

Rodgers, V.G.J. & Sparks, R.E. (1993) Effects of solution properties on polarization redevelopment and flux in pressure pulsed ultrafiltration. *Journal of Membrane Science*, **78**, 163–180.

Rowley, L. H. (1992) A screening study of 12 biocides for potential use with cellulose acetate reverse osmosis membranes. *Desalination*, **88**, 71–83.

Rucka, M., Pozniak, G., Turkiewicz, B. & Trochimczuk, W. (1996) Ultrafiltration membranes from polysulfone/aminated polysulfone blends with proteolytic activity. *Enzyme and Microbial Technology*, **18**, 477–481.

Schwinge, J., Wiley, D.E., Fane, A.G. & Guenther, R. (2000) Characterization of a zigzag spacer for ultrafiltration. *Journal of Membrane Science*, **172**, 19–31.

Smith, K.E. & Bradley Jr., R.L. (1985) Effectiveness and corrosion problems of bisulphite as a sanitiser for polysulfone ultrafiltration membranes. *Second International Conference on Fouling and Cleaning in Food Processing* (eds. D.B. Lund, B. Hallström, E.A. Plett, C. Sandu and G. Trägårdh), pp. 286–311, University of Wisconsin, Madison.

Smith, K.E. & Bradley Jr., R.L. (1987) Activity of four enzyme-based cleaners for ultrafiltration systems against proteins in skim milk and whey. *Journal of Dairy Science*, **70**, 243–251.

Smith, K.A., Colton, C.K., Merrill, E.W. & Evans, L.B. (1968) Convective transport in a batch dialyzer: determination of the true membrane permeability from a single measurement. *AIChE Symposium Series*, **64**, 45–58.

Spreer, E. (1998) Cleaning and disinfection in a dairy. *Milk and Product Dairy Technology*, pp. 423–436, Marcel Dekker, New York.

Trägårdh, G. (1989) Membrane cleaning. *Desalination*, **71**, 325–335.

Turkiewicz, B., Rucka, M., Pozniak, G. & Zboinska, E. (2006) Biofunctional membranes from poly(2,6-dimethyl-1,4-phenylene oxide) with self-cleaning properties. *Enzyme and Microbial Technology*, **39**, 527–530.

Wakeman, R.J. & Tarleton, E.S. (1991) An experimental study of electroacoustic cross-flow microfiltration. *Chemical Engineering Research and Design*, **69**, 386–397.

Walstra, P., Geurts, T.J., Noomen, A., Jellema, A. & van Boekel, M.A.J.S. (1999) Fouling, cleaning and disinfection. *Dairy Technology: Principles of Milk Properties and Processes*, pp. 363–372, Marcel Dekker, New York.

Walstra, P., Wouters, J.T.M. & Geurts, T.J. (2006) Fouling and sanitising, *Dairy Science and Technology*, 2nd edn, pp. 399–410, CRC Press (Taylor and Francis Group), Boca Raton, FL.

Walton, M. (2008) Principles of cleaning-in-place (CIP). *Cleaning-in-Place: Dairy, Food and Beverage Operations* (ed. A.Y. Tamime), 3rd edn, pp. 1–9, Blackwell Publishing, Oxford.

Wang, X.-l., Li,X.-f., Fu, X.-q., Chen, R. & Gao, B. (2005) Effect of ultrasound irradiation on polymeric microfiltration membranes. *Desalination*, **175**, 187–196.

Watkinson, W.J. (2008) Chemistry of detergents and disinfectants. *Cleaning-in-Place: Dairy, Food and Beverage Operations* (ed. A.Y. Tamime), 3rd edn, pp. 56–80, Blackwell Publishing, Oxford.

Zeman, L.J. & Zydney, A.L. (1996) *Microfiltration and Ultrafiltration: Principles and Applications*, Marcel Dekker, New York.

5 General Application for the Treatment of Effluent and Reuse of Wastewater

N.A. Milne and S.R. Gray

5.1 General wastewater quality issues

Wastewater in the dairy and beverage industries is generally considered to be similar in terms of composition to municipal wastewater. Until recently, it was generally perceived that, where appropriate, wastewater from these industries could be directed towards municipal wastewater streams with minimal treatment requirement, such as some form of basic solids removal. Where this was not possible, land application of the wastewater, effectively as irrigation water, is common after similar levels of treatment. Changes to regulations around the world are starting to see changes in how wastewater from food industries is being perceived and treated before leaving the site.

There are important differences between municipal wastewater and dairy and beverage manufacturing wastewater that are often overlooked. The most obvious is that the organic content or strength of the wastewater in manufacturing processes is significantly higher than that of municipal processes. Although a dilution effect is observed when the wastewater is added to the sewer, the impact on small sewage treatment plants in rural areas can be quite pronounced, leading to significant changes in the kinetics of treatment. Another important difference is in the distribution of solids. Whereas municipal wastewater has a significant quantity of organics as solids, dairy and beverage manufacturers typically have higher levels of fats, oils and greases and soluble organics in the form of sugars (lactose, fructose and/or glucose). The fats, oils and grease content of dairy wastewater is typically quite high at around 400–5000 mg kg^{-1} (Britz *et al.*, 2004), and leads to significant fouling in sewers. Wastewater from both the dairy and beverage industries that does not receive some form of biological pre-treatment can contain almost twice the content of soluble organics as a percentage of the total load compared with municipal wastewater. This can put greater strain on small wastewater treatment plants where these streams are disposed because of the longer kinetics required for digestion of these organics. This, coupled with charges for sewage disposal being linked to organic strength, is leading to increased attention on treatment of wastewater streams prior to sewer discharge.

Land application of many dairy and beverage wastewaters is causing concern in parts of the world where salinisation is an issue, such as Australia. Although the wastewater may contain high levels of nutrients, the concentration of sodium from clean-in-place (CIP) processes is drawing concerns arising from the sodium adsorption ratio and their

Membrane Processing: Dairy and Beverage Applications, First Edition. A. Y. Tamime.

applicability to a range of soils. In Australia, local regulators are increasing restrictions on the use of wastewater from the food manufacturing industry in irrigation for this reason. This, in turn, is driving a search for greater levels of treatment in wastewater used for this process.

Typical wastewater qualities from the dairy and beverage industries are shown in Table 5.1. As noted earlier, the focus is typically towards organic content as fats, oils and grease, biological oxygen demand (BOD) or chemical oxygen demand (COD) with salinity in the form of total dissolved solids/conductivity or sodium being a secondary issue. The organic content can vary significantly, but is sourced primarily from lost product. This may be product lost in cleaning processes, particularly the first rinse of pipes, vats and tanks, spills and leaks around processing equipment or simply wasted product. A study of dairies in the United States of America (USA) has put an estimate on product losses as 0.06–0.15% of the wastewater stream or at least 1000 L of pure product each day during operation (Danalewich *et al*., 1998). It was noted through mass balances, however, that this seemed a significant underestimate as the organic content, phosphorus and total Kjeldahl nitrogen values observed in wastewater streams are significantly higher, maybe up to 10 times higher once other sources, such as cleaning agents, had been taken into account.

As was noted earlier, salinity in the food manufacturing industry is a growing concern, and is typically sourced from CIP processes. The most commonly used chemicals in this process are caustic soda (sodium hydroxide), nitric acid and sodium hypochlorite or peracetic acid. Although other compounds have been investigated for their use in CIP, e.g. ozone (Canut & Pascual, 2008) and potassium hydroxide (Palmowski *et al*., 2005), no replacement has gained significant market traction at this point in time. Ultimately the chemicals currently in use lead to a significant increase in the salinity of the wastewater, particularly in terms of sodium content. This is compounded by the fact that, typically, caustic use far outweighs acid use in most sites, and pH adjustment of the wastewater is required (often with hydrochloric acid or nitric acid) leading to a further increase in the waste stream's salinity. This overall salinity is an issue when considering land application/irrigation with wastewater, disposal to sewers in countries where recycling of municipal wastewater is practised and in internal recycling processes where salt generally cannot be tolerated. Removal of these salts is increasingly becoming common in industry and relies almost exclusively on membrane techniques.

5.2 General wastewater treatment

Wastewater treatment in the dairy and drink manufacturing industries, due to its large organic content, occurs in similar ways to that of sewage treatment. It is broken into three basic steps: (a) physico-chemical treatment (primary treatment); (b) biological treatment (secondary treatment); and (c) disinfection (tertiary treatment). There is now also an increase in the use of a further stage of treatment in the form of desalination to promote reuse of the wastewater in other processes. In some instances, this takes the place of tertiary treatment as the membranes used in desalination will also impart some degree of disinfection and are generally unsuitable for use where oxidising chemicals like those used in disinfection are present.

Table 5.1 Typical wastewater qualities from a range of dairy and beverage manufacturing facilities.

Industry	Turbidity	COD ($mg\ kg^{-1}$)	BOD ($mg\ kg^{-1}$)	Total nitrogen ($mg\ kg^{-1}$)	Total phosphorous ($mg\ kg^{-1}$)	Conductivity ($\mu S\ cm^{-1}$)	References
Dairy	120–300 NTU	2000–10 000	1300–1500	50–65	NR	1800–2000	Koyuncu *et al*. (2000)
Dairy	15–30 NTU	1500–3000	350–600	NR	NR	TDS: 800–1200	Sarkar *et al*. (2006)
Dairy (cheese)[a]	SS: 326–3560 $mg\ kg^{-1}$	785–7619	565–5722	14.9–254	29–181	NR	Danalewich *et al*. (1998)
Dairy (milk)	SS: 360–920 $mg\ kg^{-1}$	713–1410	NR	NR	NR	NR	Samkutty & Gough (2002)
Dairy (yoghurt)	Total solids: 2750 $mg\ kg^{-1}$	4656	NR	NR	NR	NR	Strydom *et al*. (1997)
Dairy (cream)	SS: 350–1000 $mg\ kg^{-1}$	2000–6000	1200–1400	TKN: 50–60	PO_4: 20–50 as P	NR	Ince (1998)
Dairy (butter/milk powder)	NR	NR	1500	TKN: 70	35	NR	Donkin (1997)
Juice	NR	365	NR	12	2.4	NR	Noronha *et al*. (2002)
Juice	SS: 1000–7300 $mg\ kg^{-1}$	1800–6600	NR	NR	NR	NR	Chmiel *et al*. (2002)
Soft drink	SS: 0–60 $mg\ kg^{-1}$	1200–8000	600–4500	NH_3: 150–300	PO_4: 20–40	NR	Chen *et al*. (2004)
Alcohol (beer)	SS: 300 $mg\ kg^{-1}$	500	1000	40	10	NR	Ukita *et al*. (2004)
Alcohol (spirit)	SS: 300 $mg\ kg^{-1}$	200	500	20	10	NR	Ukita *et al*. (2004)
Alcohol (shochu)	SS: 13 200–17 800 $mg\ kg^{-1}$	30 700–50 100	14 200–25 900	1200–1500	140–170	NR	Nagano *et al*. (1992)
Residential sewage	NR	500	300	57	7.3	375	Britz *et al*. (2004), Milne *et al*. (2011)

NTU, nephelometric turbidity unit; COD, chemical oxygen demand; BOD, biological oxygen demand; SS, suspended solids; TDS, total dissolved solids; TKN, total Kjeldahl nitrogen; NR, not reported.

[a]Range across 14 sites.

5.2.1 *Primary treatment: solids, fats, oils and grease removal*

This stage is generally focused on removal of fats, oils and grease and some suspended solids. As a minimum, this involves a detention period where suspended solids are allowed to settle and fats and oils rise to the top of the vessel. The process can be accelerated, or its efficiency improved, through the use of coagulants. There is a tendency in recent times to move towards techniques that are assisted by the use of bubbling air. Techniques, such as dissolved air flotation and induced air flotation, are increasingly being used in food manufacturing wastewater treatment. Although investigations of hybrid processes of dissolved air flotation/membrane filtration have been investigated in mineral processing applications (al-Zoubi *et al*., 2009), they have received little attention in the food and beverage manufacturing industries to date.

Microfiltration (MF) and ultrafiltration (UF) membranes are used in some industries for the removal or concentration of fat, oil and grease from wastewater streams. A relevant example is their use in the treatment of wastewater from pre-rinse water in margarine manufacturing (Chmiel *et al*., 2002). In this case, a ceramic MF membrane with 0.2 μm nominal pore size was used to concentrate the fats and remove them from the general wastewater stream where alkaline conditions led to saponification. It had the added benefit of reducing the aeration requirement in biological treatment. The COD of the waste feed stream was quite high, being between 5000 and 10 000 mg kg^{-1}. Although a decrease in membrane permeability was witnessed, it remained above 1500 L m^{-2} h^{-1} MPa^{-1}. The retentate with a COD of 40 000 mg kg^{-1} was sent to a settling tank where the fats and oils were able to be skimmed and ultimately used in soap production. The permeate COD was kept below 250 mg kg^{-1} after entering the site's equalisation tank. In general, however, while membranes are often used in fat reduction in processing streams, they are not used in wastewater treatment for fats, oils and grease removal possibly due to the complication of fouling.

5.2.2 *Secondary treatment: biological treatment and the membrane bioreactor*

Secondary treatment focuses on the removal of the soluble organic compounds in the wastewater stream. Biological processes oxidise the carbon-based compounds aerobically to carbon dioxide, nitrogen oxides, other oxides and water, or reduce organic compounds anaerobically to methane and ammonia and under the right conditions phosphates. The processes rely on bacteria and, as such, can be considered living processes. They typically need a constant supply of wastewater with roughly consistent characteristics in order for the microbial populations to survive and continue to be effective. The batch production seen at many manufacturing sites is not suitable for many biological treatment techniques without an equalisation tank, and batch treatment in the form of sequencing batch reactors has become common for many sites. Where constant streams of wastewater can be guaranteed, however, other biological techniques are suitable.

The traditional approach to secondary treatment is the conventional activated sludge process. It uses a suspended floc of microbial populations on an organic/inorganic matrix (the activated sludge) mixed with the wastewater stream. Organic materials in the wastewater typically adsorb to the floc where they are microbially degraded. Conventional activated sludge operates as an aerobic process and air is bubbled through part of the system to provide soluble oxygen for use by the microbes. Organics in the solid phase, organic sulphur, phosphorus compounds and some ammonia may also be removed via adsorption to the floc and eventual biological conversion, but the efficiencies of this is much lower than soluble organic reduction and as such rarely occurs to any significant degree. After aeration and contact with the sludge, separation of the sludge from the treated wastewater is required. In conventional activated sludge processes this occurs through a clarifier.

Although a range of different designs can be used in conventional activated sludge systems, one important consideration is the footprint. This is typically quite large, particularly due to the clarifier and is not well suited to many industrial sites where space is one of the more important considerations when selecting treatment processes. The need to minimise the space requirements for wastewater treatment means companies are increasingly turning to membrane technologies for secondary treatment, with particular attention being paid to membrane bioreactors.

Membrane bioreactors are effectively a hybridisation of the traditional biological processes with membrane filtration technology. Where conventional activated sludge treatment would use a clarifier after sludge contact, membrane bioreactors utilise a membrane, often UF, to achieve the separation and ultimately produce a better quality effluent (Wen *et al*., 2010). Using this membrane allows for greater concentrations of biomass in the reactor, and this allows for the treatment of higher organic loadings, and ultimately allows for a reduction in the footprint of the treatment plant. The effectively complete retention of solids by the membrane means that a greater degree of entrapment and degradation of organic compounds can occur, and poorly biodegradable organics that are not commonly removed in conventional activated sludge processes can be removed in membrane bioreactors (Wang & Menon, 2009). This is important in some dairy processes where poorly biodegradable streams, such as whey wastewaters (Janczukowocz *et al*., 2008), can be more effectively treated. Importantly, this complete retention of particles also allows bacteria commonly lost in washout in conventional activated sludge plants to be retained in the system. This ultimately allows for the growth of nitrifying bacteria (Cicek, 2003) and sees some oxidation of ammonia compounds in the reactor. The bacteria retained in the membrane bioreactor are selected because of their kinetics in degradation of the wastewater alone, rather than being influenced by their settling characteristics. The prevention of washout also contributes to greater solids retention times that assist with the more efficient removal of organics. Overall, the quality of wastewater from membrane bioreactors is better than those from conventional activated sludge processes, and the high active biomass concentrations improve the robustness of membrane bioreactors to variations in feed concentrations and to shock loads.

An important operating consideration for any sludge-based process is the accumulation of sludge within the treatment system and the need to "waste" (or discharge) the sludge over time. This can become a significant cost, particularly if there is a chance of pathogenic bacteria being present in the sludge. For this reason, it is often recommended that sanitary waste is treated separately from process streams at most food processing plants (Britz *et al*., 2004). The use of membranes, leading to extended solids retention time and, consequently greater oxidation, means that lower quantities of sludge are produced using membrane bioreactors than conventional activated sludge plants (Wen *et al*., 2010). However, it is often harder to dewater this sludge leading to increased complexity and cost in their disposal (Cicek, 2003).

The main drawback in the use of membrane bioreactors is their fouling capability. The blocking of pores as a result of deposition of particulate or adsorption of dissolved matter will ultimately lead to an increase in operating pressures or loss of flux and operating capacity for the plant. Fouling in membrane bioreactors is linked to a range of phenomenon, but the most important are probably extracellular polymeric substances and soluble microbial products produced by the microbes used for treatment (Hai & Yamamoto, 2011). The wastewater quality can have an impact, but this is often indirectly – some wastewaters may alter the characteristics of a floc to increase its fouling propensity (Hai & Yamamoto, 2011). Pre-treatment has also been shown to be beneficial in some cases, particularly to reduce fouling in wastewater streams from food manufacturing (Wu *et al*., 2009; Colic *et al*., 2010). Reduction of fouling potential from an engineering point of view is focused on increasing the shear rate at the surface of the membrane. This is generally achieved through the use of high cross-flow rates or bubbling to induce turbulence at the membrane surface. Reducing the flux also aids in reducing fouling issues, but at the expense of greater membrane surface area. However, fouling cannot be entirely avoided. Although a range of options exist for removing fouling during operation (regular physical backwashes or relaxation times and occasional chemical cleaning), this has resulted in a somewhat limited uptake of membranes in practice for treatment of food wastewaters.

Whereas aerobic conditions are needed for conventional activated sludge plants, anaerobic processes can be useful in wastewater treatment where high organic loadings are present. This is due in part to the formation of the more useful methane stream that can be used as a fuel. A large number of studies into secondary treatment plants in the dairy and beverage industry utilise anaerobic treatment. The process is largely the same, but requires influents with a higher organic content and must use techniques other than bubbling to prevent fouling at the membrane surface.

The use of anaerobic or anoxic steps prior to aerobic treatment is also important in terms of nutrient reduction. Although membrane bioreactors may be more efficient than conventional processes at nitrification reactions, the ammoniacal nitrogen that is oxidised is ultimately still present as nitrates in the effluent. By using an anaerobic or anoxic pre-treatment and returning sludge from an aerobic treatment plant to the anaerobic pre-treatment system, nitrate can be reduced to nitrogen gas and complete nitrogen removal occur. Phosphorus removal can be achieved in more complex plants by adding extra anaerobic, and anoxic stages to allow biological release and uptake mechanisms to occur throughout the biological process (Ekama, 2011). It is possible to add membranes units for clarification of the final effluent from such processes. The use of submerged membrane bioreactors favours phosphorus uptake as this is maximised under aerobic conditions (Monclús *et al*., 2010b). The use of membranes, in general, allows for full retention of the suspended flocs and a separation of the solids and hydraulic retention times. This can lead to greater efficiency in terms of nitrogen removal, but phosphorus removal efficiency can be negatively impacted if the system is not well optimised and maintained (Monclús *et al*., 2010a).

There are two main configurations for membrane bioreactors. The side-stream membrane bioreactor uses a membrane filtration unit outside of the main bioreactor. A high level of suspended biomass is maintained in this reactor with the liquor being transferred from the reactor to the membrane unit for separation. There are significant benefits in using this configuration, the most important of which is the greater level of process control and optimisation that can be achieved over the unit (Wang & Menon, 2009). However, the energy requirement is often higher than submerge membrane bioreactors due the high cross-flow velocities across the membrane required to control fouling. These systems generally use hollow-fibre membranes and may be operated in inside-out mode.

The submerged membrane bioreactor has a much smaller footprint by virtue of placing the membrane within the bioreactor. Submerged membrane bioreactors were originally common in aerobic systems as they utilise the aeration of the bioreactor to provide turbulence to the membrane surface through air bubbles. This effectively reduces the potential for membrane fouling and flux losses. Anaerobic submerged membrane bioreactors are starting to appear, however, and often utilise intermittent suction or backwashing to restrict issues with fouling. While the smaller footprint may be particularly attractive, the complexity of the system is greatly increased and makes effective process control more difficult (Wang & Menon, 2009). Both flat sheet and tubular membranes may be used in submerged membrane bioreactors, and it is possible to retrofit existing tanks as submerged membrane bioreactors.

Table 5.2 gives information for selected studies on membrane bioreactor use in dairy and beverage wastewaters; it also includes studies from full-scale plants, through pilot-scale to laboratory-sized studies, and has included both aerobic and anaerobic systems. Matosic *et al*. (2009) compared the use of a conventional activated sludge plant and a membrane bioreactor for the treatment of soft drink manufacturing wastewaters. They found that membrane bioreactors were more effective and reliable for organic reduction than conventional activated sludge processes in dairy wastewaters, and that the difference was largely due to the high organic content of the wastewater stream.

In full-scale application a range of dairies, wineries, breweries, soft drink and juice manufacturers have installed membrane bioreactors or retrofitted existing systems. A distillery in Scotland uses membrane bioreactor technology for the removal of organics and copper in the wastewater stream (Wehrle Environment UK, 2012b). Two bioreactors are used at the site, the second as a polishing stage, to reduce the BOD to less than 20 mg kg^{-1} and the copper concentration to less than 0.33 mg kg^{-1}. A cross-flow tubular UF system with nominal pore size of 0.02 μm is used for the filtration, and allows the bioreactors to operate with a mixed liquor suspended solids concentration of about four times conventional systems. Flushing of the membranes occurs on a weekly basis with chemical cleaning needed every 6–8 weeks. A membrane flux of 100 L m^{-2} h^{-1} is maintained in the system using this process.

A brewery in the USA utilises a membrane bioreactor on the back of a biological nutrient reduction plant (ADI Kubota, 2012a). The plant is a submerged bioreactor that is placed after the post-anoxic section of the nutrient reduction plant. The plant is able to achieve an effluent with suspended solids concentration of less than 2 mg kg^{-1} and a reduction in BOD, total nitrogen and total phosphorus of >99.8%, >93% and >98%, respectively.

Table 5.2 Selected examples of membrane bioreactor use at full-scale, pilot-scale and in laboratory studies to treat dairy and beverages manufacturing wastewater.

Industry	Flow rate (L d^{-1})	Flux (L m^{-2} h^{-1})	Type	Organic reduction	References
Full scale					
Dairy (creamery)	1.2×10^6	100	External Aerobic UF	Effluent BOD: <9 mg kg^{-1}	Robinson (2006a)
Dairy (whey)	2×10^6	NR	External Aerobic/anoxic UF	COD: >98%	Robinson (2006b)
Dairy (creamery)	6.5×10^5	170	External Aerobic UF	COD: 95%	Wehrle Environment UK (2012a)
Distillery	1.44×10^5	100	External Aerobic (jet loop) UF	NR	Wehrle Environment UK (2012b)
Winery	1.1×10^5	15–20	Submerged Aerobic MF	COD: 95%	Bolzonella *et al.* (2010)
Winery	7.5×10^4	12.5	Submerged Aerobic MF	Effluent BOD: <40h mg kg^{-1}	Ferre *et al.* (2009)
Pilot scale					
Fruit juice	1.5×10^3 L h^{-1}	14	Submerged Aerobic UF (ceramic)	COD: >95%	Chmiel *et al.* (2002)
Winery	122	4.5	Submerged Aerobic UF	COD: 97%	Artiga *et al.* (2007)
Dairy (whey)	460	NR	External Anaerobic UF	COD: >94%	Cicek (2003)
Alcohol (shochu)	1.5×10^3 L h^{-1}	NR	External Anaerobic UF	COD: >98%	Nagano *et al.* (1992)
Beer	100–200	10–18	External Anaerobic UF	COD: >96%	Strohwald & Ross (1992)
Dairy products	Small plant	25–80	External Anaerobic UF (ceramic)	COD: 95–98%	Farizoglu & Uzuner (2011)
Dairy products	Small plant	18–52	External Batch UF	COD: 97–98%	Bae *et al.* (2003)
Laboratory scale					
Soft drinks/spring water (bottle washing)	5 L h^{-1}	5.43	Submerged Aerobic UF (0.4 μm pore)	COD: 94%	Matosic *et al.* (2009)
Food manufacturing complex	Small plant	18–40	Bio-entrapped Submerged UF	COD: >90%	Ng *et al.* (2011)

UF, ultrafiltration; MF, microfiltration; BOD, biological oxygen demand; COD, chemical oxygen demand; NR, not reported.

Anaerobic membrane bioreactors have also reached commercialisation although the examples provided in trade publications and on manufacturer websites are not in the dairy and beverage industry. One site was a salad dressing factory in the USA (ADI Kubota, 2012b). This site added membranes to an existing anaerobic bioreactor that allowed for a 60% increase in capacity. The site's wastewater was also high in COD and fats and oils (34 000 mg kg^{-1} and 1500 mg kg^{-1}, respectively) that the anaerobic membrane bioreactor was able to sufficiently treat. Reductions to 225 mg kg^{-1} COD and <5 mg kg^{-1} fats, oils and grease were achieved by the plant.

A bio-entrapped membrane reactor has been developed and investigated on a laboratory scale to reduce the effects of membrane fouling (Ng *et al.*, 2011). The sludge in the process is entrapped within small balls inside the unit, making a treatment system similar to a fixed-film bioreactor. The effectiveness of this system in dairy wastewater treatment was quite high and easily similar in terms of removal to conventional membrane bioreactor processes. The bio-entrapped system was able to run for much longer periods without chemical cleaning, reducing the cleaning requirements by about 85%.

The application of membrane-sequencing bioreactors may be of interest to facilities where batch production of wastewater cannot be easily balanced by tanks or other storage, or where an existing sequencing bioreactor is in place. It may also be useful for reduced footprint biological applications where nutrient (nitrogen and phosphorus) reduction is important. Investigation of these systems with dairy wastewater was performed by Bae *et al.* (2003) and showed significant reduction of nutrients (96% nitrogen and 80% phosphorus) when utilising a mix of anaerobic, anoxic and aerobic conditions.

Information on installed membrane bioreactors in industrial processes can be hard to come by. Most literature focuses on pilot studies with little information on whether installation of a full-scale plant occurred. However, anecdotally, the low uptake of membrane bioreactors in manufacturing processes may slowly be being reversed. A review of North American facilities has shown 10 facilities operating in food and beverage processing and one operating in dairy manufacturing (Yang *et al.*, 2006), and one supplier reports that 22 facilities have been installed in wineries in Europe (Ferre *et al.*, 2009). This is further supported by a number of manufacturers discussing installation of membrane bioreactors on their websites.

5.2.3 *Tertiary treatment: disinfection*

Tertiary treatment focuses on the disinfection of wastewater streams. This is typically performed through oxidation techniques, such as chlorination or chloramination, ozonation or ultraviolet (UV) irradiation. Membrane techniques have also been considered for their disinfection properties and some of the earliest industrial recycling projects utilised MF and reverse osmosis (RO) membranes to impart the necessary disinfection for the wastewater streams (Milne *et al.*, 2009). The main issue that can arise, however, is the need for a guarantee that the membrane is intact. This guarantee cannot be given with respect to smaller pathogens and the integrity of membranes in dairy projects has been shown in the past to be questionable (Bickers & Bhamidimarri, 1998). Subsequently, although desalination membranes may be used in treatment systems, they are not generally given a significant role in disinfection processes at the current time.

Although disinfection is important in ensuring reuse potential, it can be damaging if further treatment will be performed. The oxidising effect of most disinfection agents is damaging to membrane systems used in desalination, severely limiting membrane life and increasing membrane replacement costs. Where desalination is due to be used, disinfection is best avoided or used subsequent to the desalination step.

5.2.4 *Desalination: nanofiltration and reverse osmosis*

Whereas in the past a minimal amount of treatment was often considered sufficient for some manufacturers, increasing concern with salinity in either trade waste or water used in irrigation is starting to introduce greater demand for desalination of wastewater streams. This is coupled with increasing water shortage in many parts of the world and a resultant increase in the need to recycle water, particular at remote sites. Although many desalination techniques exist, the most common are membrane techniques: nanofiltration (NF) and reverse osmosis (RO).

RO uses pressure to drive water through a tight pore size membrane. The pores in RO membranes for brackish water desalination are on the order of 0.3 nm (Tay & Jeyaseelan, 1995), and allow easier passage of the water while rejecting ions, typically rejecting 90–95% of ions (Xie *et al*., 2004). In general, in low salinity operations 60–80% of wastewater can be recovered (Xie *et al*., 2004; Yoon & Lueptow, 2005), although it is possible to go to higher recoveries if scaling of RO membranes from sparingly soluble salts allows. To achieve this level of water recovery requires moderately high pressures in the range of 2–3 MPa (Balannec *et al*., 2005), leading to electricity requirements of 1.7 kW for every 1000 L of produced water (Vince *et al*., 2008). This represents one of the more significant costs of desalination using RO.

A more important consideration is the disposal of the concentrated brine, i.e. the remaining 20–40% of high salinity water that was not passed through the membrane. This brine stream cannot be easily reused in dairy or beverage manufacturing and, unless the site has a sewer available, disposal options are limited and costly. Groundwater injection is used in groundwater desalination in some locations, although this requires to a confined aquifer with a lower quality water to prevent contamination of fresh water sources (Mickley, 2003; Mackey & Seacord, 2008). The technique that is most reliable is evaporation ponds. Evaporation ponds need to have an impervious lining to prevent any of the water or salts leaching to nearby groundwater stores. This, coupled with the area required at some sites, makes such approaches very expensive and, in Australia, it is a significant limitation to uptake of RO for recycling projects in inland food production facilities (Barr, 2011).

NF membranes are looser, meaning they have larger nominal pore sizes of the order of 0.6–1 nm (Bowen & Doneva, 2000). Consequently, they are less selective, but allow higher fluxes of water for lower pressures. Pressures typically used for NF range from 0.5 to 1.5 MPa (Chmiel *et al*., 2002), and the energy requirements are subsequently lower than RO. The main drawback, however, is the lower rejection of ions. Polyvalent cations are generally well rejected, although this rejection depends heavily on the environment in which they are contained. In the case of milk- associated proteins, rejection is almost 100% because of the entrainment of the ions within casein micelles or their binding with other large molecules (Balannec *et al*., 2005). Monovalent cations see very poor rejection, often around 50–70% (Balannec *et al*., 2005). The requirements of the final water quality would determine whether NF may be suitable when compared with RO. As this is not often the case practically, RO dominates wastewater treatment technologies. The greater market for RO membranes than NF membranes also results in lower prices for RO membranes.

Although there have been studies investigating the use of NF or RO membranes for the direct treatment of wastewater (Akoum *et al*., 2004; Balannec *et al*., 2005; Frappart *et al*., 2006), this rarely occurs in practice due to the potential for fouling and loss of permeability in the membrane. This was highlighted by Balannec *et al*. (2005), who witnessed a drop in flux of about 50% using RO and NF membranes to concentrate a simulated wastewater with a COD of 36 000 mg kg^{-1} by three times. Unlike other UF or MF membranes, backwashing is not an option in NF or RO, and the cross-flow velocity is used to provide partial control of fouling. Nevertheless, regular chemical cleaning using acids or bases depending on the type of fouling is generally required. This adds to the cost of use and generally makes pre-treatment more effective. Secondary treatment to significantly reduce organic loading in the wastewater will reduce the potential for fouling. MF (Sarkar *et al*., 2006) or sand filtration (Miyaki *et al*., 2000) are generally used in full-scale plants in the dairy and beverage industries for removing particulates that can be problematic in RO treatments. The use of membrane bioreactors, combining biological treatment with UF or MF is probably the most economical pre-treatment in terms of size that could be utilised at this stage.

NF and RO modules can exist in two different forms: spiral wound or hollow fibres. The spiral-wound module is based on two large flat sheets of membrane separated by a permeate collection channel. The sides of these are sealed to allow flow of permeate from only one end. A spiral-wound module consists of a number of these sheets, each separated by a spacer or flow sheet, and winds them around a central permeate collection tube. The feed passes over the surface of the membranes through the turbulent conditions provided by the feed spacer. Water that permeates through the membrane is ultimately passed by the permeate channel to the central collection tube while water that does not permeate passes out of the unit as concentrate. Hollow-fibre modules use a series of extruded tubes with micro-sized diameters to pass feedwater. Bundles of these fibres are used in a single unit. Feedwater typically passes through the fibre in an inside-out configuration. Fouling can be more difficult to control in these systems and they are generally used on lower recovery systems. There are very few hollow-fibre NF membranes commercially available.

Modification to membrane modules that are currently at the laboratory scale use shear forces, in the form of vibrations (Akoum *et al*., 2004) or rotating disks (Frappart *et al*., 2006) to reduce the potential for fouling in NF or RO. Studies looking at dairy wastewater have concluded that rotating discs perform better by giving lesser permeation loses. However, the operating costs were higher than vibrating shear-enhanced filtration (Frappart *et al*., 2006). Vibration shear-enhanced filtration is also more mature, with units available with areas of 150 m^2 (Frappart *et al*., 2006).

5.3 Water reuse

The manufacture of dairy products and beverages often results in the extraction of significant amounts of water from the raw materials. Overall, it is not uncommon for plants to discharge wastewater in quantities greater than the volume of raw water they use. One of the aims of most facilities of this type, particularly in water-stressed parts of the world, is to reduce the amount of water used by increasing the volume of water recycled. The International Finance Corporation (IFC) has developed a number of water use and wastewater production targets based on European experiences (IFC, 2007a, b, c), and they are shown in Table 5.3. It is important to recognise that by increasing recycling practices these targets can often be easily reached and exceeded as has been shown in recent research by the authors (Table 5.4 and Milne *et al*., 2011). Recycling wastewater has the effect of decreasing reliance on water, decreasing wastewater disposal costs and potentially saving chemical costs where these can be effectively recovered.

Although water reuse in food manufacturing processes is possible, it is important to consider all local regulations and particularly health requirements when considering how recycled water can be used. For example, whereas Australian regulations do not prohibit the use of recycled water in production in food manufacturing (Milne *et al*., 2009), French regulations have prevented recycled water being used in any direct contact application (Balannec *et al*., 2005). Denmark on the other hand has changed its regulations in recent times to allow the use of water not at potable qualities in some processes (Casani *et al*., 2005). Having said this, there is often a perception within companies that, regardless of the level of treatment and level of microbial contamination of recycled water streams, direct contact is best avoided purely for reasons of perception. This is not necessarily the case,

Table 5.3 Water use targets as a function of production specified by the IFC (2007a, b, c) based around European datasets.

Product	Water consumption
Beer	4–7 L L^{-1} beer
Fruit juice	6.5 L kg^{-1}
Milk and fermented milk products	1.0–1.5 L L^{-1} processed milk
Cheese and whey	1.4–2.0 L L^{-1} processed milk
Milk powder, cheese and liquid products	0.8–1.7 L L^{-1} processed milk
Ice-cream	4–5 L kg^{-1} ice-cream

Table 5.4 Water use targets as a function of production specified by Milne *et al*. (2011) based around Australian data.

Product	Water consumption (L L^{-1} product)	Wastewater generated (L L^{-1} intake)
Beer	2–4	0.33–0.58
Carbonated beverages	1–2	1.36
Fruit juice	1–4	1.46
Milk	0.5–1.5	1
Ice-cream	1.4	0.82

however, as standard hazard analysis critical control point (HACCP) approaches used in the food industry can be applied in water recycling projects (Casani *et al*., 2005), and often are in municipal wastewater recycling (Milne *et al*., 2011). Subsequently, any concerns with safety should be adequately addressed where due process is followed. Consumer perception is a more complicated area, and is heavily reliant on trust and confidence (Casani *et al*., 2005; Marks *et al*., 2007). In different parts of the world trust in scientific authorities, different manufacturers and water treatment specialists will vary, as witnessed by discussions of many researchers (Casani *et al*., 2005; Po *et al*., 2005; Marks *et al*., 2007). In general, where such projects are under consideration, a high degree of communication from all parties with the general public will be required to ensure that acceptability of the reuse of recycled water is as high as possible in the given circumstances (Casani *et al*., 2005).

General discussions on wastewater recycling focus on the necessary quality of recycled water streams. Although this is predominantly focused on the end use of the water, a common guideline is to ensure the water meets drinking water standards (Mavrov & Belieres, 2000). In some instances, such as floor cleaning, this could result in water receiving more treatment than is truly required. Previous research by the authors has provided rough water treatment requirements for different end uses provided there are no problems with microbial contamination (Milne *et al*., 2009). A summary of this is provided in Tables 5.5 and 5.6. Typical uses for recycled water include in-house water replacement for boiler feed or cooling tower make-up, CIP rinse waters, floor cleaning, irrigation and toilet flushing. Of these boiler feed, cooling tower make-up and CIP water are probably the most water-intensive processes where significant quantities of water can be used.

Although mixed wastewater streams may be recycled, most studies have focused on treatment and recycling of specific streams. One of the focuses has been on the use of RO permeate or condensate from the concentration of milk and juice products. Theoretically these wastewater streams should be of high quality, but they typically contain a high level of organics transferred either through imperfections in the membranes used or carryover

Table 5.5 Approximate water quality requirements for reuse based on average water qualities.

End use	Minimum treatment level	Comments
Cooling	Tertiary	May require nitrification
Washwater (housekeeping)	Secondary	Dependent on microbial content
Dust suppression	Tertiary	NA
Transport	Secondary	May require greater filtration
Lubrication	Secondary/tertiary	NA
Fire control	Secondary/tertiary	Refer to local guidelines
Irrigation (restricted)	Secondary	Dependent on microbial content
Irrigation (unrestricted)	Tertiary	NA
Toilet flushing	Tertiary	May require colour removal for aesthetics
Boiler feed	Reverse osmosis	May require further demineralisation

Source: Adapted from Milne *et al*. (2009).
NA, not applicable.

Table 5.6 Water quality requirements for boilers and cooling towers specified in standards and by manufacturers.

Parameter	End use				
	Boiler feed (0–2 MPa)	Boiler feed (2–3 MPa)	Boiler feed (3–4 MPa)	Common boiler feed requirements[a]	Common cooling tower limits[b]
Non-volatile TOC ($mg\ kg^{-1}$)	1.0	1.0	0.5	<1	10
Oily matter ($mg\ kg^{-1}$)	1.0	1.0	0.5		
Hardness ($mg\ kg^{-1}$ as $CaCO_3$)	0.3	0.3	0.2	<0.3	30–800
Iron ($mg\ kg^{-1}$)	0.1	0.05	0.03	<0.01	
Copper ($mg\ kg^{-1}$)	0.05	0.025	0.02	<0.001	
Total nitrogen ($mg\ kg^{-1}$)				<2	
Ammonia ($mg\ kg^{-1}$ as N)				<1.9	10–25
Turbidity (NTU)				<0.1	TSS: 50 $mg\ kg^{-1}$
Conductivity ($\mu S\ cm^{-1}$)				50–120	1200
TDS ($mg\ kg^{-1}$)					1000–10 000

Source: Adapted from Milne *et al*. (2009).
TOC, total organic carbon; NTU, nephelometric turbidity unit; TDS, total dissolved solids; TSS, total suspended solids.
[a]Average of requirements used at three industrial sites.
[b]Range of minimums established by different cooling tower manufacturers.
Empty spaces, data not reported.

in thermal processes. Reported qualities for these streams are shown in Table 5.7. What is important to note is that water of this quality is not suitable for common processes such as cooling tower make-up and boiler feed. Consequently, some level of treatment is required. The most common techniques used are NF (Mavrov & Belieres, 2000; Mavrov *et al*., 2001) and RO (IDF, 1988; Vourch *et al*., 2008), although one group has also investigated the possibility of biological treatment (Bickers & Bhamidimarri, 1998).

The main focus of condensate treatment has been on membrane filtration in order to make the water suitable for boiler feed or cooling tower make-up. It has been shown that NF of condensate can produce water of a suitable quality for boiler feed, assuming a two-stage NF process is used (Mavrov *et al*., 2001). In this instance, the second stage uses a membrane with higher rejection qualities than the first (70–85% sodium chloride (NaCl) rejection versus 85–95% NaCl rejection). A 94% total yield was achieved and, over the course of 2 months, only a minor loss in membrane permeability was reported (although an unspecified CIP process was in operation during this time). In application, a dairy in Australia has installed an RO facility for the treatment of condensate from the second, third and fourth effects in the milk concentration process in order to provide make-up

Table 5.7 Water quality of dairy condensates and permeates.

Industry	Suspended solids (mg kg^{-1})	COD (mg kg^{-1})	Conductivity (μS cm^{-1})	Calcium (mg kg^{-1})	References
Dairy condensate		5–45	6–190	0.4–1	Mavrov & Belieres (2000)
Dairy condensate		2–720	7–140	0.5–1.5	Chmiel *et al*. (2000)
Skimmed milk condensate	<2	24	18		Vourch *et al*. (2008)
Whole milk condensate	2–3	40	9–15		Vourch *et al*. (2008)
Whey condensate	3	30	101		Vourch *et al*. (2008)
Skimmed milk condensate		14–88			IDF (1988)
Sweet whey condensate		34–389			IDF (1988)
Acid whey condensate		216–1053			IDF (1988)
Dairy condensate (mother liquor)		622–1117			IDF (1988)
Permeate (milk)		100–350		1–2.4	Bickers & Bhamidimarri (1998)
Permeate (sweet whey)		343–508			IDF (1988)

Empty spaces, data not reported.

water to the cooling towers (Milne *et al*., 2009). The aim was simply to decrease the COD of the wastewater to less than 10 mg kg^{-1} to reduce the potential for biological fouling of the water. The process had a very short payback period and has been operated at the site effectively since installation.

In beverage manufacturing, the most commonly separated stream for recycling is the bottle-washing stream. Membrane-based processes for the recycling of this stream are generally UF and RO systems (Tay & Jeyaseelan, 1995; Mavrov & Belieres, 2000). In both cases, removal efficiencies for organics were high (typically greater than 94%); however, the salt reduction was significantly high for RO systems as would be expected. A Singaporean study put the costs of UF and RO at US$1.10 for very 1000 L and US$1.26 for every 1000 L, respectively (Tay & Jeyaseelan, 1995). Overall, the technique used would rely not just on the costs, but the end use for the recycled water and the quality requirements of that use.

Another important focus for recycling in the dairy and beverage industries is targeted not towards water recycling, but the reuse of chemicals in the CIP process. Here, UF or NF membranes are used to allow the passage of salts used in CIP, such as sodium hydroxide (NaOH), while retaining the larger fats, proteins and suspended solids. A range of different membranes types may be used for this task, although with the extremes in pH seen in CIP, there is greater attention paid to ceramic (Bhave *et al*., 2001) and metallic membranes (Yip *et al*., 1996) over polymeric membranes. The aim of the filtration is to remove the soil associated with cleaning processes while recovering the chemicals used in cleaning at the strengths in which they were received. Although other techniques, such as centrifugation

Table 5.8 Rejection of organic content by a metallic microfiltration membrane used in caustic recovery in simulated dairy and juice effluents.

Source	Parameter (% removal)				
	Suspended solids	Fat	Protein	Carbohydrates	COD
Semi-skimmed milk (2 g fat 100 g^{-1})	96	98.9	95.4	28.2	79
Vanilla ice-cream	99	99.7	96.5	21.6	59
Chocolate milk	94	97.7	94.8		56
Juice	39	77.1	65.5	21.1	33

Source: Data compiled from Yip *et al*. (1996).
COD, chemical oxygen demand.
Empty spaces, data not reported.

and decantation are possible, it has generally been found that water quality using filtration is higher (Gesan-Guiziou *et al*., 2007) and caustic recovery is greater (Bhave *et al*., 2001). Research by Yip *et al*. (1996) into recovery in dairy product and juice manufacturing was performed using stainless steel MF membranes (pore size 0.1–0.2 μm), and a summary of results is shown in Table 5.8. They showed that the effectiveness of separation of the cleaning compounds from the soil was quite high for dairy-based products. Typically all (>95%) fats, proteins and suspended solids were removed and only carbohydrate rejection was inefficient (about 20–30%). Separation in the juice-based wastewater was, however, less with far poorer removal efficiency across organic all components.

At full scale, ceramic membranes are used at a number of manufacturing sites including juice and dairy manufacturers (Bhave *et al*., 2001). One juice site uses ceramic membranes with a nominal pore size of 50 nm for caustic recovery at 20–30 × 10^3 L d^{-1}. The result was said to reduce the site's caustic usage by 30%.

General recycling of mixed wastewater streams in food manufacturing facilities has started to gain significant attention in drought-affected areas. In Australia, this was spurred by a drought during the years 2003–2010 in the south eastern part of the country and, in particular, southern Queensland. This has led to significant improvements in water recycling ratios and in wastewater recycling technology in the area. Overall, the processes rely on traditional wastewater treatment techniques up to the secondary treatment stage. Subsequent to this, RO is used to generate a higher than potable-quality water that is used to replace potable water in almost all processes on the site. Pilot-scale studies have been highlighted in the literature, with the focus being on pre-treatment to membrane processes to avoid fouling. Sarkar *et al*. (2006) investigated the use of coagulation/flocculation followed by powdered activated carbon treatment and MF as a pre-treatment to RO. The performance at the pilot-scale plant was discussed only in terms of removal efficiencies with no discussion of fouling or cleaning requirements. A fat, oil and grease removal of 71% and BOD removal of 81% was observed after MF, suggesting the pre-treatment was successful. Overall, the process was able to produce a water that was deemed to be of acceptable quality for recycling.

A similar study by Noronha *et al*. (2002) was performed on wastewater from a fruit juice manufacturer, and the treatment process is shown in Fig. 5.1. It uses a membrane

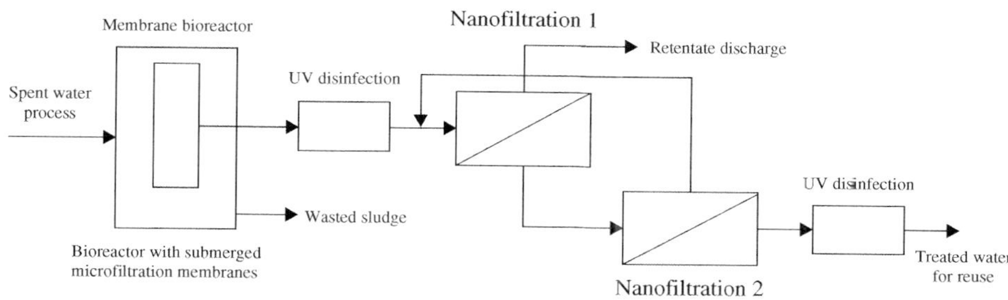

Fig. 5.1 Flow diagram for the recycling of juice manufacturing wastewater (After Noronha *et al.* (2002) and reproduced by permission of Elsevier).

bioreactor followed by UV disinfection and a two-stage NF process for the low-pressure reduction of salt content. A final UV disinfection step follows NF treatment. The plant operated for 6 months and achieved reductions for sodium of >80% and COD of >90%. Although fouling was observed in the first NF stage, the second stage proved to be resilient, and no significant loss in permeability was noted. The plant was able to meet all water quality requirements of the German Drinking Water Act except for sodium when the plant was operating at elevated recoveries.

Such a process is utilised in two breweries in Australia, leading to world's best practice with regards to water use (Apostolidis *et al*., 2011). At one site the process uses high-rate anaerobic treatment, followed by a fixed-film aerobic process, dissolved air filtration flotation, MF and RO (Apostolidis *et al*., 2011). The second brewery uses anaerobic treatment, dissolved air flotation, dissolved air filtration flotation, MF, RO, UV treatment and chlorine dioxide addition for oxidation and disinfection (Apostolidis *et al*., 2011). The processes have been reported to reduce chemical requirements at the plant, and significantly reduced the ongoing costs for water purchase and wastewater disposal. At the same time, the biogases generated during processing are used in gas-fired boilers as an alternative to natural gas. The breweries have reported a payback period on the investment in the treatment systems of 4 and 5 years, although this would be reduced based on recent increases in potable water costs and trade waste tariffs in the area (Apostolidis *et al*., 2011). Since the inception of this process at the breweries, its practice has spread through a number of food industries in Australia, including bakeries and poultry processors. In these cases, the industries were able to dispose of waste brine to sewers; however, sewers are not available at many manufacturing sites and the limitations surrounding brine disposal methods have proved to be a barrier to uptake in other parts of the country and in other industries such as dairy (Barr, 2011).

5.4 Conclusions and future applications

Increased regulation and charges for disposal of wastewaters from processing plants is driving more extensive treatment of these waters. Subsequently, membranes are finding increased application for treating these wastewaters, with MF and UF used for removal of solids, such as fats, oils and grease, membrane bioreactors for removal of organics both particulate and soluble and suspended solids, and NF and RO to desalinate brine wastes. The variable and high strength nature of the dairy and beverage wastewaters has resulted in the use of wide range of membrane materials, with ceramic membranes frequently used as well as traditional polymeric membranes. Additionally, new technologies, such as anaerobic membranes, are also finding niche markets in treatment of these waters.

The ability of membranes to deliver reliable water quality is not just enabling high treatment levels for disposal, but also for reuse. Recycled water is finding increasing application within processing plants for bottle washing, boiler feed water and other non-food contact applications, reducing water consumption and waste disposal.

References

ADI Kubota (2012a) Kubota MBR Case Study – Brewery. Available from http://www.kubota-mbr.com/industrial/ADI&Kubota_Brewery_final.pdf.

ADI Kubota (2012b) Kubota MBR Case Study – Salad Dressing Factory (Ken's Foods). Available from http://www.kubota-mbr.com/industrial/ADI&Kubota_Kens%20Foods_final.pdf.

Akoum, O., Jaffrin, M.Y., Ding, L.H. & Frappart, M. (2004) Treatment of dairy process waters using a vibrating filtration system and NF and RO membranes. *Journal of Membrane Science*, **235**, 111–122.

al-Zoubi, H., al-Thyabat, S. & al-Khatib, L. (2009) A hybrid flotation-membrane process for wastewater treatment: An overview. *Desalination and Water Treatment*, **7**, 60–70.

Apostolidis, N., Hertle, C. & Young, R. (2011) Water recycling in Australia. *Water*, **3**, 869–881.

Artiga, P., Carballa, M., Garrido, J. M. & Mendez, R. (2007) Treatment of winery wastewters in a membrane submerged bioreactor. *Water Science and Technology*, **56**, 63–69.

Bae, T.-H., Han, S.-S. & Tak, T.-M. (2003) Membrane sequencing batch reactor system for the treatment of dairy industry wastewater. *Process Biochemistry*, **39**, 221–231.

Balannec, B., Vourch, M., Rabiller-Baudry, M. & Chaufer, B. (2005) Comparative study of different nanofiltration and reverse osmosis membranes for dairy effluent treatment by dead-end filtration. *Separation and Purification Technology*, **42**, 195–200.

Barr, D. (2011) Australian Dairy Manufacturers Sustainability Council Meeting, Dairy Innovations Australia Ltd.

Bhave, R., Jung, G. & Sondhi, R. (2001) Potential Savings Through Ceramic Membrane Caustic Reclaim. Available from http://www.pall.com/pdfs/Food-and-Beverage/foo_pdf_CausticRecovery-2001.pdf.

Bickers, P.O. & Bhamidimarri, R. (1998) Aerobic treatment of reverse osmosis permeate in the dairy industry for reuse. *Water Science and Technology*, **38**, 61–67.

Bolzonella, D., Fatone, F., Pavan, P. & Cecchi, F. (2010) Application of a membrane bioreactor for winery wastewater treatment. *Water Science and Technology*, **62**, 2754–2759.

Bowen, W.R. & Doneva, T.A. (2000) Atomic force microscopy studies of nanofiltration membranes: surface morphology, pore size distribution and adhesion. *Desalination*, **129**, 163–172.

Britz, T.J., van Schalkwyk, C. & Hung, Y.-T. (2004) Treatment of Dairy Processing Wastewaters. *Waste Treatment in the Food Processing Industry* (eds. L.K. Wang, Y.-T. Hung, H. H. Lo and C. Yapijakis), pp. 1–28, CRC Press, Boca Raton.

Canut, A. & Pascual, A. (2008) Pollution prevention in food industries through both cleaning of closed equipment with ozonated water clean in place CIP systems. *WIT Transactions on Ecology and the Environment*, **111**, 615–625.

Casani, S., Rouhany, M. & Knochel, S. (2005) A discussion paper on challenges and limitation to water reuse and hygiene in the food industry. *Water Research*, **39**, 1134–1146.

Chen, J.P., Seng, S.-S. & Hung, Y.-T. (2004) Soft Drink Waste Treatment. *Waste Treatment in the Food Processing Industry* (eds. L. K. Wang, Y.-T. Hung, H. H. Lo and C. Yapijakis), pp. 255–269, CRC Press, Boca Raton.

Chmiel, H., Mavrov, V. & Belieres, E. (2000) Reuse of vapour condensate from milk processing using nanofiltration. *Filtration+Separation*, **37**, 24–27.

Chmiel, H., Kaschek, M., Bloecher, C., Noronha, M. & Mavrov, V. (2002) Concepts for the treatment of spent process water in the food and beverage industries. *Desalination*, **152**, 307–314.

Cicek, N. (2003) A review of membrane bioreactors and their potential application in the treatment of agricultural wastewater. *Canadian Biosystems Engineering*, **45**, 6.37–36.49.

Colic, M., Hogan, J. and Lechter, A. (2010) Flocculation Flotation Pretreatment Improves Operation of MBR Installed to Treat Snack Food Manufacturing Wastewater. *WEFTEC 2010*, New Orleans.

Danalewich, J.R., Papagiannis, T.G., Belyea, R.L., Tumbleson, M.E. & Raskin, L. (1998) Characterization of dairy waste streams, surrent treatment practices, and potential for biological nutrient removal. *Water Research*, **32**, 3555–3568.

Donkin, M.J. (1997) Bulking in aerobic biological systems treating dairy processing wastewaters. *International Journal of Dairy Technology*, **50**, 67–72.

Ekama, G.A. (2011) 4.14 – Biological Nutrient Removal. *Treatise on Water Science* (ed. W. Peter), pp. 409–526, Elsevier, Oxford.

Farizoglu, B. & Uzuner, S. (2011) The investigation of dairy industry wastewater treatment in a biological high performance membrane system. *Biochemical Engineering Journal*, **57**, 46–54.

Ferre, V., Trepin, A., Giminez, T. & Lluch, S. (2009) Design and performance of full-scale iMBR plants treating winery wastewater effluents in Italy and Spain. *Winery 2009 – 5th International Specialized Conference on Sustainable Viticulture*, Torino.

Frappart, M., Akoum, O., Ding, L.H. & Jaffrin, M.Y. (2006) Treatment of dairy process waters modelled by diluted milk using dynamic nanofiltration with a rotating disk module. *Journal of Membrane Science*, **282**, 465–472.

Gesan-Guiziou, G., Alvarez, N., Jacob, D. & Daufin, G. (2007) Clean-in-place coupled with membrane regeneration for re-using caustic soda solutions. *Separation and Purification Technology*, **54**, 329–339.

Hai, F.I. & Yamamoto, K. (2011) 4.16 – Membrane Biological Reactors. *Treatise on Water Science* (ed W. Peter), pp. 571–613, Elsevier, Oxford.

Ince, O. (1998) Performance of a two-phase anaerobic digestion system when treating dairy wastewater. *Water Research*, **32**, 2707–2713.

IDF (1988) *The Quality, Treatment and use of Condensate and Reverse Osmosis Permeates*, International Dairy Federation, Brussels, Belgium.

IFC (2007a) *Environmental, Health, and Safety Guidelines for Breweries*, International Finance Corporation – World Bank Group, Washington, DC.

IFC (2007b) *Environmental, Health, and Safety Guidelines for Dairy Processing*, International Finance Corporation – World Bank Group, Washington, DC.

IFC (2007c) *Environmental, Health, and Safety Guidelines for Food and Beverage Processing*, International Finance Corporation – World Bank Group, Washington, DC.

Janczukowocz, W., Zielinski, M. & Debowski, M. (2008) Biodegradability evaluation of dairy effluents originated in selected section of dairy production. *Bioresource Technology*, **99**, 4199–4205.

Koyuncu, I., Turan, M., Topacik, D. & Ates, A. (2000) Application of low pressure nanofiltration membranes for the recovery and reuse of dairy effluents. *Water Science and Technology*, **41**, 213–221.

Mackey, E.D. & Seacord, P.E. (2008) *Regional Solutions for Concentrate Management*, Water Resue Foundation, Alexandria, USA.

Marks, J.S., Martin, B. & Zadoroznyj, M. (2007) Trust in the expert system of water recycling. *Oz Water 2007 Conference and Exhibition*, Sydney.

Matosic, M., Prstec, I., Jakopovic, H.K. & Mijatovic, I. (2009) Treatment of beverage production wastewater by membrane bioreactor. *Desalination*, **246**, 285–293.

Mavrov, V. & Belieres, E. (2000) Reduction of water consumption and wastewater quantities in the food industry by water recycling using membrane processes. *Desalination*, **131**, 75–86.

Mavrov, V., Chmiel, H. & Belieres, E. (2001) Spent process water desalination and organic remocal by membranes for water reuse in the food industry. *Desalination*, **138**, 65–74.

Mickley, M. (2003) Direction in management of membrane side streams. *Membrane Technology*, **2003**(7), 5–9.

Milne, N.A., Gray, S., Tjandraatmadja, G., Diaper, C., Taylor, R., Zhang, G. & Reeve, D. (2009) *Guidance for the Use of Recycled Water by Industry*, Victoria University, CSIRO and Smart Water Fund.

Milne, N.A., Stewart, M., Reeve, D., Orbell, J. & Gray, S. (2011) *Best Practice Trade Waste Management*, Victoria University, Melbourne.

Miyaki, H., Adachi, S., Suda, K. & Kojima, Y. (2000) Water recycling by floating media filtration and nanofiltration at a soft drink factory. *Desalination*, **131**, 47–53.

Monclús, H., Sipma, J., Ferrero, G., Comas, J. & Rodriguez-Roda, I. (2010a) Optimization of biological nutrient removal in a pilot plant UCT-MBR treating municipal wastewater during start-up. *Desalination*, **250**, 592–597.

Monclús, H., Sipma, J., Ferrero, G., Rodriguez-Roda, I. & Comas, J. (2010b) Biological nutrient removal in an MBR treating municipal wastewater with special focus on biological phosphorus removal. *Bioresource Technology*, **101**, 3984–3991.

Nagano, A., Arikawa, E. & Kobayashi, H. (1992) The treatment of liquor wastewater containing high-strength suspended solids by membrane bioreactor system. *Water Science and Technology*, **26**, 887–895.

Ng, K.-K., Lin, C.-F., Panchangam, S.C., Hong, P.-K.A. & Yang, P.-Y. (2011) Reduced membrane fouling in a novel bio-entrapped membrane reactor for treatment of food and beverage processing wastewater. *Water Research*, **45**, 4269–4278.

Noronha, M., Britz, T., Mavrov, V., Janke, H.D. & Chmiel, H. (2002) Treatment of spent process water from a fruit juice company for purposes of reuse: Hybrid process concept and on-site test operation fo a pilot plant. *Desalination*, **143**, 183–196.

Palmowski, L., Baskaran, K., Wilson, H. & Watson, B. (2005) *Clean in Place – A Review of Current Technology and its Use in the Food and Beverage Industry*, Deakin University, Melbourne.

Po, M., Kaercher, J.D. and Nancarrow, B.E. (2005) *Literature Review of Factors Influencing Public Perceptions of Water Reuse*, CSIRO Land and Water, Perth.

Robinson, T. (2006a) Dairy Effluent Treatment Case Study: Dairy Crest, Davidstow Creamery, UK. Retrieved 29–2-2012, 2012, from http://www.wehrle-env.co.uk/documents/davidstow_low_res.pdf.

Robinson, T. (2006b) Dairy Effluent Treatment Case Study: Dairygold, Mitchelstown, Ireland. Available from http://www.wehrle-env.co.uk/documents/mitchelstown_low_res.pdf.

Samkutty, P. & Gough, R. (2002) Filtration treatment of dairy processing wastewater. *Journal of Environmental Science & Health, Part A: Toxic/Hazardous Substances & Environmental Engineering*, **37**, 195–199.

Sarkar, B., Chakrabarti, P.P., Vijaykumar, A. & Kale, V. (2006) Wastewater treament in dairy industries – Possibility of reuse. *Desalination*, **195**, 141–152.

Strohwald, N.K.H. and Ross, W.R. (1992) Application of the ADUF process to brewey effluent on a laboratory scale. *Water Science and Technology*, **25**, 95–105.

Strydom, J.P., Britz, T.J. & Mostert, J.F. (1997) Two-phase anaerobic digestion of three different dairy effluents using a hybrid bioreactor. *Water SA*, **23**, 151–156.

Tay, J.-H. & Jeyaseelan, S. (1995) Membrane filtration for reuse of wastewater from beverage industry. *Resources, Conservation and Recycling*, **15**, 33–40.

Ukita, M., Imai, T. & Hung, Y.-T. (2004) Food Waste Treatment. *Waste Treatment in the Food Processing Industry* (eds. L.K. Wang, Y.-T. Hung, H.H. Lo and C. Yapijakis), pp. 291–319, CRC Press, Boca Raton.

Vince, F., Aoustin, E., Breant, P. & Marechal, F. (2008) LCA tool for the environmental evaluation of potable water production. *Desalination*, **220**, 37–56.

Vourch, M., Balannec, B., Chaufer, B. & Dorange, G. (2008) Treatment of dairy industry wastewater by reverse osmosis for water reuse. *Desalination*, **219**, 190–202.

Wang, L.K. and Menon, R. (2009) Membrane Bioreactors. *Handbook of Environmental Engineering* (eds. L.K. Wang, N.K. Shammas and Y.-T. Hung), Volume 9: Advanced Biological Treatment Processes, pp. 129–156, Humana Press, New York.

Wehrle Environment UK (2012a) Dairy Wastewater Treatment Case Study: Dairy Crest, Foston, Derbyshire, UK. Available from http://www.wehrle-env.co.uk/documents/foston_low_res.pdf.

Wehrle Environment UK (2012b) Wastewater Treatment Case Study: Glenallachie Distillery, Scotland. Available from http://www.wehrle-env.co.uk/documents/glenallachie.pdf.

Wen, G., Ma, J., Zhang, L. & Yu, G. (2010) 4.07 – Membrane bioreactor in water treatment. *Comprehensive Membrane Science and Engineering* (eds. D. Enrico and G. Lidietta), pp. 195–209, Elsevier, Oxford.

Wu, Z., Zhou, Z., Wang, Z., Tian, L., Pan, Y. & Wan, X. (2009) The application of membrane bioreactor technology to the treatment of wastewater from a multifunctional supermarket. *Environmental Progress and Sustainable Energy*, **29**, 52–59.

Xie, R.J., Gomez, M.J., Xing, Y.J. & Klose, P.S. (2004) Fouling assessment in a municipal water reclamation reverse osmosis system as related to concentration factor. *Journal of Environmental Engineering and Science*, **3**, 61–72.

Yang, W., Cicek, N. & Ilg, J. (2006) State-of-the-art of membrane bioreactors: Worldwide research and commerical applications in North America. *Journal of Membrane Science*, **270**, 201–211.

Yip, V., Arntfeld, S.D. & Hydamaka, A.W. (1996) Potential for stainless steel microfiltration processing to reduce effluent from a fluid milk and ice cream processing plant. *Journal of Dairy Science*, **79**, 710–716.

Yoon, Y. & Lueptow, R.M. (2005) Reverse osmosis membrane rejection for ersatz space mission waters. *Water Research*, **39**, 3298–3308.

6 Liquid Milk Processing

G. Gésan-Guiziou

6.1 Introduction

The dairy industry may be characterised as a processing unit outside the water treatment industry which has, most successfully, been able to put membranes to work for the benefit of the industry and its customers. Nowadays, the presence of membrane equipment in dairy factories is very common, and encompasses a large variety of different processes that have been successfully implemented throughout the different manufacturing stages (i.e. from reception of raw milk to the waste-water treatment).

Membrane applications in the dairy industry dates back to the late 1960s with the concentration of milk proteins using the MMV process, named after its inventors, Maubois, Mocquot and Vassal. Simultaneously, membrane technology was applied to whey processing allowing the production of protein concentrates and, thus, transforming a waste by-product from cheese production into valuable products. Since then, the development of membrane applications in the dairy sector has been more or less tightly linked to the progress in membrane operations, such as the asymmetric organic membranes in the early 1970s, composite inorganic membranes in the early 1980s, porous ceramic membranes with multichannel configuration and uniform transmembrane pressure (UTMP) system in the early 1990s.

With regards to liquid milk processing, the main applications are the adjustment of the level of the protein (also known as standardisation of the protein content) by ultrafiltration (UF), and the removal of bacteria from milk by microfiltration (MF), which has become a worldwide success. Some applications of membranes (reverse osmosis (RO), UF) include on-farm concentration of milk, which has been successfully developed, especially when transporting milk for long distances and when large farms are involved. Recently, some studies have highlighted interesting applications devoted to the separation of milk fat globules according to their specific size, and the separation of somatic cells, but these processes have not been industrially applied yet.

6.2 On-farm concentration of milk

The concept of milk concentration on the farm, near the point of milking, dates back to the early 1970s. From that time many benefits for concentrating milk on the farm rather than at

Membrane Processing: Dairy and Beverage Applications, First Edition. A. Y. Tamime.

a separate processing plant have been claimed. The milk retentate can be separated on-site from the permeate, enabling each product to be shipped to the most profitable location. However, the most economic way to dispose of the permeate is to feed it to cows on the farm. Thus, dairy farmers can reduce their milk transportation costs because of reduction of milk volume; they are no longer paying to ship unwanted water. Concentrated milk reduces refrigeration costs because of smaller volume required for storage facilities and, once such milk reaches its destination, the heating cost is reduced due to less fluid being processed. In addition, if the concentrated milk is used for cheesemaking, the yield is increased.

On-farm milk concentration has been explored since the early part of 1980s, and the technical feasibility of the process has been shown to be easily applied. Several studies were conducted mainly by Maubois's group at the National Institute for Agricultural Research in Rennes, France, and by Cornell University (Ithaca, New York) in the United States of America (USA). In 1977, Alfa-Laval (a Swedish company and currently known as Tetra Pak) developed an on-farm UF processing unit, known as Ultratherm (Kosikowski, 1985).

The suggested process mainly used UF or RO to concentrate milk. Nowadays, nanofiltration (NF) is sometimes used. In order to reduce bacterial growth during the membrane concentration step, the initial count of bacteria in raw milk should be low because as the concentration factor of filtration is increased, the higher the concentration of the milk constituents will be achieved and also the bacterial count. Hence, in order to maintain the bacteriological quality of the product, heat treatment (thermisation, pasteurisation) is generally performed before transportation. The microbiological quality of the on-farm concentrated milk (UF, two-fold concentration factor) combined with heat treatment (72°C for 15 s) was actually shown to be acceptable (Benard *et al*., 1981). However, this heat treatment combined with a second pasteurisation performed at the cheese factory reduced the cheese yield. In the 1980s, UF was used in French farms to concentrate milk two-fold. The retentate, after thermisation and cooling, was delivered to a cheese plant for the manufacture of Emmental and Saint Paulin cheeses without any adverse effects on the yield of the products. The permeate produced at the farms was fed to cows, resulting in savings in feed (Anonymous, 1984). In another study, milk was collected from 22 farms located on a French island, it was then concentrated two-fold by UF and heat treated (72°C for 15 s). After cooling to 2°C, the concentrates were shipped to a local factory twice a week (Kosikowski, 1985). The use and economic impact of the transportation of RO milk was also evaluated in Australia (Cox & Langdon 1985). In a long study in California, a 400 cow herd was used to study the feasibility of both UF and RO on-farm (Zall, 1987). The concentration factors used in such processes ranged between 2.0 and 3.5 for UF milk and up to 2.5–2.8 for RO milk.

Despite the technical feasibility of the process, the concept of on-farm milk concentration was abandoned after several large-scale trials. This was because the regulatory concerns about the quality of raw milk and processing conditions, such as temperature, has limited UF of milk to in-plant usage. It was also generally shown that the benefits of lower transportation costs do not outweigh the capital investment and running costs (mainly attributed to membrane replacement), except in individual cases when long distances and large farms are involved. Slack *et al*. (1982) studied the economics of on-farm UF in the USA by using two- to three-fold concentrations, the economic advantages were only possible for large farms with 100 to 1000 cows.

It was not until the mid-1990s that some dairy farmers successfully used membrane technology to concentrate unpasteurised milk on the farm. The process was then extensively evaluated before the regulatory agencies approved the use of membrane filtration in Grade A dairy foods and, as a consequence, the on-farm concentration of milk became a marketplace reality. Several membrane modules were then installed on farms in the southern part of the USA, mainly in Texas, California and New Mexico, and operated exclusively on large herds with more than 1500 milking cows (Fleming, 1999). In these processes, RO, NF and mainly UF were used at farms to concentrate milk (i.e. 3.5-fold) at temperature below 8°C in order to limit the growth of bacteria in the retentate. The concentrated milk was then shipped at 4°C and transported throughout the country to several plants where it was used to standardise the protein level in milk prior to cheesemaking, or it was utilised in the manufacture of frozen dairy desserts. The total bacterial count of the concentrate was less than 300×10^3 colony forming units (cfu) mL^{-1}, and was classified as a Grade A product according to the US Food and Drug Administration (Mistry & Maubois, 2004).

6.3 Protein standardisation by ultrafiltration

For nearly 40 years, UF offered the possibility of standardising the protein content of milk. This operation makes it possible to concentrate all the milk proteins (serum proteins and caseins) and fat contents, and simultaneously allow the lactose and minerals to pass through the membrane. The change in the protein composition of milk from that which is received from the producer is called protein standardisation. The main objective of the standardisation of the proteins, one of the two most valuable components of milk, is to overcome the inherent day to day variations in milk composition, resulting in significant nutritional and technological and economical benefits. It has been reported that the average protein content of milk (3.3–3.4 g 100 g^{-1}) collected by dairy plants may show relatively wide natural variations (2.8–4.2 g 100 g^{-1}) by international comparisons (Rattray & Jelen, 1996) due to various factors, such as stage of lactation, feeding, breed, seasons of the year, weather, etc.

Depending on the objective of the standardisation, the UF process can be performed either for the concentration of protein so as to reach the prescribed protein content of milk, or for the reduction of protein content of milk by addition of ultrafiltered permeate (Fig. 6.1). Nowadays, the standardisation of proteins using the UF process is well established for the increase of the protein content of milk during the manufacture of cheese, but is not widely practised for drinking milk because it is still under debate by the regulatory bodies.

In cheese manufacture, one can note that MF using a 0.1 μm pore membrane is used more and more for the specific enrichment of milk in native casein micelles (Maubois, 1991; Gésan-Guiziou *et al*., 1999; Saboya & Maubois, 2000). The fact that less serum proteins end up in the cheese-manufacturing process is advantageous since it reduces flavour and texture defects attributed to serum proteins, and reduces also the detrimental effects of heat treatment on rennet milk coagulability.

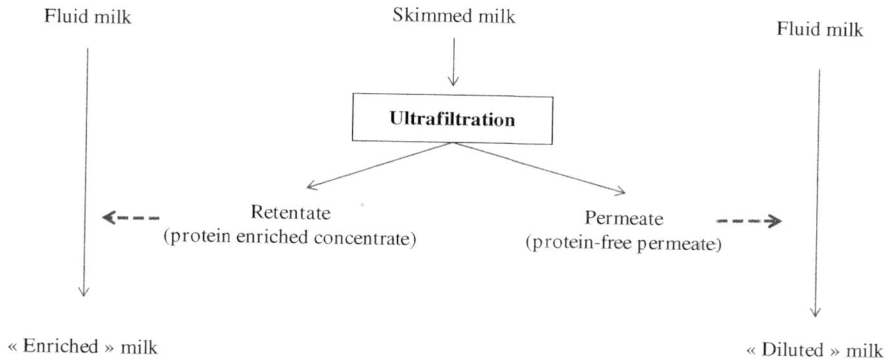

Fig. 6.1 Principles of protein standardisation process using ultrafiltration.

6.3.1 *Advantages of protein standardisation*

The standardisation of milk proteins offers several advantages, mainly for cheesemakers, and are as follows.

- It increases the output of specific equipment, mainly in cheesemaking, and lessens the need for adjustments of processing parameters by the cheesemakers. Variations in the composition of incoming milk actually result in non-efficient use of machinery in dairies and difficulties in control of the continuous cheesemaking processes (clotting rate, coagulum firmness, rate of pH drop due to buffering by caseins).
- Standardisation of protein content improves the cheese process productivity. This can be attributed to an increase in production of cheese per vat per day. A slight increase in the cheese yield is observed, and is attributed to reduced losses of fat and casein in the whey and better retention of soluble proteins in the aqueous phase of cheese (Mistry & Maubois, 2004). Caseinomacropeptide, which is classically released in the whey during cheesemaking due to the action of rennet, is also retained in the cheese curd. Moreover, the reduced variability of milk composition reduces the variability in cheese composition, and results in production of less cheese that does not meet quality standards related to functionality and body characteristics.
- Standardisation results in a constant milk composition all year round, and offers the possibility of increasing the protein content in milk without the need of adding milk powders, casein or whey protein concentrates, which is commonly practised by some cheesemakers. In addition to the improvement of process productivity, these levels of concentration provide economic benefits in terms of reduced amounts for coagulating enzymes and starter cultures compared with ordinary milk.
- Increasing the protein content of raw milk improves the physical properties of certain dairy products, and may increase their nutritional value. UF, with volume reduction ratio (VRR) of around 1.5, is, for instance, used during the manufacture of yoghurt and related fermented milk products in order to increase the protein levels without addition of milk powders (Tamime & Robinson, 2007). UF results in a better yoghurt

gel strength than those obtained with the addition of milk powders or concentration by vacuum evaporation. This process is also preferred to RO because of the reduced level of lactose in the milk base. In the case of fluid milks, higher nutritional and improved sensory properties are sometimes claimed by the manufacturer. However, in contrast to the well-established sensory effects of milk fat, no evidence in the literature confirms the sensory effects of milk proteins, i.e. protein standardisation within rather wide limits (1.5–6.4 g 100 g^{-1} for whole milk, 2.4–6.5 g 100 g^{-1} for skimmed milk) has no effect on the sensory properties of the product (Jelen, 2011). UF standardisation can also advantageously be used to bring the protein content of regular milk to a legal minimum, should such a minimum exist or be agreed upon. Present European Union (EU) legislation specifies that cow's milk shall contain not less that 2.8 g 100 mL^{-1} total protein (nitrogen × 6.38). It is thus conceivable that in some instances where the protein level in milk is low, then protein concentration may be required, especially if there were to be an international agreement on a minimum protein content in fluid milk. In the opposite scenario, in case milk contains high protein content, the UF permeate could advantageously be used to reduce the protein content of the milk. This practice is forbidden in the EU since 1997, but had led to serious profit for many ultra-high temperature (UHT) milk processors.

6.3.2 *Regulatory aspects*

Protein standardisation of milks other than drinking milks (e.g. milk used as a raw material for manufacture of products, such as cheeses, yoghurts and other dairy products) is widely being practised, mainly for technological, economical and sensory reasons. UF and MF 0.1 μm membranes can be used for this purpose.

Although protein standardisation is commonly accepted for these products and has been legally authorised for many years, the proposal to deliver consumer milk with a defined protein content is still an issue being debated in international dairy circles, such as the Codex Alimentarius Commission (Jelen, 2011). However, the UF process, which is used for standardisation that results in the reduction of protein content of milk, is unacceptable and has encountered large opposition. This could be somewhat surprising because protein content is one of the payment criteria to milk producers. The use of permeates from UF of milk to standardise the protein content in UHT milks may have been practised in the past without labelling the products as protein standardised, the dilution of milk with permeate being undetectable. To our knowledge, no country in the world has modified its legislation to allow protein standardisation (i.e. dilution) using UF permeate of regular market milk. Such a practice is expressly forbidden in the EU, even if published research has indicated undistinguishable sensory and technological properties in UHT milks obtained after blending with the UF permeate down to 3.2–2.6 g 100 g^{-1} protein (Jelen, 2011). This protein adjustment by blending UF permeate is, however, allowed in concentrated and dried dairy products and approved, in that context, by the Codex Alimentarius Commission.

Drinking milks with higher protein content than normal is now common in various countries, such as the USA and Australia (Jelen, 2011). Milk with higher than normal protein content is said to offer superior nutritional and sensory properties.

6.3.3 *Process involved*

Similar to the fat standardisation procedure in the dairy industry, the UF standardisation of protein can be accomplished either in batch mode or in continuous (in-line) mode. Typically, the milk is first skimmed and the standardisation is carried out on the skimmed milk. Then, depending on the aim of the protein standardisation process (increase or decrease in protein content, Fig. 6.1), one of the streams (retentate or permeate) is blended in varying ratios with the regular fluid milk. In a batch standardisation process, the retentate (in the case of protein concentration) or permeate (in the case of protein reduction) obtained after ultrafiltered milk is collected in a storage tank and then reblended with a batch of incoming unstandardised skimmed milk to reach the prescribed final protein content. In industrial practice, the permeate could come from a separate department or factory (Jelen, 2011). In many plants, a continuous standardisation mode is generally preferred, where the protein level is rapidly and continuously measured using special sensors as the milk is pumped through pipelines. Two options exist; first, all of the milk can be treated by the UF process and, in that case, the concentration factor is adjusted to get to the desired value. UF is then used to slightly concentrate the protein content up to a VRR of 1.7–1.8. (VRR = V_r/V_f where V_f and V_r are the volumes of feed and retentate respectively). Second, the milk is concentrated by UF up to a VRR of about 3, and the concentrate is reblended with the unstandardised milk. In both cases, the membranes used for the production of milk protein concentrates are classically of spiral-wound configurations and, to a lesser extent, tubular ceramic membranes; both membranes with a cut-off of 10–20 kg mol^{-1} operate at a transmembrane pressure of 200–300 kPa with permeation flux around 100 L h^{-1} m^{-2} at 50°C and VRR about 1.7 (Fig. 6.2).

Fig. 6.2 Commercial ultrafiltration system for the standardisation of milk (Reproduced courtesy of TIA, France; http://www.tia.fr).

UF can be performed at temperatures of around 50°C or around 10°C. At 'high' temperatures, permeation fluxes are higher, but this temperature favours both the mineral precipitation of calcium phosphate in the membrane pores and bacterial growth in the retentate. In order to limit bacterial development in spiral-wound membranes, a process duration of 6 h is generally recommended. However, at 'low' temperature (9–12°C), and especially with organic spiral-wound membranes, no calcium phosphate is observed in the membrane pores, and fouling is mainly due to the adsorption and accumulation of proteins at the membrane surface. A process duration of 10–12 h is classically observed with no or little bacterial growth. The process can last longer provided the final retentate is heat treated prior to further use.

6.4 Removal of bacteria by microfiltration

The control of bacterial growth of collected milk and the minimisation of health hazard during milk processing can be achieved through several operations. Heat treatment with various combinations of time/temperature parameters (such as thermisation, pasteurisation, UHT or sterilization by autoclaving) can be carried out to obtain the desired bactericidal effect. Although these treatments are frequently used at the industrial scale to ensure the safety of dairy products, they almost always induce irreversible modifications of milk components and, as a consequence, affect the flavour, the functionalities of the milk components, and the cheesemaking properties. Moreover, the lysed cells remain in heated milks with their potential active enzymes, resulting in alterations of milk during storage (reduction of commercial shelf-life) and processing. In cheese manufacture, centrifugation (often referred as bactofugation because the commercial equipment manufactured by Tetra Pak is marketed under the trademark of Bactofuge^{TM}) is also used for the removal of bacteria and spores from minimally pasteurised milk because it efficiently removes spores, as their density is higher than the bacteria (Gésan-Guiziou, 2010). However, the decimal reduction is generally low (~1 $\log_{10}$) and significant protein loss is observed, reaching 2.5–12 g 100 g^{-1} according to the types of machines used.

In that context, MF, which is a non-thermal preservation technology, offers an interesting alternative. Initially proposed in the mid-1990s, it has become more and more applicable for the removal of bacteria of drinking and cheese milks.

6.4.1 *Microfiltration process: operating conditions and performances*

Several studies were conducted in Sweden and France in order to optimise the original parameters described in the patent of Holm *et al*. (1986). Fig. 6.3 illustrates the MF process generally used, and it is classically and specifically applied to skimmed milk, because the size ranges of fat globules and bacteria overlap. In the case of whole milk, membrane fouling would be more pronounced, and fat would be concentrated with bacteria which can lead to difficulties for the valorisation of the retentate. It can be noted, however, that a very recent patent (Fauquant *et al*., 2009) shows that MF using a 0.8 or 1.4 μm mean pore diameter membrane can make it possible to remove bacteria and spores from milk containing fat, provided that full-fat milk has been previously homogenised in order to

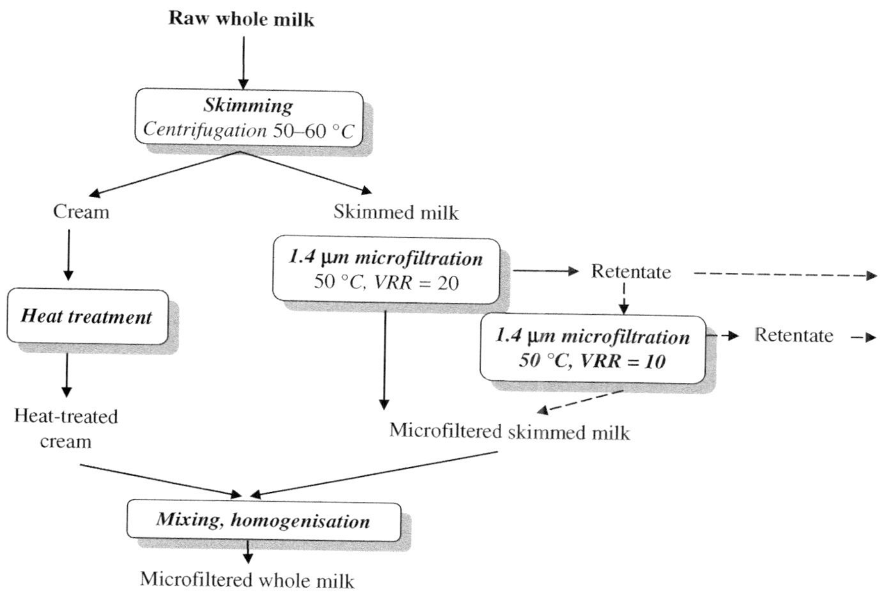

Fig. 6.3 Schematic representation of process for microfiltration of whole milk ntended for the drinking milk market. The retentate is generally discharged for other appropriate applications. VRF, volume reduction ratio.

reduce the size of the fat globules. Following the separation of the cream, the skimmed milk is treated in a MF plant. This separation is classically performed with a multichannel ceramic membrane with a pore diameter of 1.4 μm in order to retain bacteria and let the proteins pass through the membrane (Fig. 6.3); the skimmed milk is classically heated to 50°C. In order to achieve good separation, high cross-flow velocity in combination with low and UTMP are required (Gésan-Guiziou, 2010). A cross-flow velocity between 6 and 9 m s^{-1} and a TMP of about 0.05 MPa are generally applied. In order to maintain low TMP with such high cross-flow velocity and overcome membrane-fouling phenomena, most MF plants in dairies operate using either the hydraulic concept of UTMP (Sandblöm, 1974) or specific ceramic membrane with continuous variation in its hydraulic resistance. The first MF plants were equipped with UTMP systems, which were developed by Alfa-Laval Company (Bactocatch® system) in the mid-1990s. This concept involves the circulation by pumping of the co-current permeate of the retentate in order to create a pressure drop on the permeate side similar to the pressure drop on the retentate side (Sandblöm, 1974). By doing so, the TMP is homogeneous along the membrane length (Gésan-Guiziou, 2010), which results in control of the protein deposit at the membrane surface and consequently in acceptable MF performance. To create a large pressure drop on the permeate side and avoid investing in a high-capacity pump, the external space between the housing and the membrane porous media is reduced by filling the permeate compartment with plastic balls (Bactocatch® system, Alfa-Laval; Holm *et al.*, 1986).

Ceramic membranes with linear hydraulic resistance were more recently commercialised in order to obtain homogeneous filtration performance all along the membrane length.

The use of these membranes avoids a permeate circulation loop and extra investments and running costs because of the permeate pump. This new conception of membrane consists in creating an inhomogeneous membrane, having a higher hydraulic resistance (or lower permeation) at the membrane entrance where the local TMP is high, and a low resistance (or higher permeation) at the membrane outlet. The gradient of membrane resistance is established during membrane manufacture such that the membrane resistance is inversely varied according to the linear decrease in pressure in the retentate side. Two commercial membranes are proposed: (a) permeability (GP) Membralox® membranes from Pall-Exekia with a continuous variation of the porosity of the membrane support (Garcera & Toujas, 1998); and (b) Isoflux membranes from Tami-Industries with continuous variation in the thickness of the selective front layer of the membrane (Skrzypek & Burger, 2010). Both membranes are obviously constructed for well-defined hydrodynamic conditions, and consequently must be used for well-defined applications.

By combining such a system (UTMP or membrane with linear resistance) and high cross-flow velocity (6–9 m s^{-1}), it is possible to concentrate all the somatic cells up to 20 times (i.e. VRR of 20), most of the residual fat and bacteria in the MF retentate. Permeation fluxes are high, from 400 to 650 L h^{-1} m^{-2} for around 10 h of production with a volume loss of 5% at a VRR 20. In some MF industrial plants, the first retentate is concentrated 10-fold more in a second MF operation, thus leading to a total VRR of 200 and a final retentate loss of no more than 0.5% of the incoming milk. The final retentate is generally discharged separately for other suitable applications (Fig. 6.3). During this operation, the total solids and protein are largely transmitted – 99.5 and 99% of transmission, respectively (Maubois & Ollivier, 1997). Average observed decimal reduction of bacteria ranges from 2–3 $\log_{10}$ with the first ceramic membranes (Malmberg & Holm, 1988; Trouvé *et al.*, 1991), and reaches 3–4 $\log_{10}$ with the currently used 1.4 μm Sterilox or GP membranes, for milk collected in most developed dairy countries (initial total count $<200 \times 10^3$ cfu mL^{-1}). Spore-forming bacteria, which represent the main surviving species to pasteurisation, are highly retained by the MF membrane (decimal reduction $>4.5 \log_{10}$) because of their large apparent cellular volume when they are in milk. Analyses performed by the Pasteur Institute and the Institut National de la Recerche Agronomique (INRA – the English equivalent is National Institute for Agricultural Research) in France assure that microfiltered skimmed milk contains less than 1 cfu mL^{-1} of pathogenic bacteria, taking into account the usual contamination levels at farm level (Saboya & Maubois, 2000). In addition, the analyses actually show that decimal reduction of pathogenic bacteria (*Listeria monocytogenes, Brucella abortus, Salmonella typhimurium* and *Mycobacterium tuberculosis*) is 3.5–4.0 $\log_{10}$. Microfiltered milk can then be considered as safe as pasteurised milk.

Finally, in order to adjust the desired protein-to-fat ratio corresponding to the final product, the microfiltered milk was mixed with cream previously subjected to a conventional high heat treatment. Since less than 10% of the milk is heat treated at high temperature, the sensory quality of the milk is significantly improved. Therefore, MF is more efficient than bactofugation and, contrary to heat treatment procedures, it not only curbs the activity of bacteria but it physically removes bacteria, spores, lysed cells and other impurities from the milk altogether.

6.4.2 *Industrial applications*

The 1.4 μm MF process has been used increasingly for the treatment of drinking and cheese milks. For drinking milks, the objective is to produce a value-added product with a higher purity and longer shelf-life than milk treatment by conventional heat treatment processes.

Several products have been offered to consumers. France is the only country that has officially allowed the commercialisation of extended shelf-life (ESF) microfiltered raw milk. The skimmed milk is mixed with the required amount of heated cream (95°C for 20 s); the mixture is homogenised and aseptically filled. The authorised shelf-life of this milk is 3 weeks at 4–6°C. In France, the Marguerite® milk is considered to be raw milk because no pasteurisation is applied, and it has a positive phosphatase reaction. To our knowledge only one factory in France produces such milk, and annual production reached about 8 million L in 2009.

In other plants, and in most other countries throughout the world, in order to meet current regulatory requirements, whole milk produced using MF and intended for the drinking milk market undergoes a final pasteurisation step (72°C for 15 s). The shelf-life of the product is then greatly enhanced, 5 weeks at ~4°C. Owing to the improvement in flavour and storage ability of market milk, such a product has been well received in the market place, and several MF plants are currently running in Europe and in North America for the manufacture of drinking MF milk (te Giffel *et al*., 2006).

Some plants use of 1.4 μm MF as a pre-treatment in the production of UHT milk; the removal by MF of thermostable enzymes present in dead bacterial cells and in somatic cells makes it possible to decrease the intensity of heat treatment, resulting in a less cooked taste and improved storage capability of the product. However, in the drinking milk sector, MF pre-treatment of skimmed milk can be expanded to all skimmed milk used for the production of milk derivatives, such as low-heat milk powder, milk protein concentrates or casein powder.

MF could also be used with a smaller pore size of 0.5 or 0.8 μm instead of 1.4 μm, as originally proposed by Lindquist (1998). With such a low pore size, the permeation flux is around 400 L h^{-1} m^{-2} at 50°C. After mixing with UHT cream, homogenisation at 80°C, a heat treatment limited to 95°C for 6 s is applied, with the purpose of inactivating endogeneous milk enzymes only. The obtained milk, called Ultima® milk, was recognised as a sterile milk; the decimal reduction is higher than 13 $\log_{10}$ for *Clostridium botulinum*, a value that means sterility of the product (Maubois, 2011). The Ultima® milk is stable for more than 8 months at room temperature. Its organoleptic quality was judged as similar to that of a high-temperature, short-time (HTST) pasteurised milk and much better than the current UHT milk. Until now, to our knowledge, this product has not been much produced; some applications may have been developed very recently in the United Kingdom (UK).

Nowadays, in some dairy plants, the MF 1.4 μm membrane is, however, substituted by the 0.8 μm membrane pore size for the production of extended shelf-life microfiltered pasteurised milk in order to extend the duration of storage.

Another recently patented process, still using the MF process, was proposed to remove bacteria and spores from milk containing fat (Fauquant *et al*., 2009). MF using a 0.8 or

1.4 μm mean pore diameter membrane makes it possible, provided the fat was previously homogenised to reduce the size of the fat globules during the processing of whole or partially skimmed milk. In this patent, it is claimed that the bacteria are correctly retained (less than 1 and 10 cfu 10 mL^{-1} for 0.8 and 1.4 μm membranes, respectively), and the fat and protein contents are highly transmitted (95% and 90%, respectively); membrane fouling is low for a 10 h production run. The sterility of the milk obtained by 0.8 μm MF makes it possible to decrease heat treatment parameters down to 96°C for 6 s for long time conservation at ambient temperature. Compared with UHT milk (140°C for 4 s), homogenised and 0.8 μm microfiltered milks offer the following advantages: (a) lower protein denaturation and consequently possibly better nutritional and health properties; (b) better sensory properties; and (c) longer storage time (6 months at 20°C, compared with 3 months for UHT milks). The storage time of 1.4 μm microfiltered milk at 4°C is moreover longer than pasteurised milk.

The removal of bacteria with the MF process is also used for cheese milk to remove somatic cells and spores, particularly *Clostridium* spp., which survive normal pasteurisation heat treatment and cause problems mainly in semi-hard and hard cheese manufacture. Traditionally, the natural content of these spores has been controlled by addition of nitrates and other additives to prevent the 'late blowing' of semi-hard and hard cheeses, which can damage the cheeses. The large number of spores removed by MF makes it possible not to rely on the addition of nitrate and the production of natural products without additives. The MF pre-treatment of milk can then be carried out either at 50°C or at 35°C with some specific adaptations of the running parameters to give cheesemakers the chance of avoiding all detrimental effects of heat treatment on the non-fatty fraction of the milk used.

Owing to the high bacteria and spore counts in raw milk, in 2002, the French regulatory authorities permitted the provisional use of MF milk for making Protected Designation of Origin (PDO) raw milk cheese. Some cheeses are currently produced by MF, especially those using raw milk (JO, 2007). However, microfiltered milk has been described as 'too clean' by cheesemakers. The process requires knowledge of the microbial system composition that should be added to the treated milk for obtaining typical characteristics of ripened cheeses.

6.5 Fractionation of fat

MF was recently proposed to fractionate the milk fat globules according to their size. To our knowledge, no industrial application has been developed, but this process could have interesting uses in the near future. The fractionation of the native fat globule could actually be industrially attractive because its membrane has very interesting functional and health properties, mainly attributed to the phospholipid content. Although centrifugation is widely used for the separation of the fat from the milk, this process cannot separate the native milk fat globules according to their size. Separation of milk fat into small and large globules was then proposed using a ceramic MF membrane (Goudédranche *et al*., 2000; Michalski *et al*., 2006). The objective of the process consists in the fractionation of milk fat globule membranes under hydrodynamic conditions, which causes no damage to the native fat globule membrane. A disruption of the milk fat globule membrane, as observed for instance when milk is homogenised, results in a new interface between the droplet and the surrounding aqueous phase. It was shown during homogenisation that the disrupted fat globules, covered by caseins and serum proteins, promote their interactions with the cheese casein matrix, which affect the structure of the rennet gel and decrease product functionality.

It is possible using a MF operation or a cascade with or without diafiltration to prepare native fractions with different sizes of fat globule, ranging from unstable large fat globules of about 6.6 μm to very stable fat droplets of about 1.6 μm (Lopez *et al.*, 2011).

It was reported that the use of a small fat globule fraction in cheese production yields a smoother and finer texture, probably because of the interaction of the fat globule membrane with the cheese casein matrix, and the differences in triglyceride content of the fat globules according to their size (Michalski *et al.*, 2007). Small globules lead also to high water bonding capacity that can be useful in butter technology in order to improve the spreadability of the product. Moreover, it was recently shown that the size related to specific composition and structure of the fat globule membrane may affect the digestibility of milk fat: the smaller the globule, the higher the enzymatic accessibility, and the better the hydrolysis of small fat globules by both gastric and pancreatic enzymes (Berton *et al.*, 2012). Therefore, for a given concentration of fat, small globules lead to high specific surface area of the globule with the aqueous phase and enzyme accessibility, favouring fat digestibility (Lopez *et al.*, 2011). A better understanding of the specific composition and structures of the fat globules according to their size could then lead to the development of foods with improved health properties and adapted for populations such as infants or elderly persons.

6.6 Removal of somatic cells by microfiltration

Leucocytes, or somatic cells, are naturally present in milk, with size ranging from 6 to 15 μm. The number of somatic cells increases in response to pathogenic bacteria like *Staphylococcus aureus*, a cause of mastitis. Milk is considered infected when the somatic cell count (SCC) is greater than 300×10^3 cells mL^{-1}. Somatic cells contain a large number of thermoresistant enzymes (protease, lipase, catalase) and are very sensitive to mechanical stress, which results in the release of enzymes into the milk and consequently a potential impact on the quality of the manufactured dairy products (Maubois, 2002). Milks with a high SCC are known to adversely affect cheesemaking (coagulation time, yield, quality of cheese, especially flavour and texture) (te Giffel & van der Horst, 2004). Somatic cells have also been shown to have a protecting effect on *L. monocytogenes* during heat treatment (Doyle *et al.*, 1987; Griffiths, 1989).

For removal of somatic cells, the feasibility of MF was reported by many researchers (Saboya & Maubois, 2000; te Giffel & van der Horst, 2004) but, as far as we know, no industrial applications have been developed. With a 5 μm MF membrane, the somatic cells are totally retained in the retentate, and the microfiltered milk is not degraded further by the thermostable enzymes. The retention of somatic cells is also high with a 12 μm membrane, and ranges from 93% to 99% (te Giffel & van der Horst, 2004). An average pore size of 5 μm leads to milk without somatic cells, but with fat content reduced by 17 g 100 g^{-1} (Maubois, 2002). Depending on the membrane pore size (5–12 μm), the permeation fluxes range from 150 L h^{-1} to 2000 L h^{-1} for production lasting 8 h. According to Maubois (2011), this operation opens new avenues for research, for example the specific effects of a varied number of somatic cells in normal milk on milk processing, such as the stability of UHT milk, and the microstructure of cheese. Because somatic cells contain the genetic patrimony of cows, they can also be used as natural tracers of milk for identifying cows or herds which produced the milk.

6.7 Conclusions and future trends

Nowadays, membrane separation technologies offer the dairy technologist several kinds of techniques for the treatment of liquid milks. Some have largely been developed industrially both for drinking and for cheese milks (protein standardisation by UF, removal of bacteria by MF). Others are expected to grow (milk fat fractionation, removal of somatic cells), but their development will depend on the profits associated with new products and interesting functional and health properties.

In the case of protein standardisation using UF of milk, although there are no technological barriers and no real nutritional issues of major consequence, the international dairy community is still divided on the issue of general protein standardisation of milk. The adjustment of protein content by dilution of milk with ultrafiltered permeate is an economical challenge that could lead to market distortions as previously experienced with the butter surplus a few years ago. Such adjustment, however, will provide the opportunity to valorise the large surplus of UF milk permeate resulting from modern cheese manufacturing practices, which is major industrial problem. The protein standardisation of milk will certainly remain one of the highly controversial issues for the dairy industry in the future (Jelen, 2011).

In a general, membrane operations have made it possible for new and original products to be made, which was simply not possible before. The removal of bacteria using MF, which makes it possible to ensure the bacteriological quality of milk while limiting the denaturation of its constituents, is one of the best examples of such success, and would have been considered unlikely some years ago. The differences in functional and health properties between native milk fat globules obtained by MF and homogenised fat globules could become an important factor for dairy processors.

References

Anonymous (1984) L'ultrafiltration à la ferme: l'idée fait son chemin. *Technique Laitière*, **983**, 7–10.

Benard, S., Maubois, J.-L. & Tareck, A. (1981) Ultrafiltration – thermisation du lait à la production: aspects bactériologiques. *Lait*, **61**, 435–457.

Berton, A., Rouvellac, B., Robert B., Rousseau, F., Lopez, C. & Crenon, I. (2012) Effect of the size and interface composition of milk fat globules on their in vitro digestion by the human pancreatic lipase: native versus homogenized milk fat globules. *Food Hydrocolloids*, **29**, 123–134.

Cox, G.C. & Langdon, I.A. (1985) Economic evaluation of reverse osmosis for reduction in milk transport costs. *Australian Journal of Dairy Technology*, **40**, 113–117.

Doyle, M.P., Glass, K.A., Beery, J.T., Garcia, G.A., Pollard, D.J. & Schultz, R.D. (1987) Survival of *Listeria monocytogenes* in milk during high-temperature, short-time pasteurisation. *Applied and Environmental Microbiology*, **53**, 1433–1438.

Fauquant J., Robert B. & Lopez. C. (2009) Procédé pour réduire la teneur bactérienne d'un milieu alimentaire et/ou biologique d'intérêt, contenant des gouttelettes lipidiques. French Patent Application, FR 2953686 – A1.

Fleming, R. (1999) On-farm concentration of milk. *Western Dairy Management Conference*, pp. 49–54, April 8–10 1999, Las Vegas.

Garcera, D. & Toujas, E. (1998) Macroporous support with permeability gradient and it manufacturing process. European Patent Application, 0870534A1 FR 9704359 1–11.

Gésan-Guiziou, G. (2010) Removal of bacteria, spores and somatic cells by centrifugation and microfiltration techniques. *Improving the Safety and Quality of Milk : Milk Production and Processing* (ed. M.W. Griffiths), **1**, pp. 349–372, Woodhead Publishing, Cambridge.

Gésan-Guiziou, G., Boyaval, E. & Daufin, G. (1999) Critical stability conditions in cross-flow microfiltration of skimmed milk: transition to irreversible deposition. *Journal of Membrane Science*, **158**, 211–222.

te Giffel, M.C. & van der Horst, H.C. (2004) Comparison between bactofugation and microfiltration regarding efficiency of somatic cell and bacteria removal. *Bruges 2003 – Advances in Fractionation and Separation: Processes for Novel Dairy Applications*, Document No. 389, pp. 49–53, International Dairy Federation, Brussels.

te Giffel, M.C., Van Asselt, A.J. & de Jong, P. (2006) Shelf extension. *Dairy Industries International* **71**(3), 44–49.

Goudédranche, H., Fauquant, J. & Maubois J.-L. (2000) Fractionation of globular milk fat by membrane microfiltration. *Lait*, **80**, 93–98.

Griffiths, M.W. (1989) *Listeria monocytogenes*: its importance in the dairy industry. *Journal of the Science of Food and Agriculture*, **47**, 133–158.

Holm, S., Malmberg, R. & Svensson, K. (1986) Method and plant for producing milk with a low bacteria content. World Patent Application, WO 86/01687.

Jelen, P. (2011) Standardization of fat and protein content. *Encyclopedia of Dairy Sciences* (eds. J.W. Fuquay, P.F. Fox and P.L.H. McSweeney), **4**, pp. 545–549, Academic Press Elsevier, London.

JO (Official Journal of the French Republic) (2007) Décret n° 2007–628 relatif aux fromages et spécialités fromagères. *JO*, 29 April 2007, Text n°14.

Kosikowski, F.V. (1985) Ultrafiltration of milk on French farms and in the making of a new specialty cheese industry. *Journal of Dairy Science*, **68**, 2403–2410.

Lindquist, A. (1998) A method for the production of sterile skimmed milk. *PCT Patent and World Patent Application*, PCT & WO No. 57549.

Lopez, C., Briard-Bion, V., Ménard, O., Beaucher, E., Rousseau, F., Fauquant, J., Leconte, N. & Robert, B. (2011) Fat globules selected from whole milk according to their size: different compositions and structure of the biomembrane, revealing sphingomyelin-rich domains. *Food Chemistry*, **125**, 355–368.

Malmberg, R. & Holm, S. (1988) Producing low-bacteria milk by microfiltration. *North European Dairy Journal*, **54**, 30–32.

Maubois, J.-L. (1991) New applications of membrane technology in the dairy industry. *Australian Journal of Dairy Technology*, **46**, 91–95.

Maubois, J.-L. (2002) Membrane microfiltration: a tool for a new approach in dairy technology, *Australian Journal of Dairy Technology*, **57**, 92–96.

Maubois J.-L. (2011) Liquid milk products: Membrane-processed liquid milk. *Encyclopedia of Dairy Sciences* (eds. J.W. Fuquay, P.F. Fox and P.L.H. McSweeney), Vol. 3, pp. 307–309, Academic Press Elsevier, London.

Maubois, J.-L. & Ollivier, G. (1997) Extraction of milk proteins. *Foods Proteins and their Applications* (ed. S. Damodaran and A. Paraf), pp. 579–595. Marcel Dekker Inc. New York.

Michalski, M.C., Leconte, N., Briard-Bion, V., Fauquant, J., Maubois J.-L. & Goudédranche, H. (2006) Microfiltration of raw whole milk to select fractions with different fat globule size distributions: Process optimization and analysis. *Journal of Dairy Science*, **89**, 3778–3790.

Michalski, M.C., Camier, B., Gassi, J.Y., Briard-Bion, V., Leconte, N., Famelart, M.H. & Lopez, C. (2007) Functionality of smaller versus control native milk fat globules in Emmental cheeses manufactured with adapted technologies. *Food Research International*, **40**,191–202.

Mistry, V.V. & Maubois, J.-L. (2004) Application of membrane separation technology to cheese production. *Cheese Chemistry, Physics and Microbiology* (eds. P.F. Fox, P.L.H. McSweeney, T.M. Cogan and T.P. Guinee), Vol. 1, General Aspects, pp. 261–285, Elsevier, London.

Rattray, W. & Jelen, P. (1996) Protein standardization of milk and dairy products: a review. *Trends in Food Science and Technology*, **7**, 227–234.

Saboya, L.V. & Maubois, J.-L. (2000) Current developments of microfiltration technology in the dairy industry. *Lait* **80**, 541–553.

Sandblöm, R.M. (1974) Filtering process. Swedish Patent Application, No. 7 416 257.

Slack, A.W., Amundson, C.H. & Hill Jr., C.G. (1982) On-farm ultrafiltration of milk: Part – economic analysis. *Process Biochemistry*, **17**, 23–25, 30, 33.

Skrzypek, M. & Burger, M. (2010) Isoflux® ceramic membranes – practical experiences in dairy industry. *Desalination*, **250**, 1095–1100.

Tamime, A.Y. & Robinson, R.K. (2007) *Tamime & Robinson's Yoghurt: Science and Technology*, 3rd edn, pp. 13–161, Woodhead Publishing, Cambridge.

Trouvé, E., Maubois, J.-L., Piot, M., Madec, M.N., Fauquant, J., Rouault, A., Tabard, J. & Brinkman, G. (1991) Rétention de différentes espèces microbiennes lors de l'épuration du lait par microfiltration en flux tangentiel. *Lait*, **71**, 1–13.

Zall, R.R. (1987) Accumulation and quantification of on-farm ultrafiltered milk: The California experience. *Milchwissenschaft*, **42**, 98–100.

7 Membrane Processing of Fermented Milks

B. Özer and A.Y. Tamime

7.1 Introduction

Currently, there are many types of fermented milks manufactured in different parts of the world. These can be classified into the following categories: (a) gel/liquid, (b) concentrated/strained, (c) frozen or (d) dried (Campbell-Platt, 1987; Kurmann *et al*., 1992). However, an alternative scheme based on the type of starter culture used has been proposed by Robinson & Tamime (1990, 2002) to classify fermented milks into three categories: (a) lactic fermentation – mesophilic, thermophilic and probiotic; (b) yeast/lactic fermentation; and (c) mould/lactic fermentation. The manufacturing stages of all fermented milks (traditional and industrialised) are broadly similar, and the process parameters of these operations have been detailed by Tamime *et al*. (2001), Anonymous (2003), Tamime (2006), Tamime & Robinson (2007) and Tamang (2010). In general, cow's milk is usually used as the base material for the manufacture of fermented milks in the Western world, although goat, sheep and buffalo milks can also be employed, and they are mainly manufactured in the Mediterranean basin and the Indian subcontinent.

During the manufacture of most fermented milks, concentration of the milk base to increase the solids content takes place prior to the incubation stage, and the most traditional method employed in the past was boiling the milk, but this method is not practised on a large industrial scale because it is cumbersome and not economical. Currently, the solids content in the milk base is fortified by the addition of dairy powders (skimmed milk powder (SMP), whole milk powder (WMP), whey powder (WP), whey protein isolates (WPIs), whey protein concentrates (WPCs), caseinates and buttermilk powder), evaporation of the milk under vacuum or concentrating the milk using membrane filtration, such as ultrafiltration (UF) or reverse osmosis (RO). All these processes tend to improve the consistency/viscosity of fermented milks. However, in some instances, the raw milk may be processed using microfiltration (MF) to improve its microbiological quality or nanofiltration (NF) to concentrate the mineral content of the milk. In addition, in concentration of the fermentate, for example strained or concentrated yoghurt, the cloth bag method is still used for production and, in large-scale operation, the nozzle separator (i.e. more widely used industrially) or membrane filtration. In the traditional method of manufacture of these products, the curd is drained after the skimmed milk has been fermented (Bjerre & Nielsen, 1972). The whey, which is rich in lactose, ash and the whey proteins, is lost and, hence,

Membrane Processing: Dairy and Beverage Applications, First Edition. A. Y. Tamime.

the amount of milk required per kilogram of product is increased correspondingly. This eventually causes a financial loss as well as a loss of nutritive value of the end product. As mentioned earlier, the concentration of milk solids prior to fermentation offers certain advantages. For example, the elevation of milk solids improves the consistency and viscosity of fermented milks.

Nevertheless, there are different methods that can be employed for increasing the level of milk solids during the manufacture of fermented milks. Among these methods, membrane filtration has been increasingly applied during the last couple of decades, since the demand for tailor-made fermented milk products has been continuously increasing. However, membrane technologies can be applied in the fermented milks industry for the purposes of

- increasing the level of certain desired milk components (e.g. proteins and fat) in the milk base
- reducing the lactose content in milk, which may help to control post-fermentation acid development during the storage period and/or
- selectively concentrating milk components which are further used in the manufacture of fermented milks (e.g. ultrafiltered concentrated cheese whey, concentrated yoghurt made by UF, bioactive components for functional fermented milk beverages etc.).

In practice, membrane technologies can be employed in the manufacture of fermented milks, such as gel/liquid (e.g. set and stirred type) and concentrated fermented milk products. However, the known categories and the scheme proposed by Robinson & Tamime (1990) for the classification of fermented milks will be used in this review to provide an overview of the current scientific aspects of these products made using membrane technology-based production models for fermented dairy products. The processing conditions of fermented milks determine the physical, chemical, microbiological and organoleptic properties of the final products, and these aspects will be also reviewed.

7.2 Microflora of the starter cultures

The 'traditional' starter culture organisms used in fermented milks are mainly lactic acid bacteria (LAB), including probiotic microflora, acetic acid bacteria, yeasts (many different species including probiotic types, such as *Saccharomyces boulardii*) and mould (*Geotrichum candidum*), or combinations of these. Detailed characterisations of these micro-organisms have been reviewed by Tamime & Marshall (1997), Marshall & Tamime (1997), Tamime (2002), Tamime *et al*. (2006) and Tamime & Robinson (2007). The relevance genera of these micro-organisms during the manufacture of fermented milk products, including the category and fermentation temperature, are as follows:

- Lactic fermentations: (a) mesophilic species of LAB (e.g. *Lactococcus*, *Leuconostoc* and *Pediococcus*, optimal growth at ~30°C), (b) thermophilic species of LAB (e.g. *Streptococcus* and *Lactobacillus*, optimal growth at 42–45°C) and (c) therapeutic/probiotic species of LAB (*Streptococcus*, *Lactobacillus*, *Bifidobacterium* and *Enterococcus*, optimal growth at ~37°C).

- Yeast–lactic fermentations: mesophilic and thermophilic LAB (probiotic micro-organisms may be included), yeasts and miscellaneous organisms, such as *Brevibacterium linens*, *G. candidum* and acetic acid bacteria (~22°C).
- Mould–lactic fermentations: mesophilic LAB and *G. candidum* (~22°C) (Tamime & Marshall, 1997; Robinson *et al*., 2002; Tamime, 2002; Wszolek *et al*., 2006).

7.3 Patterns of production and consumption

Interest in fermented milk consumption in many countries has increased over the past few decades primarily due to: (a) masking the sharp acidic taste by the addition of fruit flavours and a sweetening agent to the product(s); and (b) enhancing the health-promoting properties of fermented milks by the addition of probiotic organisms, such as *Lactobacillus* spp., *Bifidobacterium* spp. and *Enterococcus* spp., besides the 'traditional' starter culture bacteria. As a consequence, the pattern of per capita consumption of fermented milks is still rising in many countries, a trend confirmed by the data shown in Table 7.1. Per capita consumption data for the USA were not available, but the annual production figures (×1000 tonnes) have risen from 1515 to 2525 between 2000 and 2009. In Asian countries, where fermented milks are not traditional products and were not consumed in the past, annual production figures (×1000 tonnes) have risen from 12 to 3176 between 1998 and 2009 in China, whereas in Thailand total production increased from 300 to 650 between 1998 and 2000 (IDF, 2002, 2004, 2010). In some products, such as kefir and koumiss, yeasts are blended with lactic starter culture, and mould in viili (fermented milk from Finland) (Tamime, 2006).

7.4 Manufacturing practice of gel-type (set and stirred) products

7.4.1 *Mesophilic–lactic fermentations*

The term 'mesophilic' is reserved for starter cultures with optimum growth temperatures ≤30°C. The group of products classified into this type of fermented milks consists of traditional or natural buttermilk, cultured buttermilk, Nordic/Scandinavian sour milks, Lben (in North Africa) and sour cream. To our knowledge, none of these fermented milks are made using membrane technology to fortify the solids content of the milk base with the exception of cultured buttermilk and some milk-based beverages.

In the mid-1980s, UF technology had been recommended during the manufacture of cultured buttermilk to increase the calcium content and improve the viscosity of the product (Mann, 1986), but no industrial application has been reported. However, in recent years, membrane technology has been applied during the manufacture of whey-based fermented beverages and cultured buttermilk, especially in the fermentation of liquid UF whey protein retentates where it seemed to be a feasible process during the manufacture of whey-based drinking fermented products (Özer & Kırmacı, 2010). Rodriguez-Serrano *et al*. (2002) evaluated the chemical and physical properties of a whey beverage concentrated by UF. The concentration of acetaldehyde was higher in ultrafiltered whey than in milk, and the apparent viscosity of the whey product was lower than that of milk due to lack of casein

Table 7.1 Per capita annual consumption (kg head^{-1}) of milk drinks and fermented products including yoghurt in some selected countries.

Country/consumption level	Year		
	2000	2003	2008
<10 kg head^{-1}			
Argentina	6.7	10.4	12.8
Australia	5.6	6.7	6.7
Canada	4.9	7.2	8.2
China	0.2	1.3	NR
Cyprus	NR[1]	8.0	NR
Italy	NR	7.5	8.8
Korea (Rep.)	NR	9.9	9.3
Mexico	3.0	5.1	NR
Poland	NR	7.3	7.8
South Africa	3.2	1.7	NR
Ukraine	3.2	11.0	NR
10–20 kg head^{-1}			
Austria	17.3	21.1	NR
Belgium	11.1	10.9	10.5
Czech Republic	13.8	15.7	16.3
Hungary	12.2	19.2	13.9
Norway	16.6	21.3	25.5
Slovakia	11.7	13.0	13.8
20–30 kg head^{-1}			
Germany	26.5	29.8	30.5
France	20.2	31.6	29.9
Portugal	9.7	26.2	26.6
Spain	15.7	28.0	NR
Switzerland	22.8	30.7	NR
>30 kg head^{-1}			
Denmark	35.8	45.9	NR
Finland	39.0	38.3	NR
Iceland	NR	44.3	37.9
Iran	NR	38.7	47.3
Israel[a]	28.0	22.0	NR
Netherlands	46.4	40.7	NR
Sweden	32.1	35.7	NR

Source: Data compiled from IDF (2002, 2009).
NR, not reported.
[a]Consumption data include dairy desserts and cream.

in the former. In an earlier attempt, Tratnik & Krsev (1988) investigated the production of fermented beverages from milk with demineralised whey, i.e. WPC obtained by diafiltration (DF) and by UF of demineralised whey obtained by ion exchange. WPC was mixed with the yoghurt milk base at levels varying from 0 to 50% and, as a result, the incubation time of the fermented beverage was reduced. Up to a certain level, WPC produced a firmer coagulum, a smooth structure and, in some experimental beverages, a consistency similar to cream. Incorporation of ultrafiltered WPC (more demineralised) yielded a fermented beverage with better sensory characteristics (Tratnik & Krsev, 1988). Furthermore, increase in the level of WPC in the beverage may carry a risk of unpleasant flavour and grainy consistency.

MF is primarily used in the dairy industry as an alternative to thermal processing to reduce the microbial load of raw milk. In a novel approach, Debon *et al*. (2010) studied the rheological and physico-chemical properties of MF fermented milk with added inulin (i.e. a prebiotic source). The product was prepared by MF of raw cow's milk at 45°C at a pressure of 0.3 MPa (with tangential velocity of 8 m s^{-1}), and then mixed with inulin. The authors concluded that the MF of fermented milks with or without inulin showed non-Newtonian behaviour with shear thinning, and the addition of inulin improved the rheological properties of the fermented beverage. Although the authors did not compare the MF products with a non-microfiltered (thermally processed milk) control, it was assumed that in the former products the sedimentation problem would be reduced compared with the normal industrial thermal application where heat-induced interactions between milk proteins take place, leading to an increase in the size of the colloidal particles and eventually sedimentation during storage.

Buttermilk, which is a fermented mesophilic product, fits the image of a nutritious food that can be extremely refreshing. The basic manufacturing steps of cultured buttermilk involve standardisation of the fat and solids-not-fat contents, followed by homogenisation, heating (i.e. 85°C for 30 min), fermentation by starter cultures (i.e. acid- and flavour-producing LAB), breaking the fermentate, cooling and packaging. The milk base should be standardised to a minimum of 9 g 100 g^{-1} SNF if fresh skimmed milk is used (White, 2006). The standardisation of the SNF content is usually achieved by adding SMP. In using of low-heat treated SMP, the SNF level should be increased to a minimum of 10 g 100 g^{-1} to obtain full-bodied cultured buttermilk. UF is a suitable alternative standardisation method of the SNF compared with SMP addition in the manufacture of cultured buttermilk as long as the UF processing conditions are selected carefully. Rao *et al*. (1995) and Rao (2002) demonstrated that the flux pattern of buttermilk was strongly affected by pH. Increasing the pH of buttermilk from 6.6 to 8.0 tended to reduce the initial flux rate and increase the fouling rate on the membrane surface. At higher pH values, the flux rate of buttermilk was mainly controlled by concentration polarisation and, at lower pH values, the fouling rate on the membrane surface was stimulated.

Membrane technology (especially microfiltration) is not only used to increase the SNF content in buttermilk, but also used for the fractionation of regular buttermilk and 'whey' buttermilk – the by-product from butter manufacture (Morin *et al*., 2006; Sachdeva & Buccheim, 1997). Whey buttermilk contains considerable amounts of phospholipids compared with regular buttermilk, but its protein content is lower due to the absence of caseins. According to Morin *et al*. (2006), a twofold MF concentration of regular buttermilk doubled its phospholipid content whereas that of whey buttermilk was increased by 50%.

The overall efficiency of MF processing of regular buttermilk was limited by the amount of caseins retained by the MF membrane. Morin *et al*. (2004) demonstrated that the changes in the processing temperature of MF from 7°C to 50°C did not affect the protein permeation in fresh or reconstituted buttermilk, whereas by increasing the temperature of filtration it reduced both lipid and phospholipid permeation.

7.4.2 *Thermophilic–lactic fermentations*

The term 'thermophilic' is reserved for starter cultures with optimum growth temperatures between 37°C and 45°C, and the specific micro-organisms that are often used for the production of fermented milk (mainly yoghurt and probiotic yoghurt). Several reports on the use of ultrafiltered treated milk for the manufacture of high-quality yoghurt have been published (Lankes *et al*., 1998; Tamime *et al*., 1984; Biliaderis *et al*., 1992). The formation of yoghurt gel is primarily based on heat-induced protein interactions. Therefore, increasing the protein content of the milk base above the level present in raw milk is desired. As the UF membrane is a selective method, the fat and protein contents are increased in UF yoghurts with a loss of lactose, some soluble protein fractions and minerals. Removal of milk components, which do not interfere directly with structure formation in yoghurt, is important from a practical and economical point of view. In this regard, UF technology provides some advantages to yoghurt manufacturers, such as shorter gelation time, better physical quality of the end product and extended shelf-life. When UF technology is applied to the milk base, the operation parameters should be selected carefully. In particular, the transmembrane pressure (TMP) should be kept constant throughout membrane processing of the milk so that the efficiency of the membrane is maintained. Today, a number of membrane modules are used in large-scale yoghurt production, and the input and output pressures are controlled automatically. The temperature of milk during membrane processing is also a determinative factor for the efficiency of the process. At low processing temperatures (i.e. 15–30°C), the viscosity of milk increases and the pores of the membrane are easily blocked, and the separation of low and high molecular weight components becomes difficult. At high temperatures (i.e. 50–60°C), the viscosity of milk decreases and the flux rate of milk increases and, to optimise the flux rate and membrane efficiency, the processing temperature of milk should be around 35–45°C (Attia *et al*., 1991a, b). In order to reduce membrane fouling and increase membrane efficiency, the separation of milk fat prior to membrane processing is recommended.

Membrane filtration (UF or RO) has been used to concentrate the protein content of the milk base. Two options are available depending on the daily output and capital cost of the equipment. In the first option, after the desired total solids (TS) level is achieved in the UF or RO skimmed milk retentate, cream is added to the concentrated skimmed milk to standardise the fat content in the final product (Fig. 7.1; for details of other fat standardisation units, the reader is referred to the reviews of Tamime *et al*. (2001), Anonymous (2003) and Tamime & Robinson, 2007 for more complete discussion). The other option available is to use part of the skimmed milk to be concentrated using UF or RO and, later, the skimmed milk retentate is added to the bulk of the fat standardised milk base to adjust the protein content to the desired level (e.g. ~4.5 g 100 g^{-1}). A typical illustration is shown in Fig. 7.2. Afterwards, the standardised milk base (i.e. fat and protein contents)

Fig. 7.1 An illustration of the Tetra Alfast HD3 PPT for fat standardisation of the mi k base (Reproduced by permission of Tetra Pak, Lund Sweden).

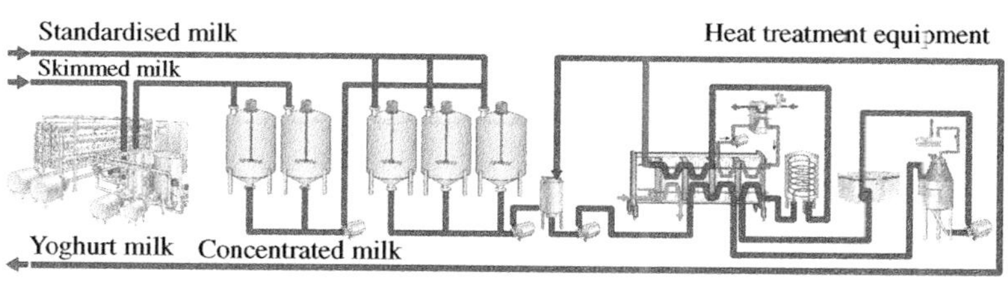

Fig. 7.2 Production line of milk base concentrated by ultrafiltration for yoghurt making (Reproduced by permission of Tetra Pak, Lund Sweden).

is processed as follows: it is pre-heated to 70°C, de-aerated, homogenised (~17.5 MPa pressure), heated to 90–95°C for up to 5 min, and cooled to the incubation temperature (e.g. 42°C for yoghurt or ~37°C for probiotic yoghurt).

In an early study, Rothwell (1982) demonstrated that yoghurt made from UF-concentrated skimmed milk and blended with whole milk to obtain a fat content of 3.2 g 100 g^{-1} in the final product gave a firmer coagulum with no whey separation. Skimmed milk concentrated by UF to 13.5 g TS 100 g^{-1} was reported to be suitable for yoghurt making (Tamime & Robinson, 2007).

It is a well-established fact that the physical properties of yoghurt made from UF-concentrated milk are far better than yoghurts made from milk concentrated by other methods (e.g. addition of SMP and other milk-based powders, and evaporation). In ultrafiltered milk base yoghurts, the risk of syneresis and excessive acid development during post-fermentation are reduced, and the consistency of the final product is improved (Marshall & El-Bagoury, 1986). On the other hand, the concentration factor (C_f) of milk by UF should be selected carefully, since concentration of milk solids by more than twofold by UF results in a very firm structure which is atypical for plain yoghurt. A good-quality yoghurt made from ultrafiltered milk base (i.e. in terms of textural properties and sensory appraisal) should have a ratio of 1.2:1.4 for the protein and lactose, which results in about 35 g 100 g^{-1} more protein in the end product than yoghurt made from RO milk base (Brazuelo *et al*., 1995). According to Renner & Eiselt-Lomb (1985), the viscosity of yoghurt made by UF ($C_f = 1.4$) was similar to yoghurt supplemented with sodium caseinate (NaCn) at a level of 2 g 100 g^{-1}, and the viscosity of the product was reported to increase by about 25% during cold storage up to 14 days.

The protein content of yoghurt is increased by about 0.7 g 100 g^{-1} and the ash content by about 0.2 g 100 g^{-1} when the UF concentration factor is raised by 0.2; therefore, a protein content of 5 g 100 g^{-1} is achieved in the milk base when applying a concentration factor of 1.4:1. The manufacture of low-lactose yoghurt and fermented milk beverages by the UF process (coupled with DF) has long been known. The manufacturing steps of these products include concentration of the milk by UF up to 20 g 100 g^{-1} protein level followed by diluting the UF retentate with water to obtain a protein level of 3.3–4.6 g 100 g^{-1} in the milk base. The low-lactose retentate is then standardised with cream to 1.4–2.6 g 100 g^{-1} fat. The average lactose content of the final product is around 0.3–0.6 g 100 g^{-1}, and it exhibits a smooth viscous texture, firm body with a clean, acid and slightly flat flavour. The major drawback of such yoghurt is the risk of development of a bitter taste after about 1 week of storage. Addition of β-galactosidase to the ultrafiltered concentrated milk is another option to produce low-lactose yoghurt (lactose level of ~0.75 g 100 g^{-1} was reported by Rasic *et al*., 1992).

Savello & Dargan (1997) demonstrated that the water-holding capacity of yoghurts made from ultrafiltered concentrated milk did not depend on different heating temperature (i.e. 110–140°C for 4–16 s and 85°C for 20 min), and the lowest syneresis was recorded in the sample concentrated by UF and heat treated at 110–130°C. The same authors failed to find a correlation between the whey protein denaturation level and syneresis in these yoghurts.

High-protein SMP (i.e. known as skimmed milk protein concentrate (SMPC)) manufactured by UF technology combined with spray drying can be used in the production of non-fat or reduced-fat yoghurts (Mistry & Hassan, 1992; Mistry, 2002). The protein composition of SMPC is similar to skimmed milk, but the former contains no lactose, which makes it particularly useful for the manufacture of yoghurt. Supplementing skimmed milk with SMPC up to 5.6 g 100 g^{-1} protein was recommended to obtain low-fat yoghurt of good quality (Mistry & Hassan, 1992). Similarly, when yoghurt was produced from spray-dried milk protein concentrates (MPCs) where the ultrafiltered retentate had been heat treated at 90°C for 30 min, the resulting product had higher viscosity and less syneresis than that made from the control powder (Butterick & Higgins, 1982).

Ordonez *et al*. (2000) investigated the possibility of using ultrafiltered concentrated milk in the manufacture of probiotic frozen yoghurt. The finished product contained high levels of protein and calcium as well as good flavour and textural quality. In addition, the culture organisms were stable in the frozen product after storage for 6 weeks. The production of bio-yoghurt made from goat's milk and blends of cow's and goat's milks (goat's to cow's milk ratios of 70:30 or 50:50) concentrated by UF or by the addition of SMP was investigated by Kavas *et al*. (2004). The UF application led to an increase in the physical parameters of the yoghurts compared with the SMP addition. However, the highest sensory scores were given to the product prepared from a cow's and goat's milk mixture (50:50) supplemented with SMP.

RO is another operation for the treatment of the milk base to increase the TS content, and will not have the same effect on the protein composition of milk as UF. The former method of concentration will result in the variation of protein/TS ratio, which is expected to influence the texture of yoghurt. Compared with the UF application, the RO treatment of milk for the manufacture of cultured milk products is rather limited. Yoghurts made from RO-treated milk have lower syneresis and higher viscosity values than yoghurts manufactured from milk fortified with SMP, but the overall quality of the yoghurts made from RO milk is lower than that of yoghurts made from ultrafiltered milk. In an early study, Davies *et al*. (1977) obtained yoghurts of higher viscosities using RO concentrated milk than those where the milk solids were fortified with SMP. Similarly, Guirguis *et al*. (1987) reported that when milk was concentrated by RO to the desired level of solids content required for yoghurt manufacture, the apparent viscosity of the product was improved by 46% and syneresis values decreased by 27% when compared with yoghurt where the milk base was fortified with SMP to the same solids content. A decrease in the syneresis of yoghurts made from RO milk was also reported by Dixon (1985).

Voutsinas *et al*. (1996) studied the physico-chemical, microbiological and physical stability changes in sheep's milk yoghurts made from RO retentate and long-term stored milk in the frozen state. The milk was thawed after 6–8 months and then made into yoghurt. Yoghurts made from frozen whole milk RO retentate had higher curd consistency and higher apparent viscosity values than the control. On the other hand, when the yoghurt was prepared from frozen skimmed milk RO retentate (i.e. similarly stored for 6–8 months in the frozen state), the overall physical and sensory qualities of the yoghurt were lower than that of the control. The former yoghurts had a thick creamy layer at their surface, a slightly grainy texture and displayed higher syneresis than the control product.

7.4.3 *Yeast–lactic fermentations*

Introduction

Alcoholic milk beverages, such as kefir, koumiss and acidophilus yeast milk, may be described as lactic/sour in taste, with ethanol content as high as 2 g 100 g^{-1}. These products have a foaming and effervescent characteristic as a result of carbon dioxide (CO_2) emission. The microflora of the starter cultures is not well defined compared with the other fermented milks described above; however, yeast and lactic acid bacteria are always present and, for detailed screening of these micro-organisms, the reader is referred to some reviews for more complete discussion (Tamime & Marshall, 1997; Tamime, 2002; Robinson *et al*., 2002; Wszolek *et al*., 2006). To our knowledge, only koumiss made from cow's milk has been processed by UF prior to heat treatment and fermentation, and will be reviewed in the subsequent section.

Koumiss

This fermented milk drink/beverage has an ancient origin, and is common in Eastern Europe and Central Asia. It is traditionally produced from mare's milk, and its highly nutritive and curative characteristics are well known. Since the cost is high and availability of mare's milk is limited, the possibility of using cow's milk instead of mare's milk for the production of koumiss has been investigated by Küçükçetin & Yaygin (1999) and Küçükçetin *et al*. (2003). Owing to the differences in the compositions of both types of milk, modification of the chemical composition of cow's milk is essential to make it suitable for koumiss production. Such a modification based on membrane filtration of cow's milk was proposed by Küçükçetin *et al*. (2003), where pasteurised skimmed milk was first ultrafiltered to produce a retentate with a protein level of 6.6 g 100 g^{-1}. The casein and whey protein fractions in the UF retentate were separated by DF in the 0.1-μm MF membrane with the ultrafiltered permeate as DF medium. After about five successive cycles, the caseins and whey proteins were fully separated from the other protein fractions, and further concentrated by means of MF and UF, respectively. To adjust the lactose content of the cow's milk similar to mare's (i.e. ~6.4–6.7 g 100 g^{-1}, Wszolek *et al*., 2006), the ultrafiltered permeate was nanofiltered and the lactose-enriched retentate was further concentrated by NF. The TS, fat, total protein and whey protein contents of the modified cow's milk were fairly similar to mare's milk. The former milk had higher titratable acidity and ash contents than mare's milk. Modified cow's milk base was processed to koumiss as shown in Fig. 7.3. A blend of LAB and yeast (*Lactobacillus delbrueckii* subsp. *bulgaricus*, *Lactobacillus acidophilus* and *Kluyveromyces marxianus* var. *lactis*) was used to ferment the modified cow's milk for the manufacture of koumiss. The viscosity of the product made from modified cow's milk was higher than that made from mare's milk, but the alcohol levels of all the koumiss samples were similar. Another alternative method of koumiss production includes mixing whole cow's milk (5 parts) with ultrafiltered cheese whey (8 parts), and the manufacturing stages are similar to those shown in Fig. 7.3 (Özer and Özer, 1999).

7.4.4 *Mould–lactic fermentations*

The only product, which is classified into this type of fermented milk, is viili made in Finland. To our knowledge, the product is not made using membrane technology to fortify the solids content of the milk base.

7.5 Manufacturing practice of concentrated products

7.5.1 *Background*

Many names are applied to concentrated yoghurt in different parts of the world, and some examples include Greek-style yoghurt (in the UK), labneh, labaneh, lebneh, labna

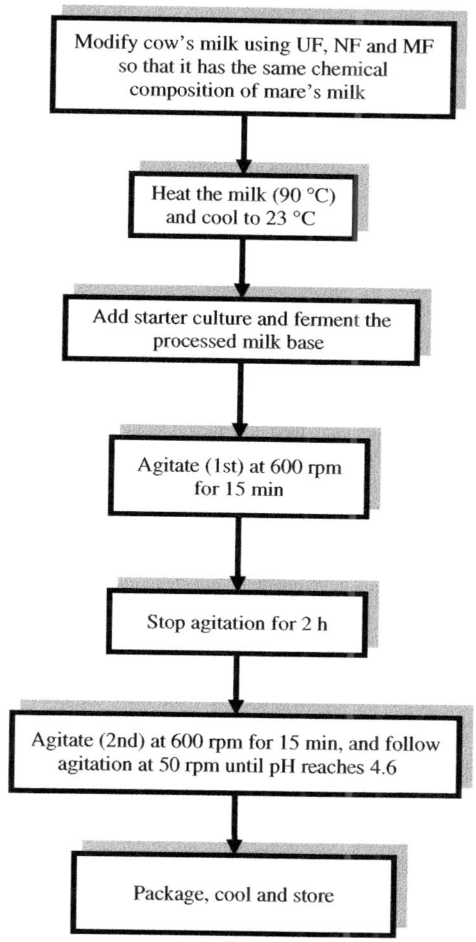

Fig. 7.3 Flow diagram of koumiss production. UF, ultrafiltration; NF, nanofiltration; MF, microfiltration; rpm, revolutions per minute (Adapted from Küçükçetin *et al.*, 2003).

(in Eastern Mediterranean countries), laban zeer, leben zeer (in Egypt and Sudan), torba, süzme (in Turkey), tan, than (in Armenia), stragisto, sakoulas (in Greece), syuzma (in Russia), matsou, mast (in Iraq and Iran), basa, zimme, kisela, mleko-slano (in Balkan countries), Ititu (in Ethiopia) and chakka or shrikhand (in India) (Tamime *et al.*, 2011).

7.5.2 *Concentrated yoghurt*

The manufacturing methods for concentrated yoghurt can vary widely from the traditional cloth bag method to more mechanised systems including nozzle separators, membrane processing and recombination (Tamime *et al.*, 2001; Nsabimana *et al.*, 2005; Özer, 2006a, b). Membrane techniques have been successfully applied in the yoghurt industry for many years, but their application to the production of concentrated yoghurt has yet to become more widespread (Özer *et al.*, 1997b).

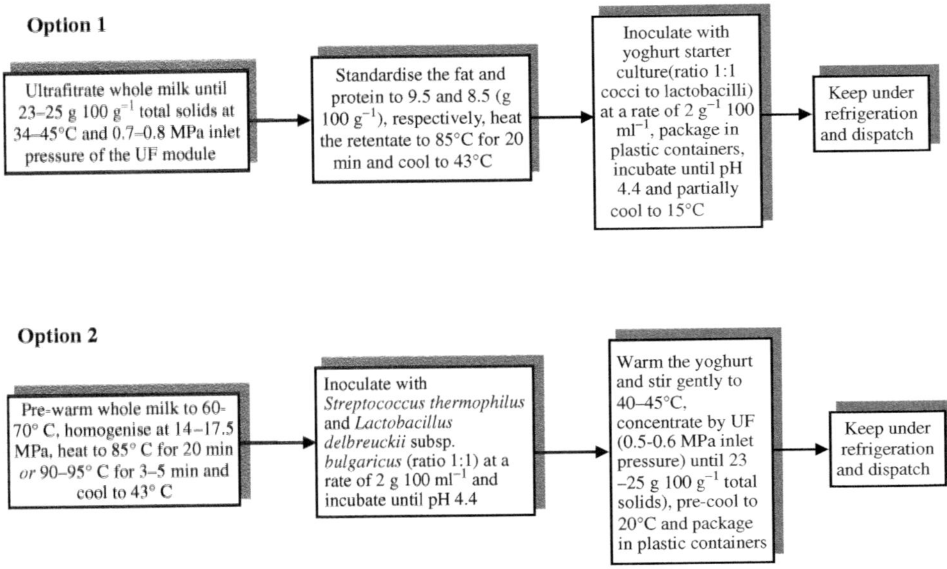

Fig. 7.4 Production of concentrated yoghurt from ultrafiltered whole milk retentate (Option 1) and by ultrafiltration (UF) of warm yoghurt (Option 2).

The different UF systems that can be employed for the manufacture of concentrated yoghurt are shown in Fig. 7.4. In option 1 (also called ultrafiltered retentate concentrated yoghurt), the milk is concentrated by UF up to the desired TS level in the final product (~23–25 g 100 g^{-1}) (Veinoglou *et al.*, 1978; Abd El-Salam & El-Alamy, 1982; El-Samragy & Zall, 1988; Hofi, 1988; Tamime *et al.*, 1989a, b; Özer *et al.*, 1997a, b, c, 1998a, b, 1999a, b). The UF module is operated at 0.7–0.8 MPa inlet pressure and at 35–45°C. The protein and fat levels of the concentrated yoghurt are adjusted to 8.5 and 9.5 g 100 g^{-1}, respectively, followed by heat treatment at 85°C for 20 min. The heat-treated retentate is further processed in a similar manner to the production of natural set yoghurt. In option 2 (also called ultrafiltered concentrated yoghurt), the natural warm yoghurt is ultrafiltered until the TS level reaches 23–25 g 100 g^{-1}. In this method, the whole milk is pre-warmed to 60–70°C, homogenised at 14.0–17.5 MPa, heated at 85°C for 20 min or 90–95°C for 3–5 min and cooled to ~43°C. The milk is inoculated with yoghurt starter cultures and fermented until pH reaches 4.4. After fermentation, the warm yoghurt is stirred gently and concentrated by a two- to four-stage UF plant with 0.5–0.6 MPa inlet pressure. The concentrated product is pre-cooled to ~20°C and packed in plastic or glass containers before transferring the product to the cold store (Tamime *et al.*, 1989b, 1991a, b, c, 2001; Özer *et al.*, 1997a, 1998a, 1999a). Although the final chemical compositions of both products are similar, the physical and organoleptic properties are remarkably different (Özer, 1997). However, a current illustration for the production of ultrafiltered concentrated yoghurt is shown in Fig. 7.5. The fermentate (i.e. pH 4.3–4.5) is agitated gently in the fermentation tank to obtain a homogeneous gel, and thermised at 55–58°C for 0–3 min to help the release of some of the whey from the protein matrix and facilitate the filtration as

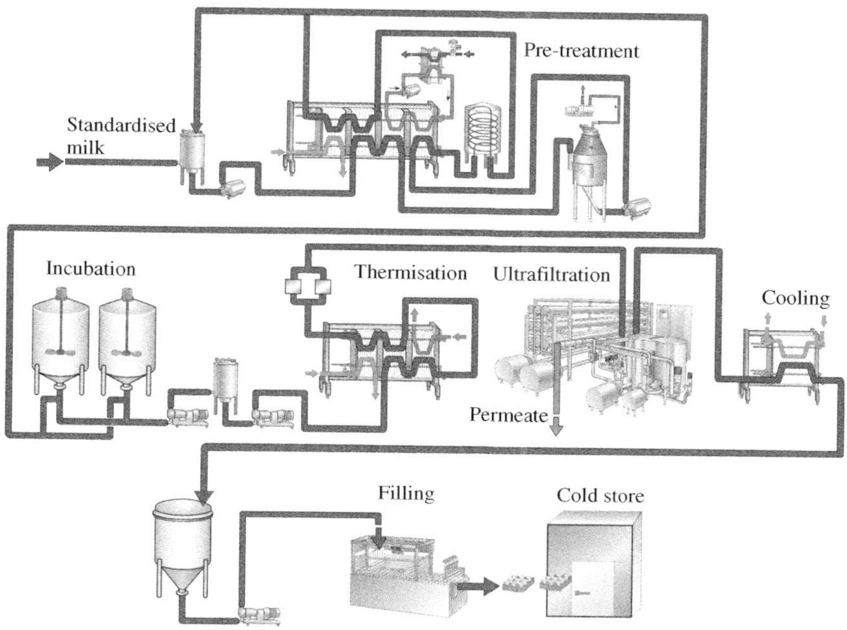

Fig. 7.5 Flow chart for the manufacture of concentrated yoghurt by ultrafiltration (Reproduced by permission of Tetra Pak, Lund Sweden).

well as to inactivate most of the bacteria in the product. The thermised yoghurt is cooled to the filtration temperature (~40°C), pumped through a duplex filter to remove any large clots and finally pumped to the UF module. Afterwards, the ultrafiltered concentrated yoghurt (i.e. retentate) is pumped to a buffer tank via a cooler where it is cooled to ~10°C and packaged. However, fruit syrup (optional) may be added and mixed with the ultrafiltered concentrated yoghurt prior to the filling machine. The SNF content of the permeate is ~6 g 100 g^{-1}, and cooled and transported to a buffer tank for further treatment, which may include heat treatment before diluting the yoghurt for the production of Ayran (i.e. low-viscosity Middle Eastern drinking yoghurt) (L.-E. Nilsson, personal communication).

Comparing the ultrafiltered concentrated yoghurt method with other production techniques (i.e. adding powders to milk, concentration of the fermentate using a nozzle separator and/or direct recombination), the main advantages of the UF system are: (a) higher yield (~10% increase); (b) shorter processing time (~25–75%); and (c) reduced whey separation during storage (Hofi, 1988; Tamime *et al.*, 1989a, 1991b; Özer *et al.*, 1997a, 1998a, 1999a; Mehaia, 2005). The reduction of the volume of milk (by ~10%), starter cultures (by ~80%) and salt (optional) (by ~50%) are additional advantages of the UF system. The body and texture of ultrafiltered concentrated yoghurt is smoother, and it has a more palatable taste than concentrated yoghurt produced by the traditional method (Özer, 1997). According to Tamime *et al.* (1989a), both concentrated products made from UF of warm yoghurt and ultrafiltered retentate should be passed through a lactic curd texturiser to obtain a smoother body, finer texture and improved sheen of the product in the absence of quick partial cooling to ~20°C of the warm concentrated yoghurt. However, Özer (1997) demonstrated that a smoother body could well be obtained in the ultrafiltered concentrated yoghurt without texturising the final product as long as the inlet and outlet pressures are kept constant throughout the UF processing of the warm yoghurt. In order to avoid the detrimental effect of shearing on the physical properties of warm yoghurt, the TMP should be 0.25–0.30 MPa and the operating temperature should be around ~45°C (Tamime *et al*., 1991a).

The use of different types of milk in the manufacture of concentrated yoghurt was examined by Tamime *et al*. (1991c). Goat's, cow's and sheep's milks were used to produce concentrated yoghurt by UF and traditional methods; The chemical compositions of all samples were 20.5–22.5 g 100 g^{-1} TS, 6.7–8.2 g 100 g^{-1} protein and 7.8–8.9 g 100 g^{-1} fat. The concentrated yoghurt made from cow's milk was found to be the firmest (1470 mN – compression response), followed by sheep's (628 mN) and goat's concentrated yoghurts (476 mN). When the samples were passed through the texturiser, the firmness of the concentrated sheep's and goat's concentrated yoghurts decreased to 157 mN and 142 mN, respectively.

7.5.3 *Shrikhand and chakka*

Shrikhand and chakka (traditional Indian fermented dairy products) produced by the UF method had similar physical properties to ultrafiltered concentrated yoghurt (Özer, 2006). However, a manufacturing protocol for ultrafiltered shrikhand has been proposed by the Indian National Dairy Research Centre and, according to this protocol, skimmed milk is heated to 85–90°C for 10–20 min, cooled to 21–22°C and inoculated with a selected blend of mesophilic starter culture until the pH reaches 4.5–4.6, which takes about 15–16 h. The fermentate (i.e. cow's or buffalo's skimmed milk) is heated to 60°C for 5 min with continuous agitation, cooled to 50°C and ultrafiltered up to about 16 g 100 g^{-1} TS. Whey is then removed from this concentrated coagulum by hanging it in a muslin cloth (eight layered) overnight at room temperature followed by mild pressing to get chakka. This UF/traditional skimmed milk chakka is pressed through a sieve for the kneading action to give smooth chakka without any visible grain (Sharma, 1998). Thermostable yoghurt bacteria can withstand high temperatures, and 60°C for 5 min is required for the UF process, which results in about 20% extra recovery of the TS. The sensory quality of chakka prepared by the UF process was reported to be slightly superior to cow's/buffalo's skimmed milk shrikhand prepared by the traditional process in terms of flavour, body and texture, colour and appearance, and overall acceptability.

7.5.4 *Ymer*

Ymer is a Danish fermented milk enriched with protein. The composition (g 100 g^{-1}) of ymer is 3.5 fat, 5.6 protein, 11 SNF and pH 4.4–4.6 (Fonden *et al*., 2006). The production of ymer can be achieved by traditional method or by UF technology. The basic manufacturing processes of traditional and ultrafiltered ymer are outlined in Fig. 7.6. The traditional production method of ymer is widely used in the industry, and today ymer production is predominantly based on using UF technology. Following the heat treatment of the milk base at 85°C for 15 s, the skimmed milk is ultrafiltered at 50–55°C to a concentration factor of 1.9, which is equivalent to 45–50% reduction of the original volume of milk.

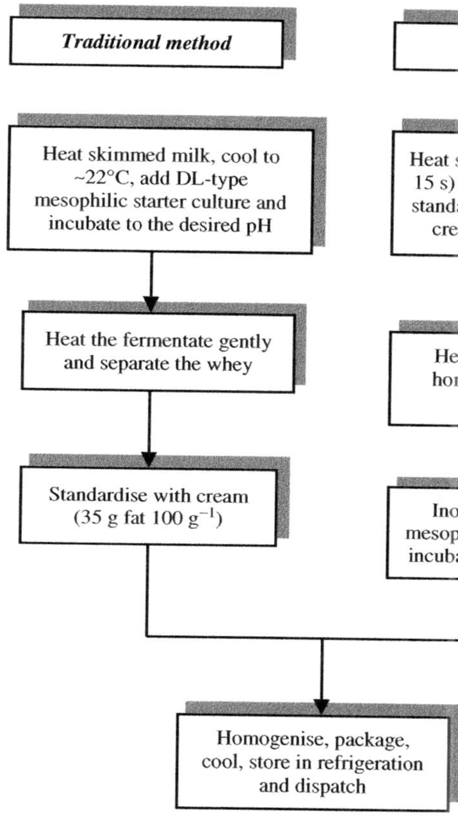

Fig. 7.6 Production methods of ymer. DL, mesophilic starter culture blend; UF, ultrafiltration.

After UF, the protein content of the milk is increased to *ca.* 6.5 g 100 g^{-1}. The fat content of the product is adjusted to the desired level by adding cream, and then a second heat treatment is applied at 85°C for 5 min. During UF, the number of micro-organisms in the milk base is likely to increase, and may cause some quality problems in ymer. Additionally, during the first heat treatment, the level of whey protein denaturation is expected to be low to reach the desired level of water-holding capacity in the product. These drawbacks are overcome by applying two-stage heat treatment to the milk base. Fermentation of ymer is carried out at 20–22°C (i.e. mesophilic lactic acid fermentation), and is normally achieved within 14–16 h. After fermentation, the pH of the product is 4.4–4.6 and, to prevent the brittle consistency in the end product, a second homogenisation is usually applied to ymer (Renner & Abd El-Salam, 1991). In ultrafiltered ymer, a major sensory problem is the development of bitterness during storage and, to overcome this, the processed milk is recommended to be cultured prior to UF treatment. Compared with traditional ymer production, ultrafiltered ymer has the vast majority of the whey proteins retained in the retentate, leading to an increase in the yield by 8–15% (Mortensen, 1984; Glover, 1985). The mineral concentration of the ultrafiltered ymer is, in general, higher than traditional ymer (Table 7.2).

Table 7.2 Levels of some minerals in traditional and ultrafiltered ymers.

Mineral	Ymer	
	Traditional	Ultrafiltration (UF)
Calcium	0.120	0.190
Magnesium	0.012	0.015
Sodium	0.041	0.045
Potassium	0.150	0.160
Phosphorous	0.120	0.130
Chloride	0.090	0.080

Source: Adapted from Renner & Abd El-Salam (1991).

7.5.5 *Skyr*

Skyr is an Icelandic concentrated fermented milk product made from skimmed milk, with added rennet, thermophilic starter culture, i.e. *Streptococcus thermophilus*, *Lb. delbrueckii* subsp. *bulgaricus* and *Lactobacillus helveticus*, and lactose-fermenting yeasts. Nozzle separators are used in large-scale production units to concentrate the thermised fermentate where the yeast floras are inactivated but not the LAB. In some commercial applications (e.g. Skyris, Iceland) *Bifidobacterium bifidum* and *Lb. acidophilus* are replaced with *Lb. delbrueckii* subsp. *bulgaricus* and *Lb. helveticus*. The application of membrane technologies in the manufacture of skyr is not common. However, to increase the yield of skyr, Gudmundsson (1987) and Wolpert (1988) recommended that skyr whey should be concentrated by UF to the same TS level of the product followed by heat treatment at 80°C for 5 min, homogenisation and cooling to 10°C. The concentrated ultrafiltered whey is then mixed with skyr and, as a consequence, the yield of the end product is increased *ca*. 30%. In some cases, skyr is mixed with ingredients including fruit preparations or other flavourings.

7.6 Quality control

7.6.1 *Compositional quality*

Viscous products

The chemical compositions of the fermented milks made from ultrafiltered retentate and ultrafiltered fermentate varied depending on the milk composition, membrane pore size and operating conditions (Table 7.3). Owing to the association of calcium and phosphorous with the casein micelle complex, the increase in the casein concentration by UF leads to an increase in the mineral content of the concentrated milks. The pore size of the membrane affects the mineral level in the end product, but the trends in the retentate composition remained similar when membranes with higher $MW_{cut\text{-}off}$ (i.e. >25 kDa) were employed (Covacevich & Kosikowski, 1977; Biliaderis *et al*., 1992).

Table 7.3 Chemical composition (g 100 g^{-1}) of some fermented milks made from ultrafiltered (UF) retentate and UF warm yoghurt and using different mammalian milks.

Product	Total solids	Fat	Protein	Ash	Titratable acidity	pH
Viscous products using different starter cultures						
Yoghurt[a]						
Sample 1	12.50	0.10	6.50	0.90	0.17	–
Sample 2	14.00	0.20	9.50	0.90	0.21	–
Sample 3	16.80	0.20	11.80	1.10	0.23	–
Yoghurt	16.30	5.90	6.10	0.60	–	–
Yoghurt (G)						
(C_f × 1.5)	14.10	4.30	4.37	0.92	–	–
(C_f × 2.0)	15.01	4.90	5.68	1.02	–	–
(C_f × 2.5)	16.73	5.80	7.35	1.12	–	–
Bioghurt (G)[b]	15.46	3.70	4.49	–	0.79	4.50
(70% G + 30% C)[b]	15.10	3.50	4.36	–	0.80	4.40
(50% G + 50% C)[b]	15.13	3.60	4.49	–	0.83	4.45
Concentrated yoghurts made from UF retentate						
Sample 1	20.10	8.75	6.36	0.96	–	4.42
Sample 2	22.70	9.00	9.00	1.35	–	–
Sample 3	22.40	10.70	8.24	–	–	4.05
Sample 4	22.60	9.00	8.30	0.81	–	4.14
Concentrated products made from UF warm yoghurt						
Sample 1	24.30	12.20	7.98	0.68	–	4.23
Sample 2	22.30	8.80	8.40	0.91	–	4.21
Sample 3	21.80	9.60	7.42	0.71	–	4.28
Probiotic products						
Sample 1 (C)[c]	23.20	9.90	8.75	0.96	0.83	4.53
Sample 2 (S)[c]	23.08	9.50	8.85	1.16	0.83	4.51
Sample 3 (C)[b]	23.17	9.09	8.15	0.92	0.97	4.26
Sample 4 (S)[c]	23.41	9.07	8.55	1.00	1.11	4.30
Sample 5 (C)[d]	23.30	9.79	8.56	1.07	0.89	4.56
Sample 6 (S)[d]	23.40	9.75	8.56	1.17	1.00	4.53
Concentrated bioghurt						
(C)[e]	23.50	9.80	8.89	1.13	0.85	4.63
(S)[e]	22.97	9.40	8.89	1.13	0.89	4.61

Source: Data compiled from Biliaderis *et al*. (1992), Akin (1999), Özer *et al*. (2002), Domagala (2012) and Özer (2000a). C' cow's milk; G' goat's milk; S' sheep's milk.

[a]Cow's skimmed milk was used; the calcium and phosphorous contents ranged between 0.27–0.44 and 0.15–0.26, respectively.

[b]Starter culture used consisted of *Streptococcus thermophilus* and *Lactobacillus acidophilus*.

[c]Starter culture used consisted of *Bifidobacterium animalis* subsp. *lactis*.

[d]Starter culture used consisted of *Lb. acidophilus*.

[e]Starter culture used consisted of *B. animalis* subsp. *lactis*, *Lb. delbrueckii* subsp. *bulgaricus*, and *St. thermophilus*.

Enriching the milk base with ultrafiltered buttermilk powder (UF BMP) yields a high protein/TS ratio (0.47), and less smoothness of the yoghurt, compared with that of a product with the same TS level, but fortified with non-ultrafiltered product which gives a protein/TS ratio of 0.33 (Trachoo & Mistry, 1998). Supplementing yoghurt milk base with UF BMP also results in a very dense matrix and low whey separation during storage. Tamime *et al.* (1984) demonstrated that the texture of yoghurt made from ultrafiltered concentrated milk was smoother than yoghurt manufactured from milk supplemented with sodium caseinate (NaCn). Overall, the texture of yoghurt is determined by the method of production used. Although the yoghurt gel is based on the protein–protein interactions, increasing the protein level of the yoghurt milk base does not always result in the same textural properties in the end product. In all the TS fortification systems, apart from the protein level in milk (or, in narrower terms, protein-to-TS ratio), the number and distribution of both strong and weak (reversible) bonds are determinative on the physical properties of the yoghurt (Özer *et al.*, 1998a). Lankes *et al.* (1998) compared three yoghurt samples fortified by different methods including UF, vacuum evaporation (EV) and SMP addition in terms of their dynamic rheological properties by small amplitude rheometry. The results indicated that yoghurt made from ultrafiltered milk had a higher elastic modulus (G') than the other products and a higher complex viscosity (G^*) and a lower phase shift (tan δ). Similar results were obtained by Becker & Puhan (1981) and Savello & Dargan (1995), who reported an increase in Brookfield and Posthumus viscosities by 118% and 33%, respectively, in yoghurt made from ultrafiltered milk. Cheng *et al.* (2000) showed that yoghurt made from ultrafiltered milk gave higher gel firmness than yoghurt made from milk supplemented with SMP. According to Tamime *et al.* (1984), the firmness of yoghurts decreased in the RO, UF, EV and SMP products, respectively.

Many factors affect the physical characteristics of the yoghurt gel (Roefs & van Vliet, 1990; Dickinson & McClement, 1996) and, while the importance of post-incubation conditions, such as in-plant shear and storage temperature, cannot be underestimated, it is believed that the rheological characteristics of yoghurt-like fermented products are determined mainly by the gelation process. As a result of the increase in protein and, therefore, the buffering capacity of milk, the gelation profiles of the membrane-concentrated milks differed from those made from unconcentrated milk (especially in G', η' and phase shift angle values). In high protein systems, like the ultrafiltered milk, flocculation kinetics may change so that, in effect, the systems enter into a 'metastable' state in which the strength of the gel can continue to increase (Schulze *et al.*, 1991). Overall, the number and/or distribution of the strong and weak bonds within the gel increase with time (Dickinson, 1994). The concentration model influences the number and distribution of the bonds over the gel network. Table 7.4 shows the gel development rate and gel onset time (min) of the yoghurts made from ultrafiltered milk and yoghurts concentrated by UF or RO. In general, fermented milks made from ultrafiltered milk exhibit shorter onset time and greater phase angle values than unconcentrated milks. Biliaderis *et al.* (1992) demonstrated that the sudden increase in the G' values and a decrease in the tan δ values occurred within the pH range of 5.2–5.7, where some casein leaves the micelle and causes dissociation of weakly bound β-casein and κ-casein. This change in the micellar structure makes it more susceptible to ionic calcium and, thereby, the rapid development of the α-casein network. In general, during the early stages of fermentation (i.e. 0–90 min), no viscoelasticity is evident in the

Table 7.4 Apparent gelation rate constant, onset of gelation and rheological parameters of yoghurt and concentrated yoghurt using different membrane filtration.

Parameters	Yoghurt[a]		Concentrated yoghurt[b]	
	Total solids (14 g 100 g^{-1})	Total solids (16 g 100 g^{-1})	UF – total solids (22.2 g 100 g^{-1})	RO – total solids (23.2 g 100 g^{-1})
Gel development rate[c]	ND	ND	4.88	8.66
Gel onset point[c]	90	85	90	120
G′[d]	1220	2010	2685	1093

Source: Data compiled from Biliaderis *et al*. (1992) and Özer (1997).
UF, ultrafiltration; RO, reverse osmosis; ND, not determined.
[a]Milk base was concentrated by UF.
[b]Warm fermentate was concentrated by using different membrane filtrations.
[c]Time in minute to attain a G′ of 1 Pa.
[d]Elastic modulus (Pa) at the end of incubation period.

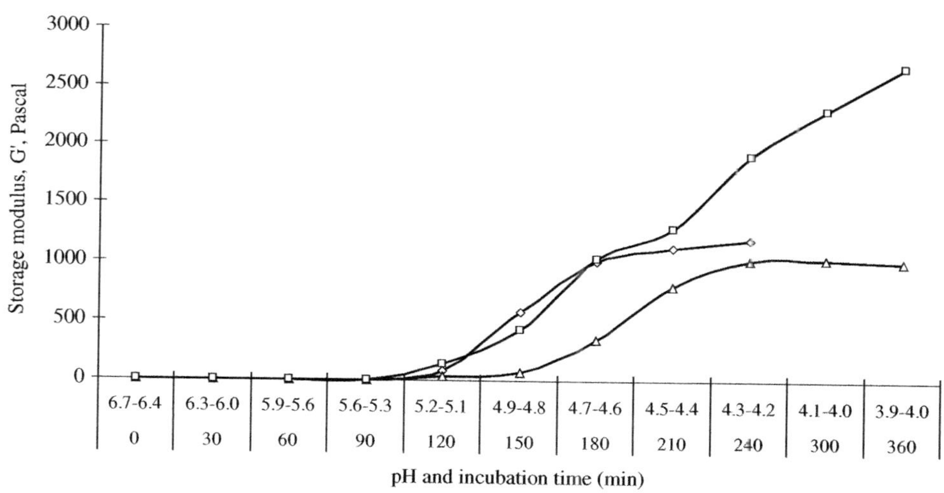

Fig. 7.7 Variation in the storage modulus (G′) values of the concentrated yoghurts as function of time and pH. Experimental conditions: amplitude range 10^{-3}–10^{-2} mNm at 0.25 Hz; temperature 25°C; parallel-plate geometry, plate radius 10 mm; gap setting 1 mm. Results are the average of eight separate runs. Standard errors are less than symbol dimensions. ◇, Control yoghurt (16 g 100 g^{-1} TS); □, fermentation of ultrafiltered retentate (~23 g 100 g^{-1} TS); Δ, fermentation of reverse osmosis retentate (~23 g 100 g^{-1} TS) (Adapted from Özer *et al.*, 1998a. Reproduced by permission of Elsevier).

milk bases (both ultrafiltered and unconcentrated milks). The storage modulus (G′) values then show a rapid increase followed by a plateau region. However, the time needed for the ultrafiltered milk to reach the plateau value is much longer than that of unconcentrated milk (Fig. 7.7). This can be attributed to the much lower rate of rearrangement of the casein network because of higher buffering capacity in the milk concentrated by UF. The tan δ values of the ultrafiltered milk remain similar, suggesting the same interactive forces are involved in gel formation, but to different degrees (Özer *et al.*, 1998b).

The rheological properties of the yoghurt gel are also affected by the heat-induced thiol–disulphide interchange reactions taking place mainly between κ-casein and β-lactoglobulin. The UF treatment influences the level of total and reactive thiol groups in concentrated milk. Özer *et al*. (2002) compared the levels of total and reactive thiol groups in the yoghurt samples produced from milks concentrated to *ca*. 16 g 100 g^{-1} by UF or RO. The highest value for reactive thiol groups was obtained in yoghurt made from ultrafiltered milk (5.66 μmol SH protein g^{-1}), followed by a similar product made from RO milk (4.36 μmol SH protein g^{-1}) and the control (12 g 100 g^{-1} TS, 2.47 μmol SH protein g^{-1}). The ratio of reactive thiol groups to total thiol groups was found to be independent of the increase in the protein content and, while 81% of the total thiol groups in the control sample were activated by heat treatment at 85°C for 20 min, this figure was 74% and 66% in samples made from ultrafiltered milk and RO milk, respectively. It is well known that the casein micelles have a heterogeneous structure and, if an increase in the protein content of milk leads to an increase in the heterogeneity, there might well be an inverse relationship between protein content and the activation of thiol groups. The lower levels of reactive thiol groups in yoghurts made from RO milk may be attributed to higher lactose content in this sample. Sugars can affect the conformation, and may change the heat denaturation/coagulation kinetics of proteins (Garret *et al*., 1988).

The choice of membrane type used also affects the physical quality of the fermented milks. Domagala & Kupiec (2003) found that yoghurt produced from ultrafiltered goat's milk with a 10 kDa molecular cut-off membrane was characterised by higher viscosity and hardness than that produced using membranes with 30 kDa and 100 kDa molecular cut-off. The same authors also showed a significant difference in the cohesiveness values of yoghurts made from milk concentrated through UF membranes with 10 kDa and 30 kDa molecular cut-off. Yoghurts produced from concentrated milk with different membranes did not vary in gumminess. Also, resilience of yoghurts produced from untreated milk and from ultrafiltered milk did not vary significantly between the samples. In contrast to the physical parameters, the yoghurts manufactured from milk concentrated by UF with molecular cut-off 10 kDa and 100 kDa received lower sensory scores than those produced with a 30 kDa molecular cut-off membrane. This could be a result of the fact that milk concentrated with the 10 kDa membrane had a high protein content, and yoghurt made from this milk had a very compact gel structure in addition to a lower dry matter content in 100 kDa caused by migration of some high molecular weight substances to the permeate. Furthermore, the time of UF of milk was prolonged when the 10 kDa membrane was used, and significant losses in milk dry matter were evident when the 100 kDa membrane was employed. Therefore, Domagala & Kupiec (2003) concluded that the membrane of 30 kDa pore size seems to be the most suitable for concentrating goat's milk used for the manufacture of yoghurt. Other researchers reported that it could be possible to obtain improved quality yoghurt made from ultrafiltered cow's milk by using membranes having 10–20 kDa pore sizes (Biliaderis *et al*., 1992; Savello & Dargan, 1995, 1997).

According to Biliaderis *et al*. (1992), ultrafiltered milk base yoghurts ($MW_{cut\text{-}off}$ 10 kDa at 25°C) had higher levels of acetaldehyde, pyruvic acid, acetic acid and lactic acid, and lower concentrations of orotic and citric acids than SMP fortified yoghurt. Lactose is one of the major routes for acetaldehyde formation by the yoghurt starter cultures and, despite lower concentration of lactose in the ultrafiltered milk, the acetaldehyde production in the

Table 7.5 Textural properties of goat's milk yoghurt.

Sample	Apparent viscosity (Pa s^{-1})	Hardness (N)	Adhesiveness ($\times 10^3$ N m^{-1})	Extrusion force (N)	Syneresis (%)	Sensory evaluation[a]
Non-concentrated	0.48	0.15	0.14	1.20	32.0	3.83
UF (C_f × 1.5)	1.33	0.26	0.47	2.54	12.0	4.54
(C_f × 2.0)	2.23	0.37	0.73	3.52	14.0	4.40
(C_f × 2.5)	2.47	0.49	1.23	5.05	6.0	4.18

Source: Data compiled from Domagala (2012).
UF, ultrafiltration.
[a]Five-point rating scale was used.

respective yoghurt was not impaired (Biliaderis *et al*., 1992). This possibly indicates that the other metabolic pathways of acetaldehyde production (i.e. tyrosine, methionine, valine and acetyl phosphate metabolisms) were not negatively affected by the membrane concentration process. This assumption was further supported by Alizadeh *et al*. (2008), who established a positive correlation between the acetaldehyde level and the protein content in the yoghurt made from ultrafiltered milk. Marshall & El-Bagoury (1986) compared lactic acid and acetaldehyde production by the yoghurt starter cultures in goat's milk fortified by UF, RO and the addition of goat's milk powder. Starter culture activity was impaired when milk was concentrated by RO, and aroma was poor in milk fortified by addition of powder, whereas yoghurts made from ultrafiltered retentate had satisfactory acidity and aroma. Threonine (0.1 mg g^{-1}) addition to unfortified milk and milk fortified by addition of powder or by UF of the milk produced yoghurts with high levels of acetaldehyde. In contrast, Karademir *et al*. (2002) found no significant differences between the yoghurts manufactured from goat's milk fortified by UF, EV or SMP with regard to the acetaldehyde content, consistency and viscosity values. Goat's milk yoghurt is, in general, characterised by a softer body and more delicate structure than cow's milk yoghurt. Domagala (2012) proposed that concentration of goat's milk by UF was a satisfactory way of improving the textural and rheological properties of the yoghurt. The same author compared the textural parameters and sensory scores of goat's milk yoghurts made with three different concentration factors (C_f) (i.e. 1.5×, 2.0× and 2.5×) (Table 7.5). Overall, with an increase in the C_f from 1.5× to 2.5×, the physical properties of the goat's milk yoghurt increased accordingly. Concentration of goat's milk by UF influenced the microstructure of yoghurt as well. Domagala (2012) showed that the yoghurts made from ultrafiltered milk had a more compact microstructure composed of bigger micelle clusters than its non-concentrated counterpart. With the increase in concentration, while the size of agglomerates became bigger, the void spaces over the protein network were smaller. An opposite finding was reported by Tamime *et al*. (1984), who demonstrated that yoghurt made from ultrafiltered cow's milk had a more open microstructure than yoghurt made with SMP added to the milk base.

The potential use of UF technology in the manufacture of viscous bioghurt was evaluated by Kavas *et al*. (2004). A slow development of acidity was noted in the bioghurts made from ultrafiltered milk compared with products made with the addition of SMP to the milk base. Interestingly, the compression response values (i.e. probe penetration into the gel) of the ultrafiltered samples were lower than those made with added SMP, despite the fact that the former sample had higher TS levels. The same authors also demonstrated that the consistency of the bioghurts increased throughout storage for 14 days. This finding was further supported by the organoleptic evaluations.

Concentrated products

Concentrated yoghurt is a typical weak viscoelastic gel whose storage modulus (G′) is greater than its loss modulus (G″) within the linear viscoelastic region. In spite of the fact that both concentrated yoghurts made by fermenting ultrafiltered milk (fermenting UF retentate) and after fermentation (UF of warm yoghurt) had more or less similar chemical composition, the former product was stiffer than the latter product. The RO treatment caused a much weaker structure with fairly low G′ values than the ultrafiltered-treated samples, Figs. 7.8 and 7.9 show the structural breakdown in the concentrated yoghurts produced by different membrane applications at higher amplitudes, which suggests that the overall domain of the protein may be reduced and that different methods of manufacture may be producing materials that have different space occupancy in the gel material (Özer *et al.*, 1999b). The RO treatment of warm yoghurt led to a very weak and sponge-like structure, giving it an atypical appearance for concentrated yoghurt. The rheology of concentrated yoghurt is related to the protein content and the severity of mechanical agitation during

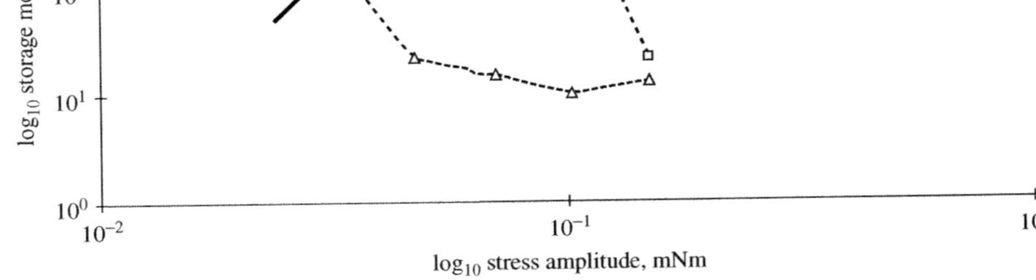

Fig. 7.8 The storage modulus (G′) of concentrated warm yoghurt using membrane filtration as a function of stress sweep. Torque range was between 1.5×10^{-2} and 1.5×10^{-1} mNm at 0.25 Hz frequency. Measuring temperature was 25°C; parallel plate geometry with 10 mm plate radius was employed at 1 mm fixed gap setting. Results are the average of eight separate runs each done five times. Standard errors are less than symbol dimensions. ◇, Traditional concentrated yoghurt (cloth bag method) (~23 g 100 g^{-1} TS); □, concentrated yoghurt made by ultrafiltration of the warm fermentate (~23 g 100 g^{-1} TS); Δ, concentrated yoghurt made by reverse osmosis of the warm fermentate (~23 g 100 g^{-1} TS). Points indicated represent the strain value at the gel breaking point (Adapted from Özer *et al.*, 1998b, and reproduced by permission of John Wiley and Sons).

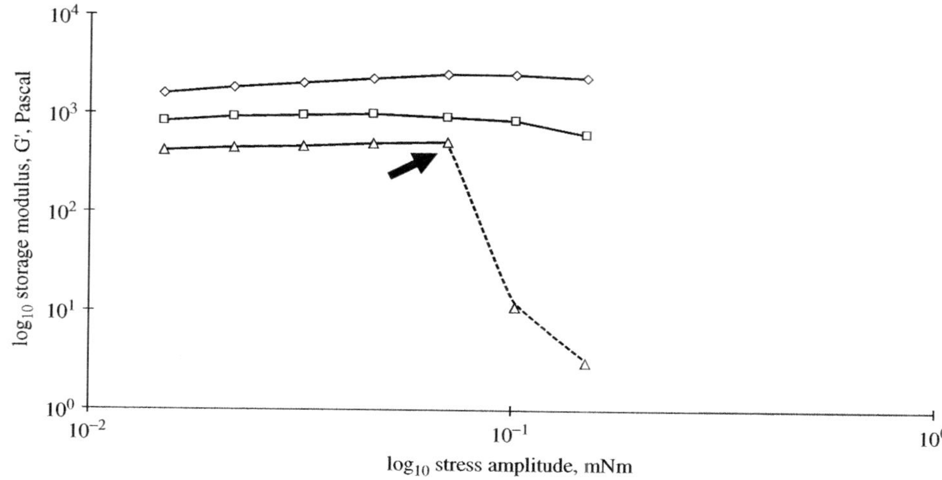

Fig. 7.9 The storage modulus (G′) of samples concentrated prior to fermentation as a function of stress sweep. Torque range was between 1.5×10^{-2} and 1.5×10^{-1} mNm at 0.25 Hz frequency; measuring temperature was 25°C; parallel plate geometry with 10 mm plate radius was employed at 1 mm fixed gap setting. Results are the average of eight separate runs each done five times. Standard errors are less than symbol dimensions. ◇, Traditional concentrated yoghurt (cloth bag method) (~23 g 100 g^{-1} TS); □, concentrated yoghurt made by ultrafiltration of the warm fermentate (~23 g 100 g^{-1} TS); Δ, concentrated yoghurt made by reverse osmosis of the warm fermentate (~23 g 100 g^{-1} TS). Points indicated represents the strain values at the gel breaking point (Adapted from Özer *et al.*, 1998b), and reproduced by permission of John Wiley and Sons).

membrane processing. The minimum protein concentration at which a gel can be formed is ~3 g 100 g^{-1}, which is presumably due to the lower number of protein contacts in the larger serum volume (Özer *et al.*, 1999b). Overall, yoghurts with high protein levels have greater gel strengths. The dependence of complex modulus (G*) on the protein concentration is non-linear, indicating that the number of stress-carrying strands is not proportional to volume fractions of casein particles, and so the network formed is very heterogeneous. Özer (2006a, b) postulated that UF of warm yoghurt at low inlet pressures (i.e. 0.7–0.8 MPa) caused a more homogeneous structure in the gel material, bringing about an increase in the number of smaller conglomerates over the gel network and, thus, the number of stress-carrying effective bonds.

Tamime *et al.* (1989b, 1991a) and Özer *et al.* (1999b) showed that concentrated yoghurts made from milk concentrated by UF prior to fermentation displayed continuity in structure. The slight discontinuity in the concentrated yoghurt made by UF of warm yoghurt was also evident, indicating the limited detrimental effect of membrane application on the weak gel structure of concentrated yoghurt. Both types of fermented milk (e.g. natural plain yoghurt and concentrated yoghurt) made by membrane techniques have similar microstructures in which the protein particles are adsorbed onto minute fat globules (Tamime *et al.*, 1989b, 1991a).

Özer (1997) employed confocal scanning laser microscopy (CSLM) to study the distribution of protein networks in the serum phase in concentrated yoghurt gel produced by UF or RO, and to find out the relationship between structure and rheology of the resulting gels

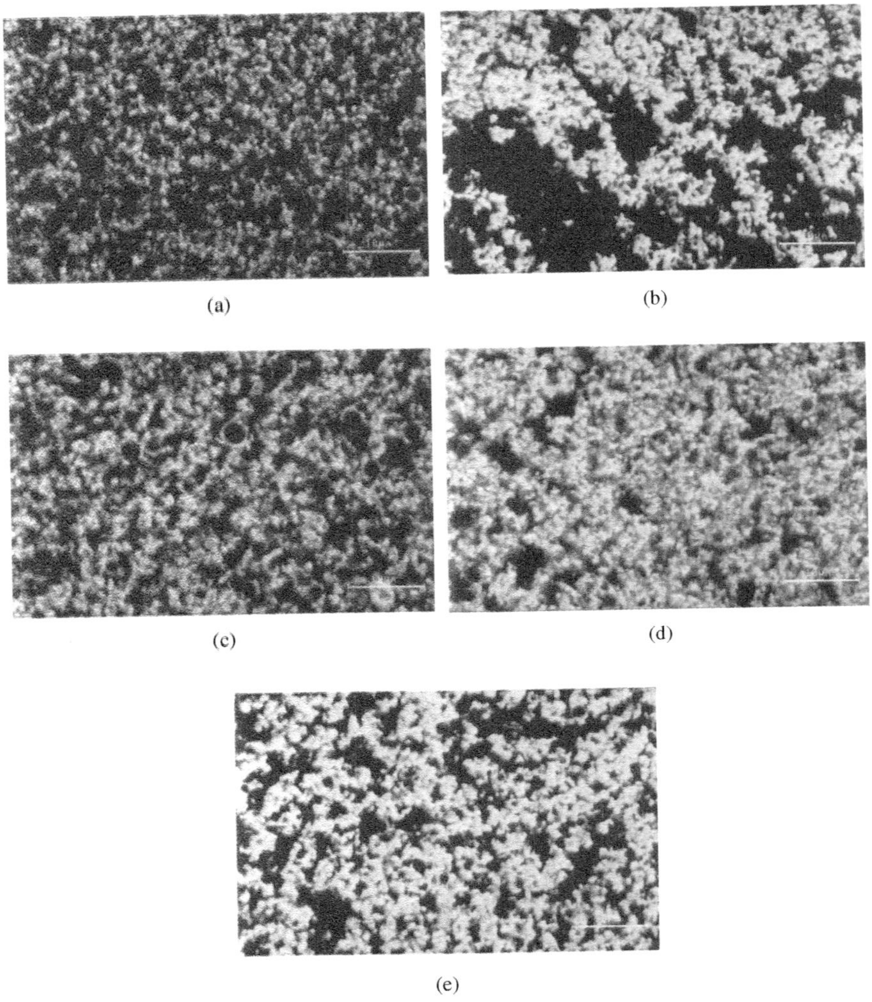

Fig. 7.10 Microstructure (confocal laser scanning microscopy) of concentrated yoghurts. (a) Concentrated yoghurt made by reverse osmosis (RO) of the warm fermentate; (b) concentrated yoghurt made from fermenting RO retentate; (c) concentrated yoghurt made by ultrafiltration (UF) of the warm fermentate; (d) concentrated yoghurt made from fermenting UF retentate; (e) traditional concentrated yoghurt (i.e. cloth bag method). The total solids contents of the samples were ~23 g 100 g^{-1}. Bar, 10 μm (Adapted from Özer, 1997).

(Fig. 7.10). Much denser structures were observed in the concentrated yoghurts with higher protein contents (fermented before or after UF application) than the samples with lower protein levels (fermented before or after RO application). It was strikingly evident that the yoghurt concentrated by RO after the fermentation stage had smaller void spaces in its gel network than yoghurt produced from RO retentate. Apart from this peculiar structure in the former sample, there was a close relationship between protein content, gel strength and the dimensions of void spaces in the protein matrix. It has been suggested that, during

RO application of warm yoghurt at relatively high operating pressure (2 MPa), the protein matrix was broken down and the broken particles were distributed over the gel network more evenly. However, the much stronger gel structure of ultrafiltered concentrated yoghurt suggests that, whereas UF may have caused a homogenisation effect, similar to RO, the result of the lower pressure during the UF process was an even distribution of a relatively less disturbed protein network. These observations were further confirmed by scanning electron microscopy (SEM) and transmission electron microscopy (TEM) (Özer, 1997). Both SEM and TEM micrographs revealed that RO application to warm yoghurt caused structural damage in the casein strands, whereas much finer strands were evident in yoghurt concentrated by UF.

However, Akin (1999) found that the firmness of concentrated bioghurt made from ultrafiltered cow's milk was lower than the product made from added SMP to the milk base (69 g *vs*. 81 g) but, when goat's milk was used in the manufacture of concentrated bioghurt, the firmness of the ultrafiltered products was slightly higher than that made from goat's milk fortified with SMP (64 g *vs*. 53 g). In addition, Akin (1999) demonstrated that the firmness values of concentrated probiotic milk drinks and concentrated probiotic yoghurt made from ultrafiltered cow's milk were lower than the products made with added SMP to the goat's milk base. Also, the effect of type of milk on the firmness and the concentration of lactic acid isomers of the concentrated products were limited (Akin, 1999). In all concentrated bioghurts made from cow's, goat's or sheep's milk, the L(+) lactic acid isomer was predominantly present irrespective of the milk type; however, the method of concentration of fermented milk affected the level of lactic acid isomer.

7.6.2 *Microbiological quality*

The increase in protein content of milk with UF leads to an increase in the buffering capacity of the milk (Alvarez *et al*., 1998). Therefore, the negative effect of reduction in pH on the growth of the yoghurt starter cultures (*St. thermophilus* and *Lb. delbrueckii* subsp. *bulgaricus*) is prevented to some extent (Mistry, 2002; de Castro-Morel & Harper, 2003). The growth of the starter cultures is stimulated in yoghurts made from milk concentrated by UF due to the presence of the greater amount of nutrients available. If the ultrafiltered milk base is heat treated at 85–90°C, the growth of *St. thermophilus* is particularly stimulated because of reduced concentration of oxygen, which exhibits an inhibitory effect on this organism. However, the proteolytic activities of the bacteria become weakened and the release of low molecular weight peptides is hindered (Nakazawa *et al*., 1991). The same authors also demonstrated that the ratio of the total free amino acids to protein was 0.0375, and the UF treatment did not affect the particle size of low molecular weight peptides in yoghurt. Overall, the growth trend of the yoghurt starter cultures in ultrafiltered milk retentate is similar to that in unconcentrated milk. Since the optimum temperature range for the growth of *St. thermophilus* is around 40–45°C, UF of warm yoghurt at this temperature stimulates the growth of this organism. Özer & Robinson (1999b) investigated the behaviour of the starter cultures in concentrated yoghurts produced by fermenting ultrafiltered and RO milk retentates, and by removing whey from batches of natural yoghurt using the same techniques. The growth of *St. thermophilus* was similar in all samples, with the exponential growth phase ending around 180 min of incubation. The stationary phase of growth of *St.*

thermophilus coincided with the exponential growth phase of *Lb. delbrueckii* subsp. *bulgaricus*. During concentration of warm yoghurt by UF or RO, the growth of *St. thermophilus* and, to a more limited extent, *Lb. delbrueckii* subsp. *bulgaricus* was reported to continue (Özer & Robinson, 1999b). In concentrated yoghurts made from milk retentates (i.e. UF and RO), the counts of both yoghurt organisms were lower than in that manufactured by concentrating warm yoghurt. The ratio of 1:1 between *St. thermophilus* and *Lb. delbrueckii* subsp. *bulgaricus* is extensively quoted in the literature as the optimum ratio for a good-quality yoghurt (Tamime & Robinson, 2007). Such a ratio is maintained in concentrated yoghurt provided it is made from yoghurt with a similar ratio (Salji, 1991). This trait is of critical importance for establishing the characteristic aroma and flavour of the end product, and improving the shelf-life of concentrated yoghurt without adversely affecting its quality.

Although the total counts of yoghurt starter cultures are expected to be similar in concentrated yoghurts made from milks concentrated by different membrane techniques, it is expected to observe higher counts for *St. thermophilus* colonies than *Lb. delbrueckii* subsp. *bulgaricus*; this could be attributed to the concentration of stimulatory factors, such as whey proteins, retained in the membrane treated milk (Hickey *et al*., 1983). Özer & Robinson (1999b) calculated higher maximum generation times (G_{max}) (~1.15–1.23 h) for the mixed cultures in concentrated yoghurt systems than that in unconcentrated yoghurts (~2.13–2.20 h). The level of acidity is one of the major determinative factors for the viability of the yoghurt organisms. Sinha (1991) and Yamani & Ibrahim (1996) found a strong relationship between the viability of *St. thermophilus* and the level of acidity in the final product. In general, in unconcentrated milks the acidity has a limited effect on the growth of *Lb. delbrueckii* subsp. *bulgaricus* but, when the milk is concentrated by UF to 23–24 g 100 g^{-1} TS, the effect of acidity on the growth of this organism becomes relatively more important (Özer & Robinson, 1999b). This may be explained by the prolonged exposure of the cells of *Lb. delbrueckii* subsp. *bulgaricus* to levels of acidity during the longer incubation periods used with the concentrated milk.

7.6.3 *Organoleptic properties*

Biliaderis *et al*. (1992) reported that by increasing the TS content of plain yoghurt from 12 to 16 g 100 g^{-1} using UF, the sensory scores of perceived thickness and graininess became more sensitive to increasing TS than in yoghurts made by fortifying the milk base with SMP. The same authors found no marginal difference between yoghurts made from ultrafiltered milk or fortified with SMP addition regarding the perceived sourness.

Abrahamsen & Holmen (1980) reported that the major sensory problem in the yoghurts made from ultrafiltered milk was 'too dry' and 'pudding-like' texture. A remarkable decrease in the firmness of the yoghurts was also noted during storage for 14 days. Additionally, the flavour of low-lactose yoghurt made from diluted ultrafiltered milk was flat, thus the addition of BMP and/or sodium citrate to stimulate lactic acid fermentation was recommended (Kosikowski, 1979).

In general, the overall characteristics of traditional concentrated yoghurt (with TS levels of 23–25 g 100 g^{-1}) could be summarised as being slightly creamy in colour, short in body and with a lumpy texture, which makes this product difficult to spread. This problem can be overcome by increasing the initial TS level in the milk base used for the

Table 7.6 Comparison of the sensory scores[a] of concentrated yoghurt awarded by the Egyptian and American panellists.

Sensory attributes	Storage period (days)	American panellist	Egyptian panellist
Appearance	1	4.07	4.24
	28	3.43	3.88
Flavour	1	3.43	3.88
	128	3.06	3.83
Sourness	1	3.36	3.53
	28	4.06	3.39
Consistency	1	3.76	3.65
	28	3.44	3.72
Overall acceptability	1	3.50	3.88
	28	3.00	3.33

Source: Data Adapted from El-Samragy & Zall (1988).
[a]Rating scale used was 1, dislike very much ... 5, like extremely.

manufacture of concentrated yoghurt. Regarding the texture and appearance, concentrated yoghurt made from warm fermentate concentrated by UF is superior to either traditional concentrated yoghurt or concentrated yoghurt made from fermented ultrafiltered retentate. UF of warm yoghurt results in a smoother body and creamy texture with a more shiny appearance (Özer, 1997). The concentrated yoghurt produced from ultrafiltered retentate is, in general, characterised by a crumbly structure, a coarse texture and wheying-off after the coagulum is broken. In cases of using a structuriser to obtain a smoother body and texture in concentrated yoghurt made by removing whey from warm yoghurt by UF, the choice of structuriser head is important. Tamime *et al*. (1991b) compared two different structuriser heads, causing low and high turbulence, in terms of sensory properties of the concentrated yoghurt. The unsmoothed samples were dull in appearance, granular and very rough in texture, possibly due to the effect of slight agglomeration of the casein particles at low pH. Creating high turbulence during structurising resulted in a smoother body in concentrated yoghurt than using low turbulence but, as a consequence, the firmness of the product was lower.

The organoleptic preferences of consumers vary from one society to another. El-Samragy & Zall (1988) compared the organoleptic preferences of Egyptian and American panels of concentrated yoghurt (23.18 g 100 g^{-1} TS) (Table 7.6). The Egyptian panellists, who were more familiar with such concentrated products, i.e. flavour attribute, gave slightly higher scores for the product stored for 28 days than the American panellists, with the exception of the scores awarded for the sourness attribute.

7.7 Conclusion

Over the past few decades, great developments in membrane technology have taken place, and their possible applications during the manufacture of gel (set and stirred) and viscous (i.e. concentrated) yoghurts have been successful, at least in semi-industrial trials. It is evident that the physico-chemical aspects, rheological properties and sensory profiling of yoghurt made by UF of the milk base are superior (i.e. firmer and minimum syneresis during storage) to similar products made using SMP to increase the protein content in the product. Although the latter approach could be more economical during the manufacture of yoghurt, the improvements in the overall characteristics of yoghurt made from ultrafiltered milk should not be overlooked. In addition, the use of membrane filtration of the milk base can allow the manufacturer of yoghurt to produce different product characteristics, such as low-lactose yoghurt.

Nozzle separators are widely used world-wide for the manufacture of concentrated yoghurt, but it is feasible to produce a similar product using membrane technology. It could be argued, however, that UF of warm fermentate for the production of concentrated yoghurt may be slightly more expensive in terms of initial capital investment and possible replacement of membranes once every 2 years depending on ware and tear, but the increase in yield of the product is in favour of the membrane system of production. This is proven during the manufacture of skyr (see Section 7.5.5). Furthermore, the permeate produced from UF of warm yoghurt is much 'clearer' (i.e. containing only lactose, minerals and a very low amount of soluble nitrogen) than the whey from the nozzle separator; the permeate is even colourless when using a Carbosep membrane (i.e. carbon-based composites), which can be easily used in beverage drinks, or during the production of low-viscous drinking yoghurt (ayran) in the Middle East.

References

Abd-El Salam, M.H. & El-Alamy, A.H. (1982) Production and properties of yoghurt and concentrated yoghurt (Labneh) from ultrafiltered recombined milk. *Research Bulletin*, No. **1803**, pp. 1–11, Ain Shams University, Cairo, Egypt.

Abrahamsen, R.K. & Holmen, T.B. (1980) Yogurt from hyperfiltrated, ultrafiltered and evaporated milk and from milk with added milk powder. *Milchwissenschaft*, **35**, 399–402.

Akin, N. (1999) [Inek ve koyun sutunden uretilen bazi konsantre fermente sut urunlerinin sertligi ve duyusal ozellikleri]. *Turkish Journal of Veterinary and Animal Sciences*, **23**, 583–590.

Alizadeh, A., Ehsani, R.M. & Homayouni, A. (2008) Acetaldehyde production rate in yoghurt made from ultrafiltered skim milk. *Asian Journal of Chemistry* **20**, 6529–6534.

Alvarez, F., Arguello, M., Cabero, M., Riera, F.A., Alvarez, R., Iglesias, J.R. & Granada, J. (1998) Fermentation of concentrated skim-milk. Effects of diffeent protein/lactose ratios obtained by ultrafiltration-diafiltration. *Journal of the Science of Food and Agriculture*, **76**, 10–16.

Anonymous (2003) *Dairy Processing Handbook*, 2nd and revised edn of G. Bylund (1995), Tetra Pak Processing Systems AB, Lund.

Attia, H., Bennasar, M. & de la Fuente, T. (1991a) Study of the fouling of inorganic membranes by acidified milks using scanning electron microscopy and eletrophoresis. I. Membrane with pore diameter 0.2 μm. *Journal of Dairy Research*, **58**, 39–50.

Attia, H., Bennasar, M. & de la Fuente, T. (1991b) Study of the fouling of inorganic membranes by acidified milks using scanning electron microscopy and eletrophoresis. II. Membrane with pore diameter 0.8 μm. *Journal of Dairy Research*, **58**, 61–65.

Becker, T. & Puhan, Z. (1981) Effect of different processes to increase the milk solids non fat content on the rheological properties of yogurt. *Milchwissenschaft*, **44**, 626–629.

Biliaderis, C.G., Khan, M.M. & Blank, G. (1992) Rheological and sensory properties of yogurt from skim milk and ultrafiltered retentates. *International Dairy Journal*, **2**, 311–323.

Bjerre, P. & Nielsen, P.S. (1972) Application of ultrafiltration in the manufacture of cultured milk products. *19th International Dairy Congress, IE, 738*.

Brazuelo, A., Suarez, E., Riera, F.A., Alvarez, R., Iglesia, J.R. & Granda, J. (1995) Protein-enriched yogurt by ultrafiltration of skim-milk. *Journal of the Science of Food and Agriculture*, **69**, 283–290.

Butterick, J.K. & Higgins, J.J. (1982) Functional properties of milk powders from ultrafiltered skim-milk. *International Dairy Congress*, **1**, Book 2, 439.

Campbell-Platt, G. (1987) *Fermented Foods of the World*, Butterworth, London.

Cheng, L.J., Augustin, A.M. & Clarke, P.T. (2000) Yoghurts from skim milk-whey protein concentrate blends. *Australian Journal of Dairy Technology*, **55**, 110.

Covacevich, H.R. & Kosikowski, F.V. (1977) Skim milk concentration for cheese-making by alternative ultrafiltration procedures. *Journal of Food Science*, **42**, 1359–1361, 1379.

Davies, F.L., Shankar, P.A. & Underwood, H.M. (1977) The use of milk concentrated by reverse osmosis for the manufacture of yogurt. *Journal of the Society of Dairy Technology*, **30**, 23–28.

Debon, J., Prudencio, E.S. & Petrus, J.C.C. (2010) Rheological and physico-chemical characterization of prebiotic microfiltered fermented milk. *Journal of Food Engineering*, **99**, 128–135.

de Castro-Morel, M. & Harper, W.J. (2003) Effect of retentate heat treatment and spray dryer inlet temperature on the properties of milk protein concentrates (MPCs). *Milchwissenschaft*, **58**, 13–15.

Dickinson, E. (1994) Computer simulation of particle gel formation. *Journal of Chemistry Society Faraday Transaction*, **90**, 173–180.

Dickinson, E. & McClement, D.J. (1996) *Advances in Food Colloids*, Blackie Academic and Professionals Publishers, Glasgow.

Dixon, B.D. (1985) Dairy products prepared from reverse osmosis concentrate-market milk products, butter, skimmilk powder and yoghurt. *Australian Journal of Dairy Technology*, **40**, 91–95.

Domagala, J. (2012) Instrumental texture, syneresis, and microstructure of yogurts prepared from ultrafiltered goat milk: Effect of degree of concentration. *International Journal of Food Properties*, **15**, 558–568.

Domagala, J. & Kupiec, B.E. (2003) Changes in texture of yoghurt from ultrafiltrated goat's milk as influenced by different membrane types. *Electronic Journal of Polish Agricultural Universities* **6** (1).

El-Samragy, Y.A. & Zall, R.R. (1988) Organoleptic properties of yoghurt-cheese labneh manufactured using ultrafiltration. *Dairy Industries International*, **53** (3), 27–28.

Fonden, R., Leporanta, K. & Svensson, U. (2006) Nordic/Scandinavian fermented milk products. *Fermented Milks*, (ed. A.Y. Tamime), pp. 156–173. Blackwell Publishing, Oxford.

Garret, J. M., Stairs, R. A. & Annett, R. G. (1988) Thermal denaturation and coagulation of whey proteins: effect of sugars. *Journal of Dairy Science*, **71**, 10–16.

Glover, F. A. (1985) *Ultrafiltration and Reverse Osmosis for the Dairy Industry*, Technical Bulletin 5, The National Institute for Research in Dairying, Reading, UK.

Gudmundsson, B. (1987) Skyr. *Scandinavian Dairy Journal*, **1** (4), 240–242.

Guirguis, N., Versteeg, K. & Hickey, M.W. (1987) The manufacture of yoghurt using reverse osmosis concentrated milk. *Australian Journal of Dairy Technology*, **42**, 7–10.

Hickey, M.W., Roginski, H. & Broome, M.C. (1983) Growth and acid production of group N streptococci in ultrafiltered milk. *The Australian Journal of Dairy Technology*, **38**, 138–143.

Hofi, M.A. (1988) Labneh (concentrated yoghurt) from ultrafiltered milk. *Scandinavian Dairy Industry*, **2** (1), 50–52.

IDF (2002) *World Dairy Situation 2002*, Document No. 378, pp. 60, International Dairy Federation, Brussels.

IDF (2004) *World Dairy Situation 2004*, Document No. 391, pp. 70, International Dairy Federation, Brussels.

IDF (2009) *World Dairy Situation 2009*, Document No. 438, pp. 98, International Dairy Federation, Brussels.

IDF (2010) *World Dairy Situation 2010*, Document No. 446, pp. 176–177, International Dairy Federation, Brussels.

Karademir, E., Atamer, M., Tamuçay, B. & Yaman, S. (2002) Some properties of goat milk yogurts produced by different fortification methods. *Milchwissenschaft* **57**, 261–263.

Kavas, G., Uysal, H., Kilic, S., Akbulut, N. & Kesenkas, H. (2004) Production and selected properties of bioghurt made from goat milk and cow milk mixtures by ultrafiltration and addition of skim milk powder. *International Journal of Food Properties*, **7**, 473–482.

Kosikowski, F.V. (1979) Low lactose yogurt and milk beverage by ultrafiltration. *Journal of Dairy Science*, **62**, 41–46.

Küçükçetin, A. & Yaygin, H. (1999) Studies on the properties of the koumiss made from original mare's milk and whey powder added cow's and goat's. *Special Issue for the 11th Congress of KUKEM-Biotechnology*, Isparta, *Turkey*, **23**:29–35.

Küçükçetin, A., Yaygin, H., Hinrichs, J. & Kulozik, U. (2003) Adaptation of bovine milk towards mares' milk composition by means of membrane technology for koumiss manufacture. *International Dairy Journal*, **13**, 945–951.

Kurmann, J.A., Rasic, J.L. & Kroger, M. (1992) *Encyclopedia of Fermented Fresh Milk Products*, Van Nostrand Reinhold, New York.

Lankes, H., Özer, B.H. & Robinson, R.K. (1998) The effect of elevated milk solids and incubation temperature on the physical properties of natural yogurt. *Milchwissenschaft*, **53**, 510–513.

Mann, E.J. (1986) Buttermilk. *Dairy Industries International*, **51** (10), 9–10.

Marshall, V.E. & El-Bagoury, E. (1986) Use of ultrafiltration and reverse osmosis to improve goats' milk yogurt. *Journal of the Society of Dairy Technology*, **39**, 65–66.

Marshall, V.M.E. & Tamime, A.Y. (1997) Physiology and biochemistry of fermented milks. *Microbiology and Biochemistry of Cheese and Fermented Milk* (ed. B.A. Law), 2nd ed., pp. 153–192, Blackie Academic & Professional, Glasgow.

Mehaia, M.A. (2005) Manufacture of fresh labneh from goat's milk using ultrafiltration process. *Journal of Food Technology*, **3**, 24–29.

Mistry, V.V. (2002) Manufacture and application of high milk protein powder. *Lait*, **82**, 515–522.

Mistry, V.V. & Hassan, H.N. (1992) Manufacture of nonfat yogurt from a high milk protein powder. *Journal of Dairy Science*, **75**, 947–957.

Morin, P., Jimenez-Flores, R. & Pouliot, Y. (2004) Effect of temperature and pore size on the fractionation of fresh and reconstituted buttermilk by microfiltration. *Journal of Dairy Science*, **87**, 267–273.

Morin, P., Pouliot, Y. & Jimenez-Flores, R. (2006) A comparative study of the fractionation of regular buttermilk and whey buttermilk by microfiltration. *Journal of Dairy Science*, **90**, 2132–2140.

Mortensen, H. (1995) Delta's Olympian aspirations. *Scandinavian Dairy Information*, **9** (4), 8–10.

Nakazawa, Y., Furusawa, M., Hohno, H. & Shida, T. (1991) Manufacture and proteolytic properties of yogurt from milk concentrated by ultrafiltration. *Lebensmittel-Wissenschaft and Technologie*, **24**, 491–494.

Nsabimana, C., Jiang, B. & Kossah, R. (2005) Manufacturing, properties and shelf-life of labneh: a review. *International Journal of Dairy Technology*, **58**, 129–137.

Ordonez, G.A., Jeon, I.J. & Roberts, H.A. (2000) Manufacture of frozen yogurt with ultrafiltered milk and probiotic lactic acid bacteria. *Journal of Food Processing and Preservation*, **24**, 161–176.

Özer, B.H. (1997) *Rheological Properties of Labneh (Concentrated Yoghurt)*, PhD. Thesis, The University of Reading, Reading, UK.

Özer, B.H. (2006a) Production of concentrated products. *Fermented Milks* (ed. A.Y. Tamime), pp. 128–155, Blackwell Publishing, Oxford.

Özer, B.H. (2006b) [*Yoğurt Bilimi ve Teknolojisi*.] *Sidas Yayıncılık*, Izmir, Turkey.

Özer, B.H. & Kırmacı, H.A. (2010) Functional milks and dairy beverages. *International Journal of Dairy Technology*, **63**, 1–15.

Özer, D. & Özer, B.H. (1999a) Fermented products of Eastern Europe and Asia: koumiss, kefir and tarhana. *Encyclopedia of Food Microbiology*. (eds. R.K. Robinson, Batt, C. & Patel, P.), pp. 798–805, Academic Press, London.

Özer, B.H & Robinson, R.K. (1999b) The behaviour of starter cultures in concentrated yoghurt (labneh) produced by different techniques. *Lebensmittel- Wissenschaft und-Technologie*, **32**, 391–395.

Özer, B.H., Robinson, R.K, Grandison, A.S. & Bell, A.E. (1997a) Rheological characteristics of labneh (concentrated yoghurt) produced by various concentration techniques. *Textural Properties of Fermented Milks and Dairy Desserts*, Special Issue **9802**, pp.181–185, International Dairy Federation, Brussels.

Özer, B.H., Robinson, R.K, Grandison, A.S. & Bell, A.E. (1997b) Comparison of techniques for measuring the rheological properties of labneh (concentrated yoghurt). *International Journal of Dairy Technology*, **50**, 129–134.

Özer, B.H., Grandison, A.S., Robinson, R.K. & Lewis, M.J. (1997c) Changes in the physical properties of labneh (concentrated yoghurt) during storage. *Proceedings of IFT Annual Meeting*, pp. 117–118.

Özer, B.H., Robinson, R.K., Grandison, A.S. & Bell, A.E. (1998a) Gelation properties of milk concentrated by different techniques. *International Dairy Journal*, **8**, 793–799.

Özer, B.H., Bell, A.E., Grandison, A.S. & Robinson, R.K. (1998b) Rheological properties of concentrated yoghurt (labneh). *Journal of Texture Studies*. **29**, 67–79.

Özer, B.H., Stenning, R., Grandison, A.S. & Robinson, R.K. (1999a) Effect of protein concentration and distribution on the rheology of concentrated yoghurt. *International Journal of Dairy Technology*, **52**, 135–138.

Özer, B.H., Stenning, R., Grandison, A.S. & Robinson, R.K. (1999b) Rheology and microstructure of labneh (concentrated yoghurt). *Journal of Dairy Science*, **82**, 682–689.

Özer, B.H., Robinson, R.K. & Grandison, A.S. (2002) Effect of elevation of total solids by ultrafiltration and reverse osmosis on thiol groups in milk. *Milchwissenschaft*, **57**, 609–611.

Rao, H.G.R. (2002) Mechanisms of flux decline during ultrafiltration of dairy products and influence of pH on flux rates of whey and buttermilk. *Desalination*, **144**, 319–324.

Rao, H.G.R, Lewis, M.J. & Grandison, A.S. (1995) Effect of pH on flux during ultrafiltration of sweet whey and buttermilk. *Journal of Dairy Research*, **62**, 441–449.

Rasic, J.L., Kosikowski, F.V. & Bozic, Z. (1992) Nutient yogurt from low lactose milk using lactase-UF retentate procedure. *Milchwissenschaft*, **47**, 32–35.

Renner, E. & Eiselt-Lomb, U. (1985) Investigation on the protein fortification of yogurt – 2. effect of the product composition. *Milchwissenschaft*, **40**, 462–463.

Renner, E. & Abd El-Salam, M.H. (1991) *Application of Ultrafiltration in the Dairy Industry*. Elsevier Applied Science, London.

Robinson, R.K., Tamime, A.Y. & Wszolek, M. (2002) Microbiology of fermented milks. *Dairy Microbiology* (ed. R.K. Robinson), 3rd edn, pp. 367–430, John Wiley & Sons Inc., New York.

Robinson, R.K. & Tamime, A.Y. (1990) Microbiology of fermented milks. *Dairy Microbiology – The Microbiology of Milk Products*, (eds. R.K. Robinson), 2nd edn, pp. 291–343, Elsevier Applied Science, London.

Rodriguez-Serrano, G. M., Perez-Hernandez, G., Gallardo, F., Gomez-Ruiz, L. & Garcia-Garibay, M. (2002) Performance of yogurt cultures during the fermentation of whey concentrated by ultrafiltration for the elaboration of a fermented beverage. *Milchwissenschaft*, **57**, 540–543.

Roefs, S.P.F.M. & van Vliet, T. (1990) Structure of acid casein gels. 2. Dynamic measurements and type of interaction forces. *Colloids and Surfaces*, **50**, 161–175.

Rothwell, J. (1982) Dairy products. *Dairy Ingredients in Food Products*, Bulletin No. **147**, pp. 77–81, International Dairy Federation, Brussels.

Sachdeva, S. & Buccheim, W. (1997) Recovery of phospholipids from buttermilk using membrane processing. *Kieler Milchwirtsschaftliche Forschungsberichte*, **49**, 47–68.

Salji, J.P. (1991) Concentrated yoghurt: a challenge to our food industry. *Food Science and Technology Today*, **5**, 18–19.

Savello, P.A. & Dargan, R. A. (1995) Improved yogurt physical properties using ultrafiltration and very high temperature heating. *Milchwissenschaft*, **50**, 86–90.

Savello, P.A. & Dargan, R.A. (1997) Reduced yogurt syneresis using ultrafiltration and very-high temperature heating. *Milchwissenschaft*, **52**, 573–577.

Schulze, K., Raeuber, H.J. & Guthy, K. (1991) Viscoelastic properties of milk during yogurt fermentation. *Rheology*, **1**, 160–169.

Sharma, D.K. (1998) Ultrafiltration for manufacture of indigenous milk products: Channa and shrikhand. *Indian Dairyman*, **50**, 33–37.

Sinha, R.P. (1991) Effect of carbohydrate on the viability of *Streptococcus thermophilus*. *Journal of Food Protection*, **54**, 537–541.

Tamang, J.P. (2010) *Himalayan Fermented Foods*, CRC Press (Taylor & Francis Group), Boca Raton, FL.

Tamime, A.Y., (2002) Microbiology of starter cultures. *Dairy Microbiology* (eds. R.K. Robinson), 3rd edn, pp. 261–366, John Wiley & Sons Inc., New York.

Tamime, A.Y. (ed.) (2006) *Fermented Milks*, Blackwell Publishing, Oxford.

Tamime, A.Y. & Marshall, V.M.E. (1997) Microbiology and technology of fermented milks. *Microbiology and Biochemistry of Cheese and Fermented Milk* (ed. B.A. Law), 2nd edn, pp. 57–152, Blackie Academic & Professional, Glasgow.

Tamime, A.Y. & Robinson, R. K. (2007) *Tamime and Robinson's Yoghurt: Science and Technology*, 3rd edn, Woodhead Publishing, Cambridge.

Tamime, A.Y., Kalab, M. & Davies, G. (1984) Microstructure of set style yogurt manufacture from cow's milk fortified by various methods. *Food Microstructure*, **3**, 83–92.

Tamime, A.Y., Davies, G., Chehade, A.S. & Mahdi, H.A. (1989a) The production of 'labneh' by ultrafiltration: a new technology. *Journal of the Society of Dairy Technology*, **42**, 35–39.

Tamime, A.Y., Kalab, M. & Davies, G. (1989b) Rheology and microstructure of strained yoghurt (labneh) made from cow's milk by three different methods. *Food Microstructure*, **8**, 125–135.

Tamime, A.Y., Kalab, M. & Davies, G. (1991a) The effect of processing temperatures on the microstructure and firmness of labneh made from cow's milk by the traditional method or by ultrafiltration. *Food Structure*, **10**, 345–352.

Tamime, A.Y., Davies, G., Chehade, A.S. & Mahdi, H.A. (1991b) The effect of processing temperatures on the quality of labneh made by ultrafiltration. *Journal of the Society of Dairy Technology*, **44**, 99–103.

Tamime, A.Y., Kalab, M., Davies, G. & Mahdi, H.A. (1991c) Microstructure and firmness of labneh (high solids yogurt) made from cows, goats and sheeps milks by traditional method or by ultrafiltration. *Food Structure*, **10**, 37–44.

Tamime, A.Y., Robinson, R.K. & Latrille, E. (2001) Yoghurt and other fermented milks. *Mechanisation and Automation in Dairy Technology* (eds. A.Y. Tamime & B.A. Law), pp. 152–203, Sheffield Academic Press, Sheffield.

Tamime, A.Y., Skriver, A. & Nilsson, L.-E. (2006) Starter culture. *Fermented Milks*, (ed. A.Y. Tamime), pp. 11–52, Blackwell Publishing, Oxford.

Tamime, A.Y., Wszolek, M., Božanić, R. & Özer, B. (2011) Popular ovine and caprine fermented milks. *Small Ruminant Research*, **101**, 2–16.

Trachoo, N. & Mistry, V.V. (1988) Application of ultrafiltrated sweet buttermilk and sweet buttermilk powder in the manufacture of non-fat and low fat yogurts. *Journal of Dairy Sciences*, **81**, 3163–3171.

Tratnik, L. & Krsev, L. (1988) Production of fermented beverages from milk with demineralized whey. *Milchwissenschaft*, **43**, 695–698.

Veinoglou, B., Anifantakis, E. & Stiakakis, I. (1978) Production of strained yoghurt from ultrafiltered cow's milk. *XX International Dairy Congress*, **IE**, 831.

Voutsinas, L.P., Katsiari, M.C., Pappas, C.P. & Mallatou, H. (1996). Production of yogurt from sheep's milk which had been concentrated by reverse osmosis and stored frozen. 2. Compositional, microbiological, sensory and physical characteristics of yogurt. *Food Research International*, **29**, 411–416.

White, C. (2006) Cultured buttermilk. *Manufacturing Yogurt and Fermented Milks* (ed. R. Chandan), pp. 279–284, Blackwell Publishing, Iowa.

Wolpert, V.M. (1988) Skyr and skyr whey. *Dairy Industries International*, **90**, 29.

Wszolek, M., Kupiec-Teahan, B., Skov Gulard, H. & Tamime, A.Y. (2006) Production of kefir, koumiss and other related products. *Fermented Milks* (ed. A.Y. Tamime), pp. 174–216, Blackwell Publishing, Oxford.

Yamani, M.I. & Ibrahim, S.A. (1996) The differential enumeration of *Lactobacillus delbrueckii* subspecies *bulgaricus* and *Streptococcus salivarius* subspecies *thermophilus* in yoghurt and labneh using an improved medium. *Journal of the Society of Dairy Technology*, **49**, 103–108.

8 Cheese

V.V. Mistry

8.1 Background

Broadly speaking, cheesemaking involves the removal of water from milk in a defined manner using a combination of acid, temperature control and mechanical means. Depending on how these steps are employed, the net result includes products with a diverse range of sensory, nutritional and functional characteristics. For the cheese manufacturer, in addition to the quality of final product, economics of production is also of great importance. Thus, the yield of cheese, the cost of production per unit of cheese, etc., are monitored and procedures are sought that will maximise efficiency. Methods used to maximise efficiency include the application of different processes, such as co-precipitation for incorporating whey proteins in milk, addition of skimmed milk powder (SMP) or use of thermally concentrated milk within allowable guidelines and improvement in equipment design. Newer cheesemaking equipment offers the capability of significantly increased recovery of fat from milk relative to older equipment. The cheesemaking industry has been revolutionised by membrane processing in many interesting ways that have led to significant new processes, product development and, to some extent, new varieties. This phenomenon was inspired by unique technical developments in France in the late 1960s with the invention of the MMV process (Maubois *et al.*, 1969; Maubois & Mocquot, 1971). The MMV process, named after its inventors (Maubois, Mocquot and Vassal), led to significant advances in cheesemaking, improvements in plant efficiency, and increases in cheese yield. It also led to the development of continuous cheesemaking and new cheese varieties. As a result of evolution in the engineering of membrane processing and strengthened understanding of the properties of milk and cheese, commercial applications of membrane processing for cheesemaking have continued to evolve (Korolczuk *et al.*, 1986; Maubois, 2002).

Beneficiaries of these processes now include fresh, soft, semi-hard and hard cheeses as well as supplementary applications, such as brine recovery (Kosikowski & Mistry, 1997). Membrane applications for cheesemaking now encompass reverse osmosis (RO), nanofiltration (NF), ultrafiltration (UF) and microfiltration (MF). RO removes water from milk and generates characteristics similar to thermal evaporation, NF selectively removes small minerals, UF removes lactose and small minerals, and MF is used to separate bacteria from cheese milk (Saboya & Maubois, 2000) and casein enrichment of cheese milk through the separation of whey proteins (Maubois *et al.*, 2001).

Membrane Processing: Dairy and Beverage Applications, First Edition. A. Y. Tamime.

This range of concentration and separation techniques provides capabilities for selectively employing one or more of the techniques for cheesemaking. In this chapter, the application of membrane processing for manufacturing soft and hard cheeses will be discussed.

8.2 Properties of membrane processed concentrates

Concentration of milk with UF, RO or MF alters its composition in a manner that will have an impact on its properties. These alterations are important to understand as they have a direct influence on cheesemaking. Three specific characteristics of concern are buffering capacity, rennet coagulation and rheological characteristics.

8.2.1 *Buffering capacity*

When milk is ultrafiltered, there is an increase in buffer capacity because mineral salts (Ca, Mg, P) bound to casein micelles are concentrated in the same proportion as proteins. The net outcome is a modification in acidification kinetics by lactic acid bacteria (LAB), the final pH of cheese, rennet coagulation kinetics and rheological characteristics of the curd, activity of ripening enzymes, lysis of mesophilic LAB during ripening (Saboya *et al*., 2001), growth and rate of survival of spoilage flora (Rash & Kosikowski, 1982) and water-holding capacity of the cheese mass during ripening. This phenomenon also occurs when milk is microfiltered to enrich micellar casein.

Thus, as milk is concentrated a higher production of lactic acid by starter cultures is required to obtain the end pH in cheese, which may be ~5.2 in hard cheeses and 4.6 in soft and fresh cheeses (Mistry & Kosikowski, 1985). This can result in acid-tasting cheeses (Maubois, 1979). This acid will also release a large quantity of calcium salts into the aqueous phase of cheese curd during acidification and modify casein micelle aggregation. As a result, cheese texture can become crumbly or sandy (Brulé *et al*., 1975), and spreadability and stretching properties are poor (Green & Grandison, 1987).

Buffering also leads to higher numbers of starter cultures in curd and resulting cheese than in non-ultrafiltered curd and cheese (Mistry & Kosikowski, 1985), and can cause bitterness in some cheeses, such as Quarg (Mortensen, 1985). This increase in starter culture count(s) can also be used for producing a new bulk starter (Mistry & Kosikowski, 1986a). This starter is manufactured by fermenting whole or skimmed milk retentate containing 12 g 100 g^{-1} protein with a mesophilic starter culture at 22°C for 12–15 h. It has a built-in internal pH control mechanism due to the buffering capacity of the ultrafiltered retentate, which maintains the pH of the starter steady at 5–5.2. It, therefore, has a greater activity than traditional bulk starters made from high heat-treated milk and maintains its activity for 10–12 h at room temperature. The ultrafiltered fermentate also has a high protein content, which contributes to increased cheese yield (Mistry & Kosikowski, 1986b), making it suitable for either traditional or ultrafiltered cheesemaking.

Rash & Kosikowski (1982) showed that increased buffering capacity of pH 6.7 of ultrafiltered milk could also offer a favourable environment for the growth and survival of certain bacteria, such as enteropathogenic *Escherichia coli*. It is important, therefore, to adjust the mineral salts content of ultrafiltered retentates depending on cheese variety. This can be achieved by a reduction in milk pH before or during UF either by the activity of the starter culture or by the addition of an acidifying agent, such as glucono-δ-lactone (GDL) or organic acids in some countries. Acidification solubilises the colloidal calcium and magnesium phosphate salts and eliminates them. This can also be achieved by the addition of sodium chloride (NaC1) (0.5–0.9 g 100 g^{-1}) to ultrafiltered retentate during or after UF. The increase in ionic strength resulting from NaC1 addition reduces the ionisation of casein phosphoseryl groups and, consequently, leads to solubilisation of colloidal calcium in the permeate or in the aqueous phase of ultrafiltered retentate (up to 15–18 g 100 g^{-1} depending on the pH and amount of NaC1 added) (Brulé *et al*., 1975).

8.2.2 *Rheology of concentrated milks*

As the protein content of milk is increased by UF and/or the temperature is lowered, the milk will possess pseudoplastic behaviour (Culioli *et al*., 1974). This behaviour has an impact on equipment design of membrane systems. Thus, viscosity of ultrafiltered milk increases with protein content. The manufacture of semi-hard or related cheeses from these highly viscous ultrafiltered retentates requires removal of all dissolved gases which are entrapped in the liquid and the use of special mixing devices (static and dynamic) to enable thorough blending of rennet and starter cultures in the retentate (Maubois, 1987). If dissolved gases are not removed by the application of a vacuum, a spongy curd is obtained and the appearance and taste of the cheese are poor. If rennet is not mixed satisfactorily, the resulting curd will be flaky because of localised coagulation.

8.2.3 *Rennet coagulation*

Rennet coagulation in cheesemaking is a complex phenomenon that is dependent on several factors, including temperature, protein concentration, enzyme strength, pH, etc. If the same amount of rennet is added to equal volumes of milk or ultrafiltered retentate, the rennet clotting time is not affected by the increase in protein content, but the time from clotting to cutting is reduced (Dalgleish, 1980; Maubois & Mocquot, 1971; Upreti, *et al*., 2011). The impact of these phenomena has been discussed at length by Mistry & Maubois (2004).

8.3 Applications of ultrafiltration in cheesemaking

Cheesemaking using ultrafiltered milk can be conducted in three main ways: (a) protein standardisation of milk; (b) use of intermediate or medium concentrated retentates; and (c) use of liquid pre-cheeses.

8.3.1 *Protein standardisation*

This is probably the most widely used method of UF for cheesemaking because it can be adapted to most cheese varieties with the potential of economic benefits. Milk is ultrafiltered to a concentration of up to two-fold (on a protein basis) and used for various types of cheeses (Govindasamy-Lucey *et al*., 2011). Generally, a concentrate containing 3.7–4.5 g 100 g^{-1} protein is employed while using conventional cheesemaking equipment. This method enables uniformity in the composition of milk, hence the term 'protein standardisation' and is used for cheeses, such as Camembert, Cheddar, Mozzarella and many other cheese varieties. Advantages of using this procedure include

- uniformity in milk composition from day to day
- firm rennet curd and, therefore, lower losses of casein to whey
- increased cheese yield (~6 g 100 g^{-1} on protein basis)
- improved cheesemaking efficiency (more cheese per vat)
- importantly, no need for additional specialised cheesemaking equipment other than a UF equipment.

Increased cheese yield is attributed to higher fat and protein recovery and retention of some whey proteins.

For hard cheeses, such as Cheddar, a concentration of up to 1.6- to 1.7-fold is used. At higher levels, rennet curd is extremely firm and difficult to handle, and fat losses to whey may be high. The moisture content of Cheddar cheese made with this process tends to decrease with protein content in milk. Reasons suggested include rapid syneresis because of coarser networks of protein gels. Lowering of cooking temperature and use of homogenised cream can increase cheese moisture. Homogenisation can also be used to further increase the yield of Cheddar cheese made by ultrafiltered milk (Oommen *et al*., 2000). Fat recovery in Cheddar cheese made from milk ultrafiltered to 4.6 g 100 g^{-1} protein and without cream homogenisation was 94.7%, whereas with homogenised cream it was 96.8%.

For Cottage cheese, industrial trials have demonstrated that for optimum yield, flavour and body characteristics, ultrafiltered retentate of 1.2- to 1.7-fold concentration is desirable (Kealey & Kosikowski, 1986). Firm texture and flat flavour are produced above these levels. If thermisation of milk is conducted prior to UF, Cottage cheese yield is increased (Zall & Chen, 1986).

In the late 1970s, Maubois (1979) in France developed the concept of on-farm UF of milk for cheesemaking. This method involved the UF of milk to less than two-fold concentration on the farm prior to transportation to the cheese factory. The permeate was fed to cows at the farm. The objective was to lower the cost of transportation of milk and to increase cheese yield. The economics of the process should also take into account the disposal of permeate that is produced on the farm. Subsequent studies in the United States of America (USA) suggested that this process would be economical for farms with 100–1000 cows. This method of on-farm UF of milk is now being used in the USA, where milk is ultrafiltered cold to 3.5-fold at a collection centre and then transported long distances (>500 km) to cheese factories. At these cheese factories the retentate is used to raise the total solids content of the cheese milk to 13.5–15.0 g 100 g^{-1}.

8.3.2 *Medium or intermediate concentrated retentates*

Milk is concentrated to two- to five-fold prior to cheesemaking. Diafiltration (DF) may be adopted to adjust the mineral content and buffering capacity of the ultrafiltered retentate.

Much higher quantities of whey proteins are retained in the cheese, and the yield is also higher than with the protein standardisation method. There are significant changes in physico-chemical properties, requiring specially designed equipment. Rennet curd, for example, is very firm and difficult to handle with conventional equipment.

The most notable example of use of this method for commercial applications was the APV-SiroCurd process for Cheddar cheese (Tamime, 1993). This method was developed in Australia and involved continuous rennet coagulation of ultrafiltered milk of 40–45 g 100 g^{-1} total solids. A small portion of the ultrafiltered milk was pre-fermented with starter culture and used as bulk starter at the rate of 10–12 mL 100 mL^{-1}. The rennet curd was cut with specially designed wire knives and cubed curd pieces transferred into a rotating drum where syneresis took place during heating to 38°C for 30–40 min. This was followed by automated cheddaring at the optimum pH and milling and salting. Yield increases of 6–8% were claimed but, after several years of operation, this process was abandoned because of technical difficulties and poor economics.

Another cheese that enjoyed commercial success for a period of time was structured Feta-type cheese. This product was developed in Denmark in response to demand from people that desired cheeses having an appearance and texture (presence of mechanical holes) similar to those of traditional products, characteristics they did not find in UF Feta-type cheese made from liquid pre-cheeses (Mortensen, 1985). Pasteurised, fat-standardised milk, generally homogenised at 18 MPa and 60°C, is ultrafiltered at 50°C. The final concentrate contains 28.5 g 100 g^{-1} total solids, which corresponds to a concentration factor of 3:1. Lipase, starter culture or GDL is added to the ultrafiltered retentate, previously homogenised at 5 MPa at 65°C, heat treated to 80°C for 60 s and cooled to 34°C. After a short storage period in a tank, the retentate is pumped to specially- designed Alfa-Laval (currently known as Tetra Pak) Alcurd or APV-Pasilac systems/coagulators. Rennet is added in-line and, in both types of systems, it is mixed thoroughly and the retentate coagulates in the tubes of the equipment. The resulting coagulum is removed from the tubes, cut into cubes, moulded and drained (16–24 h at 10–14°C) until the pH has decreased to 4.8 (Skovhauge, 1987). The product of this process is virtually indistinguishable from the traditional product. A yield increase of about 14% on a solids basis is claimed over the traditional process (Mortensen, 1985), a far smaller value than that obtained with the process using liquid pre-cheese (30%). The use of this process too has diminished in favour of standard Feta cheese, which now is a protected name in Greece.

8.3.3 *Liquid pre-cheeses concept*

In this process, conventional cheesemaking equipment is not used because milk is ultra-filtered to a concentration equal to the composition of the cheese being manufactured. After setting with rennet, acid development takes place followed by additional treatments required for the specific cheese variety. This method has the highest yield potential because of minimal whey separation and consequent maximum whey protein retention in the cheese.

Whereas the protein standardisation technique can be adapted to most cheese varieties, the liquid pre-cheese concept is more limited in its applicability because it is not possible to achieve the composition of all cheeses by ultrafiltration. The process developed for this method was originally for Camembert cheese through the MMV process mentioned earlier. It has also been applied to Feta-type cheese and Pavé d'Affinois of France.

For Camembert cheesemaking using this method, pasteurised milk is ultrafiltered to five-fold concentration and, at 20°C, a mesophilic starter culture (2 g or mL 100 mL^{-1}) and salt (0.75 g 100 mL^{-1}) are added. After a pH of 5.5 is reached, rennet is added and the mixture transferred into moulds where curd develops. At the proper firmness, curd wheels are removed, brined for 30 min, sprayed with *Penicillium camemberti* spores and ripened at 11–12°C for 12 days under high humidity for mould growth on the surface of the cheese wheels. This process results in a yield increase of 12–15%.

A high level of success with the liquid pre-cheese concept has also been achieved with Feta-type cheese. In this Danish procedure (UF Cast Feta-type cheese), five-fold ultrafiltered whole milk (i.e. homogenised, blended with starter culture, salt and lipase–rennet mixture) is coagulated in 20 kg tins where curd forms. The metal can is filled to one-third its capacity when the dosing machine delivers the right amount of retentate. The partially filled container travel along a conveyor belt (~10–20 min) and, by the time the metal container returns to the dosing/filling machine, the retentate has already coagulated. This operation is repeated three times, so that the coagulum is formed in three layers. Because the filling/coagulation of milk is done intermittently, a horizontal layer/division of the Feta-type cheese is formed without cutting. After the third layer of ultrafiltered retentate has coagulated in the container, the cans pass through a sequence of cutting stages, thus resulting in 27 portions of retail cheese. The curd is then covered with a parchment paper before the dry salt is added; the container is then sealed and turned upside down to allow the whey seeping from the coagulated curd to dissolve the salt (i.e. forming the brine) and held for ripening (Tamime & Kirkegaard, 1991). Afterwards, the cheese portions are removed from the 20 kg cans, packaged and retailed.

According to Tamime *et al*. (2006)

> "a similar process is known as Tetra Tebel Casofill® Bacteriologically Acidified Feta-type (BAF) (Kiers, 2001), which is suitable for the manufacture of brined cheeses (i.e. Feta-type), Queso Fresco and cream cheese (Bredahl *et al*., 1995). As shown in Fig. 8.1, the process line consists of the following.
>
> *Pre-treatment*
>
> The raw milk is pre-warmed to ~60°C, separated, and the skimmed milk is microfilterd or bactofugated (optional treatment). The retentate from the MF and/or the bactofugate containing the bacteria could be discharged to waste depending on the process configuration that has been installed, and the permeate (i.e. the bulk of the skimmed milk) is standardised automatically to 1:1, fat-to-protein ratio. The principle of the in-line direct standardisation system (i.e. two streams – one for the skimmed milk and the other for the cream containing ~40 g fat 100 g^{-1}) consists of control valves, flow density meters and a computerised control loop to adjust the fat content in the skimmed milk to a desired value. Descriptions of the in-line fat standardisation systems available on the market have been reported by Tamime & Robinson (1999 (2007)), Muir & Tamime (2001), Tamime *et al*. (2001) and Anonymous (2003). Alternatively, recombined skimmed milk could be used

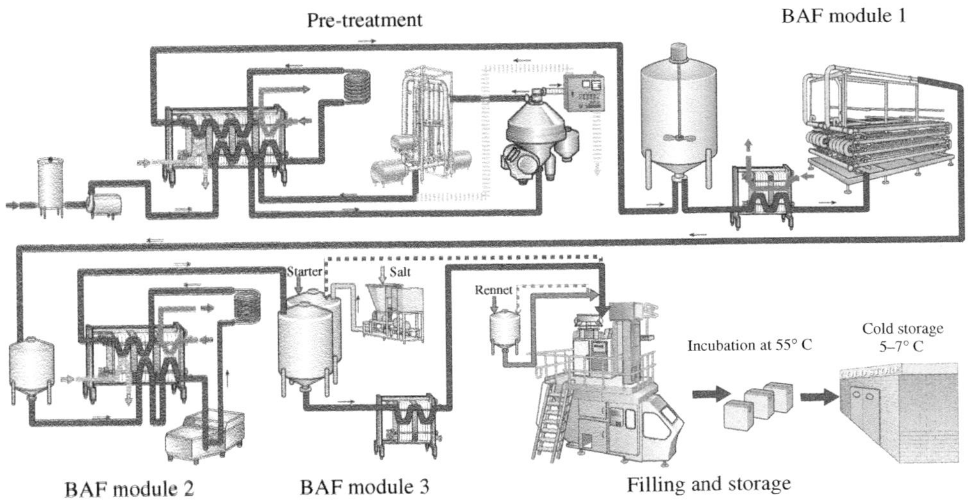

Fig. 8.1 Flow diagram of the cheesemaking process using ultrafiltered milk. BAF, bacteriologically acidified feta-type (After Tamime *et al*., 2006, reproduced with permission of Tetra Pak B.V. Leeuwarden, The Netherlands and Wiley-Blackwell).

instead of raw milk, but the addition of melted anhydrous milk fat (AMF) could either be added during the rehydration stage of the process or metered into the retentate before the homogenisation stage (Anonymous, 2003).

Afterwards, the standardised milk is pasteurised at 75°C for 15 s to ensure the inactivation of pathogenic micro-organisms, cooled to 5–7°C and stored for further processing.

Module 1

The cold stored processed milk is pre-warmed to 50°C and pumped to the UF module to concentrate the milk total solids (TS) to ~38.5 g 100 g^{-1} (i.e. crude protein 16.40, fat 17.12, lactose 3.22, ash 1.67 – this is equivalent to 5.5-fold concentration of the milk solids); the permeate may be discarded (optional). The UF plant (e.g. Tetra Alcross® US/UC) could be supplied with spiral-wound or ceramic membranes (Fig. 8.1). The capacity of such UF modules ranges between 5 000 and 10 000 L h^{-1}. In addition, the permeate is pumped into the regeneration section of the plate heat exchanger (PHE) to heat the incoming cold milk from 5–7°C to the UF temperature. If the permeate is not discarded, it is cooled to <7°C for further processing, for example, the manufacture of drinking-type fermented milks or concentrated using the RO technique and the retentate is added to the UF retentate in order to increase the cheese yield.

Module 2

In this section of the BAF process (Fig. 8.1), the concentrated milk is heated to 75–80°C for 30 s, homogenised using a two-stage homogeniser at 4.5 and 0.5 MPa, respectively, and cooled to the incubation temperature of the starter culture. The primary objectives of the heat treatment and the homogenisation of the UF cheese milk are: (a) to provide a stable fat-in-water emulsion of the milk fat; (b) to induce physico-chemical changes in the milk; and (c) to inactivate any micro-organisms present in the milk.

In addition, high heat treatment (e.g. 90°C) and high homogenisation pressure (i.e. >18 MPa) of the retentate as previously reported by Tamime & Kirkegaard (1991) are not applied in the BAF process. Nevertheless, the retentate may be heated at ~150°C for a few seconds for ultra-high temperature (UHT) sterilisation during the manufacture of Queso Fresco cheeses (i.e. not salted) that are very popular in the Iberian peninsula and South America.

Module 3

In this part of the production line (see Fig. 8.1), the processed milk is cooled 30°C, and inoculated with a mesophilic starter culture (i.e. direct-to-vat inoculation (DVI) or direct-to-vat set (DVS) type). Typical examples are cultures R-703 or R-704, which can be obtained from Chr. Hansen A/S. These are blends of *Lactococcus lactis* subsp. *lactis* and *Lactococcus lactis* subsp. *cremoris* that do not produce CO_2, categorised as fast acid-producers, and inoculated at a rate of 1 000 units 10 000 L^{-1} of retentate. The inoculated milk is fermented between 31 and 37.8°C in an intermediate tank for 12–15 h or pH <5.0. Incidentally, dry salt (2–3 g 100 mL^{-1}), which is metered into the retentate, is added at two stages; *firstly*, 0.5 g salt 100 mL^{-1} is added with the starter culture before the incubation period, and *secondly*, the remainder of the salt (i.e. 1.5–2.5 g 100 mL^{-1}) is added to the fermentate at pH <5.0, and the time required to mix the bulk of the salt in the fermentate is ~90 min. Afterwards, the fermentate is heated to 45–50°C, mixed with the coagulant (e.g. Chy-Max® (rennet powder) from Chr. Hansen A/S – 2080 international milk clotting units (IMCU) g^{-1}, which is rehydrated at a rate of 1.5–2.5 g L^{-1} of good potable water free from chlorine – this corresponds to 3120–5200 IMCU L^{-1}), and metered in-line to the fermented retentate at a rate of 1 mL 100 mL^{-1} at the point of filling into the retail container (see Fig. 8.1). Alternatively, Standard Rennet (i.e. before dilution with water) could be used at a rate of 20–30 mL 100 kg^{-1} of retentate (E. Høeier, Personal communication). The coagulation time is within 15–20 min, and the containers are stored at ambient temperature (ca. 25°C) overnight before they are transferred to the cold store (5–7°C). The cheese can be consumed direct after cooling or can be stored for up to twelve months at refrigerated temperature.

Other coagulants that can be used during the manufacture of Feta-type cheese have been reported by many researchers. Saldamli & Kaytanli (1998) concluded that Maxiren® 50 (from *Kluyveromycece lactis* by recombinant technology) was more suited to Turkish white cheese than parallel products made with Fromase® 46T (from *Mucor miehei*) or Rennilase® 150 L (from *Rhizomucor miehei*), and the quality of the cheese was similar to those made with standard rennet. Whilst Kandarakis *et al*. (1999) concluded that the quality and yield of Feta cheese made with Chymax® (from *Escherichia coli*) were similar to the product made with Ha-La® (from calf rennet) (see also Anifantakis & Kandarakis, 1983; Alichanidis *et al*., 1984).

A Tetra Brik aseptic machine (models TBA/8 and TBA/19) is used to fill the fermentate including the added salt and the coagulant, and the packaging throughput ranges between 6 000 and 7 000 pack h^{-1} depending on the capacity of the container (e.g. 250 or 500 mL). However, technical information regarding the formation of the packaging container, sterilisation of the packaging material and filling procedure has been reported by Tamime & Robinson (1999 (2007)) and Muir & Tamime (2001). The viscosity of the UF fermentate is greater that milk and, as a consequence, a large diameter nozzle on the filling head should be used to facilitate easy filling of the cartons (see Fig. 8.1)."

A variation of this concept is applicable to the manufacture of processed cheese. This type of cheese is generally manufactured from base natural, hard cheese and may involve the inclusion of different ingredients, such as emulsifiers, dried milk products (e.g. whey powder, SMP), butter, etc. (Kosikowski & Mistry, 1997). Therefore, manufacturers often formulate such cheese taking into consideration the cost of ingredients and target least cost for optimum quality. Such cheeses offer flexibility in formulation by selecting a specific ingredient and combining it with appropriate other ingredients that will result in a product of a specific functionality. For example, cheese protein can be replaced by an ultrafiltered milk product containing a high level of protein. A cheese base made by UF can satisfactorily replace the young cheese component in the manufacture of conventional processed cheese. An Australian process (Ernstrom *et al*., 1980) was commercialised in the USA. In this process, whole milk or whole milk acidified to pH 5.7 is ultrafiltered to 40% of its original weight and then diafiltered to 20% of its original weight. The product was fermented with starter cultures for 16 h at 30°C, and then vacuum evaporated to 64 g 100 g^{-1} total solids. This product is then included in a processed cheese formulation.

8.3.4 *Application of ultrafiltration for fresh and soft cheeses*

The application of membrane processing, particularly UF, for fresh cheeses is challenging because of the low pH requirements of such cheese and the high buffering capacity discussed above (Mistry & Maubois, 2004). Such cheeses tend to be very acidic because of excessive acid production due to the high buffer capacity. The manufacture of fresh acid-type cheeses, such as Cream, Quark and Ricotta, was particularly challenging until mineral and ceramic membranes became available. These membranes made it possible to ultrafilter acid curd to high levels with few fouling problems. The general principle is to ferment high heat-treated milk to pH 4.6–4.8, and then ultrafilter the curd to the desired concentration (Mahaut *et al*., 1982). Traditionally for Quark, a centrifugal separator is used to separate the curd and whey. The advantage of the UF procedure is that whey proteins are retained and, therefore, cheese yield is increased. For cream cheese, a Cornell University procedure involves the blending of heavy cream with 27.5 g 100 g^{-1} solids skimmed milk retentate to achieve the composition of cream cheese. This mixture is pasteurised and homogenised, then fermented and pasteurised again along with stabilisers. In a patented process for cream cheese using UF is used for a portion of the mix and the process generates very little whey (Weissman, 1982). Other examples of soft cheeses that use UF include the French Pavé d'Affinois that was developed in 1982 and subsequently increased in production. For making this cheese, milk is ultrafiltered fourfold and fermented with a thermophilic starter culture. Rennet is added and the mixture is poured into rectangular plastic trays that are 5 cm high. The trays are set in an incubator at 43°C for 6 h to allow acidification and coagulation to occur. After cooling to room temperature, the cheese slabs are removed from the trays and cut with an automatic dividing knife in 96 pieces, each having the size and form of a small rectangular paving stone (~7 × 5 × 5 cm). The resulting fresh cheeses are ripened for 10 days, as for Camembert (Korolczuk *et al*., 1986).

8.4 Cheese quality

While whey proteins retention is advantageous for cheese yield, whey proteins generally are inert filler materials and undergo very little proteolysis during aging. Flavour development is, therefore, slow. Retarded proteolysis along with the high water-binding capacity of whey proteins also influences the texture of the cheese. Furthermore, the high buffering capacity of such cheese retards autolysis of starter cultures cells and breakdown of casein. These effects become more pronounced as whey protein concentration in cheese is increased (protein standardisation versus liquid pre-cheese concept). Texture defects of cheeses made from ultrafiltered milk, such as sandiness, firmness or crumbliness, can be attributed to higher Ca salts content of the retentates (pH 6.7) that are used for cheesemaking. It is important to recognise calcium retention and develop ways to adjust it in order to maintain optimum texture (Kindstedt & Guo 1998). The mineral content may be adjusted by pre-acidification or diafiltration as described earlier.

The quantity of whey proteins retained depends on the variety and on the degree of UF concentration. If all the whey proteins of milk are retained, they will represent approximately 20 g 100 g^{-1} of the total protein in the cheese. Lower quantities will be retained when a low concentration of UF is used. The water-binding capacity of whey proteins is much higher than that of casein, and ultrafiltered cheeses are less susceptible to drying during retailing than traditional cheeses.

Because of whey protein retention in ultrafiltered cheeses, they generally ripen at a slower rate. The larger the amount of whey proteins incorporated, the slower the flavour development. The effect of whey proteins on flavour development is less pronounced in cheeses manufactured with the protein-standardisation process because of the smaller quantities of whey proteins present, but is more significant in cheeses made from higher concentrations and those that are ripened for long periods.

The retarded maturation could be due to several reasons. The high content of the whey protein (mainly β-lactoglobulin) in ultrafiltered cheeses could inhibit to some extent the general proteolytic activity of rennet (Creamer *et al*., 1987) and plasmin (Visser, 1981). Undenatured whey proteins found in ultrafiltered cheeses are resistant to proteolysis by these proteases and by starter-derived enzymes. The high buffering capacity of ultrafiltered cheeses prepared from retentate (pH 6.7) is the most probable cause. The high buffer capacity retards or even completely inhibits the rate of mesophilic starter cultures autolysis (Saboya *et al*., 2001) and, consequently, hydrolysis of the casein network. The rate of α_{s1}-casein breakdown, as well as of β-casein, has been found to be reduced in ultrafiltered cheese (Creamer *et al*., 1987). The rate of flavour development of ultrafiltered cheese may be improved by the addition of flavour-producing enzymes (Spangler *et al*., 1990) or cell extracts (Saboya *et al*., 2001), which are totally retained in the retentate contrary to what happens in traditional cheesemaking where 80–90% of the added enzyme is lost in the whey. Non-starter bacteria or slow-acidifying lactic micro-organisms may be used for their proteolytic and flavour production potential because cheesemaking by UF permits separate management of acidification, drainage and ripening flora.

The impact of high mineral retention on cheese functionality and flavour is also of concern. Excessive calcium retention makes it difficult to obtain optimal functionality in cheeses, such as Mozzarella, and may lead to bitterness in fresh acid curd cheeses. The latter is also because of increased buffering, which leads to high levels of starter cells. Employing pre-acidification during ultrafiltration can control mineral content of cheese.

8.5 Applications of microfiltration in cheesemaking

MF of milk for cheesemaking is a relatively new concept, but is rapidly gaining commercial acceptance because of the potential to employ a wide range of membrane pore sizes (0.05–10 μm). This flexibility makes it possible to achieve the desired specific separation as well as fractionation of milk constituents. Ceramic as well as polysulphone MF membranes are used. Applications of MF for cheesemaking include processes for removal of bacteria and casein standardisation of cheese milk; both these approaches are commercialised.

8.5.1 *Removal of bacteria*

Reducing the microbial load and eliminating any existing pathogens of milk prior to cheesemaking by processes, such as pasteurisation, are a common practice and in some cases even mandatory. On the other hand, heat treatment of milk, such as pasteurisation, is believed to alter the cheesemaking characteristics of milk and flavour characteristics of cheese. High-speed centrifugation of milk (bactofugation) was developed to remove bacteria and particularly spores of *Clostridium tyrobutyricum* from milk (Kosikowski & Mistry, 1997), but the process is not as efficient as newly developed MF procedures.

The MF process of Tetra Pak for removing bacteria and spores is called the Tetra Alcross Bactocatch® process. Raw skimmed milk is microfiltered using a membrane of pore size 1.4 μm at 35–50°C. The retentate (i.e. smaller portion which is equivalent to 10% of the original volume of milk) contains bacteria and the permeate is the bacteria-free milk. This bacteria-free milk can be blended with heated cream for standardisation of the fat content. Bacteria removal efficiencies of 99.6–99.98% are reported, i.e. relatively sterile milk is obtained. The retentate and cream are mixed and heated to ~120°C for a few seconds, cooled and blended with permeate. Hence, bacteria are removed and milk components are recovered.

This process is particularly suitable for fluid milk production, but is attractive for manufacturing cheeses, such as Swiss, because of the possibility of removal of spores of *C. tyrobutyricum* without using nitrates or excessively high heat. On the other hand, French researchers have demonstrated that under normal circumstances microfiltered milk is not ideal for eye formation in Swiss cheese because of the removal of non-starter lactic acid bacteria by MF. Satisfactory distribution of eyes in Emmental cheese made from microfiltered milk requires incorporation of specific non-starter lactic acid bacteria, such as heterolactic strains, along with mesophilic lactic, thermophilic lactic and propionic starters added to this milk at the beginning of cheesemaking (Maubois, 2002). Similarly, the development of typical flavour in Camembert cheese requires the addition of *Hafnia alvei*, a bacterial species able to produce volatile sulphur compounds from methionine in the cheese (Cousin, 1994).

8.5.2 *Casein standardisation*

With MF membrane of 0.1 μm pore size separation of casein and whey proteins can be accomplished (Hurt & Barbano, 2010). With this process, the casein content of milk is increased from 2.5 to 3.5 g 100 g^{-1} or more and, hence, cheese yield is increased. Furthermore, the microfiltrate (permeate) that is generated from this process contains whey proteins, but no glycomacropeptide, which is normally found in whey from conventional cheesemaking procedures. With this method, it is, therefore, possible to standardise the casein content of cheese milk while producing an 'ideal whey' that has better functional characteristics than whey with glycomacropeptides.

In recent work, French workers (Garem *et al*., 2000) have used this approach in combination with UF to improve the cheesemaking properties and dried milk. In this patented process (Quiblier *et al*., 1992), whey proteins are completely or partially removed from milk by MF as mentioned above. The permeate of MF, which contains the whey proteins, is ultrafiltered and the permeate of UF, which contains lactose, minerals and water, is blended with the retentate of MF. The retentate of MF contains casein, hence with the last blending step milk with low or no whey protein is obtained. This milk is evaporated and spray dried. The spray-dried product can be used for cheesemaking after reconstitution without any typical difficulties encountered with conventional powder, which contains denatured whey proteins because of the high heat treatments employed during manufacture.

8.5.3 *α_s-/β-casein ratio adjustment by microfiltration*

By using a combination of MF and physico-chemical attributes of the casein micelle, it is possible to adjust the ratios of the various casein fractions in milk for cheesemaking (Mistry & Maubois, 2004). It is known that hydrophobic binding entraps most of the β-casein in the micelles. When milk is cooled to a temperature lower than 5°C, β-casein is solubilised. It can be separated by 0.2 μm pore size MF membranes (Terré *et al*., 1987) or by 100 000 daltons cut-off UF membranes (Murphy & Fox, 1991a). In this process β-casein is in the permeate and the retentates have a casein content with an increased α_s-/β-casein ratio (Terré *et al*., 1987). Such milk can be favourable for the manufacture of certain cheeses (Terré *et al*., 1987; Murphy & Fox, 1991b). For example, it is known that degradation of β-casein during the cheese-ripening process often results in bitter peptides. With β-casein-reduced milk, it may be possible to reduce these effects. Similarly, cheeses made from the milk of goats and sheep have unique characteristics, partly because of differences in their casein fractions. Thus, adjustment of the ratios of casein fractions could lead to the development of new cheese as well.

8.5.4 *Recovery of fat and brine*

MF also offers additional opportunities in the cheese industry not directly related to cheesemaking, but to related processes. For example, MF can be used to recover 99% of the fat from the cooker and stretcher water during Mozzarella cheesemaking (Kosikowski & Mistry, 1997). A two-stage 0.05 μm pore size ceramic membrane with total membrane surface of 15 m^2 is used, treating ~1100 kg of cooker water h^{-1}. Fat is retained as retentate, and salts and lactose are removed in the permeate. This permeate can be mixed into the whey stream for further treatment. The final product contains 40 g 100 g^{-1} fat and 55 g 100 g^{-1} total solids. Fat recovery by this method is far greater than that obtained by centrifugal separation of the cooker water, which is ~60%. In another application, membrane systems are employed for cleaning and recovery of brining systems in cheeses, such as Mozarella (Kosikowski & Mistry, 1997). Typically, brine salting of cheese usually takes place at 5–10°C with pH adjusted to the pH of the cheese being brined. For example, for Mozzarella cheese, the pH of brine is adjusted to ~5.3 with a food grade acid, such as acetic acid. It is also desirable to add calcium in the form of calcium chloride to the brine, and the amount added depends on the level of calcium in the cheese. For Mozzarella and Swiss cheeses ~0.1 g 100 g^{-1} calcium in the brine is sufficient. Calcium addition and pH adjustment help attain optimum body characteristics of the products. If calcium is not added to the brine, leaching of calcium from the cheese would occur and lead to softening of the body and texture of the cheese. This defect would make the cheese difficult to shred for application in pizza making.

When brine is repeatedly used in a cheesemaking operation, there will be build-up of cheese solids, such as proteins, fat, minerals and even fines in the brine. It is important that these be removed at regular intervals or else yeast and bacterial growth can occur in the brine and eventually in the cheese. Such brine can be purified with membranes employing MF or UF. In large automated brining systems, the brine canals are directly connected to fines savers to remove solid cheese particles. This is followed by purification in a membrane system where milk solids, such as fat and proteins, and also bacteria are removed. The purified brine (permeate) is then chilled and pumped to a brine buffer tank where the pH, and salt and calcium content are adjusted. From here, the rejuvenated brine is returned to the canals. This procedure allows for an economical maintenance of the brining system, and makes it unnecessary to frequently discard corrosive brine solutions.

8.6 Nanofiltration

Dairy applications of NF are very recent, and major use currently lies in the area of whey processing. Such use includes demineralisation and concentration of whey and reduction of salt from salty whey. High permeability of monovalent ions (40–90%) and low permeability of multivalent ions (5–20%) typically characterise NF. Consequently, it is possible to concentrate milk with NF to obtain an altered mineral balance. Experimental work suggests that this would offer potential for soft cheese manufacture.

8.7 Milk protein concentrates

To produce milk protein concentrates (MPCs), UF, MF or a combination of various concentration technologies are utilised (Garem, *et al.*, 2000). These products present interesting new technical possibilities and have, therefore, become important products for cheesemaking (Mistry, 2002). There are no specific standards of identity in any country for MPC, and cover a wide range of compositional parameters and functional characteristics. In the dried product, the milk protein content may range from 35 to over 85 g 100 g^{-1}. Unlike WPCs, MPCs contain both major milk protein groups in proportions similar to milk.

Manufacturing technologies for MPC include UF, diafiltration and spray drying, if the end product is to be in the dry form (Mistry & Hassan, 1991). It is essential to have raw milk of good quality (low total and spore counts). Skimmed milk is ultrafiltered and diafiltered to ~21 g 100 g^{-1} total protein. With diafiltration, the lactose content is reduced such that the final product has less than 0.5 g 100 g^{-1} lactose. The diafiltered product is then spray dried to less than 5 g 100 g^{-1} moisture. This dried or liquid product can be used to supplement milk for cheesemaking using techniques described earlier. MPCs have introduced new possibilities in cheesemaking (Bhaskar *et al.*, 2001; Bhaskar *et al.*, 2001; Moran *et al.*, 2001a, 2001b). Moran *et al.* (2001a) have developed a continuous method for manufacturing processed cheese in which acidified milk is ultrafiltered and diafiltered to a concentration factor of 4–7, followed by evaporation up to 70 g 100 g^{-1} solids to give reduced-calcium pre-cheese. Processed cheese is made from this pre-cheese in the traditional manner using flavouring agents and emulsifiers. No cheese base is required, but, instead, high protein concentrates are used as the base material along with flavouring agents.

Concentrates with high micellar casein content have been developed using MF (Schuck *et al.*, 1994) followed by drying. The casein content of such powder is ~90 g 100 g^{-1} and forms almost 96 g 100 g^{-1} of total protein (Saboya & Maubois, 2000). When used in cheesemaking, increases in cheese yield have been reported (Caron *et al.*, 1997). Such powders also provide flexibility in usage regardless of the extent of heat treatment because the κ-casein/β-lactoglobulin complex typically found in skimmed powders does not exist due to the removal of the β-lactoglobulin during MF.

8.8 Future potential

Membrane processing of milk has made possible the introduction of many innovations for cheesemaking. It has evolved from processes requiring highly specialised equipment to those that have now become an integral part of the cheesemaking operation. Cheesemaking efficiency has improved and new cheese varieties have been developed. The process that led the development of membrane applications in cheesemaking, i.e. the MMV process for Camembert cheese, is no longer used for this cheese because of difficulties in meeting consumer expectations for appearance of the cheese, but it inspired the development of applications for other cheeses, such as Feta-type, Pavé d'Affinois, Le Petit Moulé, La Roche (blue cheese), and other varieties. Innovations in membrane and process design have also assisted with these applications. The challenge of standards of identity pertaining to membrane-processed milk have not been fully resolved in many countries, but Codex Alimentarius regulations do permit use of such milk and progress will continue. Membrane processing makes it possible to separate and concentrate milk in a well-defined manner, allowing the use of resulting products in unique ways with useful functional and nutritional properties.

References

Alichanidis, E., Anifantakis, E.M., Polychroniadou, A. & Nanou, M. (1984) Suitability of some microbial coagulants for Feta cheese manufacture. *Journal of Dairy Research*, **51**, 141–147.

Anifantakis, E.M. & Kandarakis, J.G. (1983) [Utilisation de la pepsin bovine en fabrication de fromage Feta fait à partir du lait de brebis.] *Lait*, **63**, 416–424.

Anonymous (2003) *Dairy Processing Handbook*, 2nd and revised edn. of G. Bylund (1995), Tetra Pak Processing Systems AB, Lund.

Bhaskar G.V., Singh H. & Blazey N.D. (2001) Milk protein products and processes. World Patent Application No. WO 01/41578 A1.

Blazey N.D., Dybing S.T., Knights R.J. & Huang I.-Lo. (2001) Methods for producing cheese and cheese products. United States of America Patent Application No. 6177128.

Bredahl, B., Borgudd, L. & Nielsen, J. (1995) A method of producing ready-to-distribute soft/hard cheese. The Patent Cooperation Treaty (PCT) Application No. WO 95/09537.

Brulé, G., Maubois, J.L., Vandeweghe, J., Fauquant, J. & Goudedranche, H. (1975). Utilisation de l'ultrafiltration sur membrane pour la fabrication de fromages de type pâtes fraîches. *Revue Laitiere Francaise*, **328**, 117–122.

Caron A., St-Gelais, D. & Pouliot Y. (1997) Coagulation of milk enriched with ultrafiltered or diafiltered microfiltered milk retentate powder. *International Dairy Journal*, **7**, 445–451.

Cousin, O. (1994) *Réensemencement des Laits Microfiltrés en vue de leur Transformation en Fromage Camembert*, Ph.D. thesis Ensar/INRA – Rennes.

Creamer, L.K., Iyer, M. & Lelievre, J. (1987) Effect of various levels of rennet addition on characteristics of Cheddar cheese made from ultrafiltered milk. *New Zealand Journal of Dairy Science and Technology*, **22**, 205–214.

Culioli, J., Bon, J.P. and Maubois, J.-L., Goudedranche, H. and Piot, M. (1974). Etude de la viscosité des rétentats et des préfromages obtenus après traitement du lait par ultrafiltration sur membrane. *Lait*, **54**, 481–500.

Dalgleish, D.G. (1980) Effect of milk concentration on the rennet coagulation time. *Journal of Dairy Research*, **47**, 231–235.

Ernstrom, C.A., Sutherland, B.J. & Jameson, G.W. (1980) Cheese base for processing. A high yield product from whole milk by ultrafiltration. *Journal of Dairy Science*, **63**, 228–234.

Garem, A., Schuck, P. & Maubois, J.-L. (2000) Cheesemaking properties of a new dairy-based powder made by a combination of microfiltration and ultrafiltration. *Lait*, **80**, 25–32.

Govindasamy-Lucey, S., Jaeggi, J.J., Martinelli, C., Johnson, M.E. & Lucey, J.A. (2011) Standardization of milk using cold ultrafiltration retentates for the manufacture of Swiss cheese: Effect of altering coagulation conditions on yield and cheese quality. *Journal of Dairy Science*, **94**, 2719–2730.

Green, M.L. & Grandison, A.S. (1987) Secondary (non-enzyme) phase of rennet coagulation and post-coagulation phenomena. *Cheese: Chemistry, Physics and Microbiology* (ed. P.F. Fox), Vol. **1**, pp. 97–134, Elsevier Science Publishing Co., London.

Hurt, E. & Barbano, D.M. (2010) Processing factors that influence casein and serum protein separation by microfiltration. *Journal of Dairy Science*, **93**, 4928–4941.

Kandarakis, I., Moschopoulou, E. & Anifantakis, E.M. (1999) Use of fermentation produced chymosin from *E. coli* in the manufacture of Feta cheese. *Milchwissenschaft*, **54**, 24–26.

Kealey, K.S. & Kosikowski, F.V. (1986) Cottage cheese from ultrafiltered skim milk retentates in industrial cheese making. *Journal of Dairy Science*, **69**, 1479–1483.

Kiers, G. (2001) Filling cheese as a liquid. *Scandinavian Dairy Information*, **15**, 20–21.

Kindstedt, P.S. & Guo, M.R. (1998) A physico-chemical approach to the structure and function of Mozzarella cheese. *Australian Journal of Dairy Technology*, **53**, 70–73.

Korolczuk, J., Maubois, J.-L. & Fauquant, J. (1986) Mécanisation en fromagerie de pâtes molles. *Proceedings of XXII International Dairy Congress*, pp. 123–128, Reidel D. Publishing Company, The Hague.

Kosikowski, F.V. & Mistry, V.V. (1997) *Cheese and Fermented Milk Foods*, 3rd edition Vol. **1**, Origins and Principles, F.V. Kosikowski LLC., Great Falls.

Mahaut, M., Maubois, J.-L., Zink, A., Pannetier, R. & Veyre, R. (1982) Eléments de fabrication de fromages frais par ultrafiltration sur membrane de coagulum de lait. *Technique Laitiére*, **961**, 9–13.

Maubois, J.-L. (1979) Application des techniques à membrane dans l'industrie fromagère. *Génie Rural*, **1**, 15–19.

Maubois, J.-L. (1987) Industrial development in ultrafiltration. *Proceeding Symposium on Cheese Biotechnology and International Food Development*, October 1987, pp. 24, Cornell University, Ithaca.

Maubois, J.-L. (2002) Membrane microfiltration: a tool for a new approach in dairy technology. *Australian Journal of Dairy Technology*, **57**, 92–96.

Maubois, J-L. & Mocquot, G. (1971) Préparation de fromage à partir de préfromage liquide obtenu par ultrafiltration du lait. *Lait*, **51**, 495–533.

Maubois, J.-L., Mocquot, G. & Vassal, L. (1969) Procédé de traitement du lait et de sous produits laitiers. **French Patent Application** No. 2 052 121.

Maubois, J.-L., Fauquant, J., Famelart, M.H. & Caussin, F. (2001) Milk microfiltrate, a convenient starting material for fractionation of whey proteins and derivatives. *The Importance of Whey and Whey Components in Food and Nutrition*, Proceedings of 3rd International Whey Conference, pp. 59–72, B. Behr's Verlag, Hamburg.

Mistry, V.V. (2002) Manufacture and application of high milk protein powder. *Lait*, **82**, 515–522.

Mistry, V.V. & Hassan, H.N. (1991) Delactosed, high milk protein powder. 1. Manufacture and composition. *Journal of Dairy Science*, **74**, 1163–1169.

Mistry, V.V. & Kosikowski, F.V. (1985) Fermentation of ultrafiltered skim milk retentates with mesophilic lactic cheese starters. *Journal of Dairy Science*, **68**, 1613–1617.

Mistry, V.V. & Kosikowski, F.V. (1986a) A naturally buffered bulk retentate starter from ultrafiltered milk. *Journal of Dairy Science*, **69**, 945–950.

Mistry, V.V. & Kosikowski, F.V. (1986b) Application of retentate starter produced from ultrafiltered milk to the manufacture of Cheddar cheese. *Journal of Dairy Science*, **69**, 1484–1490.

Mistry, V.V. & Maubois, J.-L. (2004) Application of membrane separation technology to cheese production. *Cheese: Chemistry, Physics and Microbiology* (eds. P.F. Fox, P.L.H. McSweeney, T.M. Cogan and T.P. Guinee), 3rd edn, Vol. **1**., pp. 261–286, Elsevier, London.

Moran J.W., Dever H.A., Miller A.M., Silver R.S. & Hyde M.A. (2001a) Continuous on-demand manufacture of process cheese. United States of America Patent Application No. 6183804.

Moran J.W., Trecker G.W. & Monckton S.P. (2001b) Continuous manufacture of process cheese. United States of America Patent Application No. 6183805.

Mortensen, B.K. (1985) Recent developments in the utilization of milk proteins in dairy products. *Milk Proteins '84'*, Proceedings of the International Congress on 'Milk Proteins' Luxembourg (eds. T.E. Galesloot and B.J. Tinbergen), pp. 109–119, PUDOC, Wageningen.

Muir, D.D. & Tamime, A.Y. (2001) Liquid milk. *Mechanisation and Automation in Dairy Technology*, (eds. A.Y. Tamime & B.A. Law), pp. 53–94, Sheffield Academic Press, Sheffield.

Murphy, J.M. & Fox, P.F. (1991a) Fractionation of sodium caseinate by ultrafiltration. *Food Chemistry*, **39**, 27–38.

Murphy, J.M. and Fox, P.F. (1991b) Functional properties of α_s-rich/κ-rich or β-rich casein fractions. *Food Chemistry*, **39**, 211–228.

Oommen, B.S., Mistry, V.V. & Nair, M.G. (2000) Effect of homogenization of cream on composition, yield and functionality of Cheddar cheese made from milk supplemented with ultrafiltered milk. *Lait*, **80**, 77–91.

Quiblier, J.P., Ferron-Baumy, C., Garric, G. and Maubois, J.L. (1992) Procédé de traitement des laits permettant au moins de conserver leur aptitude fromagère. European Patent No. 0542 583 B1

Rash, K.E. & Kosikowski, F.V. (1982) Behavior of enteropathogenic *Escherichia coli* in Camembert cheese made from ultrafiltered milk. *Journal of Food Science*, **47**, 728–732.

Saboya, L.V., Goudédranche, H., Maubois, J.-L., Lerayer, A.L.S. & Lortal, S. (2001). Impact of broken cells of lactococci and propionibacteria on the ripening of Saint-Paulin UF cheeses: extent of proteolysis and GC-MS profiles. *Lait*, **81**, 699–713.

Saboya, L.V. & Maubois, J.-L. (2000) Current developments of microfiltration technology in the dairy industry. *Lait*, **80**, 541–553.

Schuck, P., Piot, M., Mejean, S., Le Graet, Y., Fauquant, J., Brulé, G. & Maubois, J.-L. (1994) Déshydratation par atomisation de phosphocaséinate natif obtenu par microfiltration sur membrane. *Lait*, **74**, 375–388.

Skovhauge, E. (1987) Production of cheese with open structure by means of ultrafiltration technique. *North European Dairy Journal*, **53**, 61–67.

Saldamli, I. & Kaytanli, M. (1998) Utilisation of Fromase, Maxiren and Rennilase as alternative coagulating enzymes to rennet in Turkish white cheese production. *Milchwissenschaft*, **53**, 22–25.

Spangler, P.L., Jensens, L.A., Amundson, C.H., Olsons, N.F. & Hill, C.G. (1990) Gouda cheese made from ultrafiltered milk, effects of concentration factor, rennet concentration, and coagulation temperature. *Journal of Dairy Science*, **73**, 1420–1428.

Tamime, A.Y. & Kirkegaard, J. (1991) Manufacture of Feta cheese – industrial. *Feta and Related Cheeses* (eds. R.K. Robinson and A.Y. Tamime), pp. 70–143, Ellis Horwood, Chichester.

Tamime, A.Y. (1993) Modern cheesemaking: hard cheeses. *Modern Dairy Technology* (ed. R.K. Robinson), Vol. 2 – Advances in Milk Products, pp. 49–220, Elsivier Applied Science, London.

Tamime, A.Y. & Robinson, R.K. (1999 (2007)) *Tamime and Robinson's – Yoghurt Science and Technology*, 3rd edn, pp. 19–23, 143–147 & 217–220, Woodhead Publishing, Cambridge.

Tamime, A.Y., Robinson, R.K. & Latrille, E. (2001) Yoghurt and other fermented milks. *Mechanisation and Automation in Dairy Technology*, (eds. A.Y. Tamime & B.A. Law), pp. 152–203, Sheffield Academic Press, Sheffield.

Tamime, A.Y., Robinson, R.K. & Kiers, G.L.J. (2006) Industrial manufacture of Feta-type cheese. *Brined Cheeses* (ed. A.Y. Tamime), pp. 77–116, Blackwell Publishing, Oxford.

Terré, E., Maubois, J.L., Brulé, G. & Pierre, A. (1987) Procédé d'obtention d'une matière enrichie en caséine β, appareillage pour la mise en oeuvre de ce procédé et application des produits obtenus par ce procédé comme aliments, compléments alimentaires ou additifs en industrie alimentaire et pharmaceutique ou dans la préparation de peptides à activité physiologique. French Patent Application No. 2 592 769.

Upreti, P., Mistry, V.V. & Acharya, M.R. (2011) Characterization of rennet coagulation of milk concentrated by vacuum condensing and ultrafiltration. *Dairy Science & Technology*, **91**, 383–395.

Visser, S. (1981). Proteolytic enzymes and their action on milk proteins: A review. *Netherlands Milk and Dairy Journal*, **35**, 65–88.

Weissman, B.J. (1982) Ultrafiltration process for the preparation of cream cheese. United States of America Patent Application No. 4341801, 27 July.

Zall, R.R. & Chen, J.H. (1986) Thermalizing milk as opposed to milk concentrate in a UF system affects cheese yield. *Milchwissenschaft*, **41**, 217–218.

9 Whey Processing

L. Ramchandran and T. Vasiljevic

9.1 Introduction

Whey, once known as a major pollutant of the natural waterways, is now recognised as a pool of valuable ingredients. This by-product of cheese and caseinate manufacturing is liquid gold if its potential can be harnessed completely. Its increased utilisation in myriad products has been aided by superior nutritional quality, powerful biological properties and unique functional characteristics of whey proteins. Fig. 9.1 highlights the potential functionality of whey that governs its utility.

Increasing knowledge on the biological functions of whey constituents has drawn interest of researchers, manufacturers and consumers alike, and has transgressed whey utilisation from an ingredient for animal feed to high-value ingredients for functional or health-promoting foods. This realisation has been instrumental in boosting the demand and thereby the market for whey products and ingredients. As a consequence, processing whey to recover its beneficial constituents became more acceptable for optimal utilisation of its nutritional and functional properties rather than treating whey for disposal. Advances in processing technology have propelled whey proteins as important nutritional and functional ingredients in the special niche of food for health in the food industry.

9.2 Whey: components, their functionality and uses

Whey is the greenish-yellow liquid produced as a result of milk coagulation. Two types of whey are commonly produced depending on the type of cheese or caseinate. Sweet whey (pH not less than 5.6) is produced by enzymatic coagulation of milk, such as rennet coagulation during Cheddar cheesemaking. Acid whey (pH not exceeding 5.1) is produced during the manufacturing of acid-coagulated milk products, such as cottage cheese and casein (Jelen 2003). All types of whey, irrespective of their origin, are composed of proteins – mainly, α-lactalbumin (α-La), β-lactoglobulin (β-Lg), bovine serum albumin and minor proteins, such as lactoferrin, immunoglobulins (Ig), growth factors (GFs); lactose and salts, but their concentrations vary with the type of whey. These constituents of whey influence the extent of processing it can undergo and the quality, functionality and utility of the end product. The various components of whey and their functionality are shown in Table 9.1.

Membrane Processing: Dairy and Beverage Applications, First Edition. A. Y. Tamime.

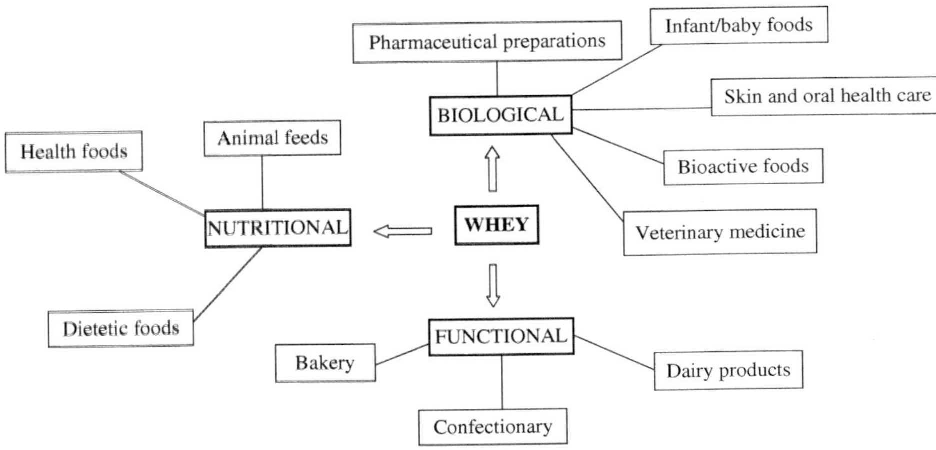

Fig. 9.1 Potential functions and uses of whey.

Table 9.1 Whey constituents – their functionality and uses.

Class of constituent	Individual constituents	Properties	Intended uses
Major proteins	β-lactoglobulin α-lactalbumin Bovine serum albumin Immunoglobulins	Foam stabiliser Gelling properties High emulsifying properties High whippability Amino acid composition Affinity for glycosylated receptors Immunomodulatory effect	Confectionery, Infant formula and neutraceuticals Meringue like products Functional foods
Minor proteins	Lactoferrin Lactoperoxidase Growth factors (GFs)	Antimicrobial activity Promotes iron utilisation Cancer prevention Antiviral activity Immune system, Insulin-like GF Transforming GF Epidermal GF Basic fibroblast GF	Bioactive ingredient in supplement foods Skin care and oral health care products Infant formulae Pharmaceutical and neutraceutical applications
Carbohydrates	Lactose Oligosaccharides	Low calorific value glycaemic index and cariogenity Infant health	Food and drug industries Baby/infant formulations Health care products
Salts	Ca, Mg, P, Na	Bone health Weight control	Dietetic products Health products
Hydrolysed constituents	Glycomacropeptide Caseinomacropeptide Bioactive peptides	Therapeutic benefits involving digestive antimicrobial Immunomodulatory Antithrombotic Antihypertensive	Baby foods Functional foods

Preliminary utilisation of whey was in the form of whey powder. Developments in engineering and technology led to production of whey protein concentrate (WPC) and whey protein isolate (WPI). Traditionally, processing of whey involved heating and drying that altered whey protein structure and consequently functionality and applicability. High mineral content of whey and the fat content of certain types of whey were also found to affect whey protein functionality, limiting their utilisation in food and feed products (Brans *et al*., 2004). Developments in membrane science and technology proved instrumental in introducing membranes as a less intensive and less detrimental method of whey processing. The understanding that whey proteins should maintain their native structure intact in order to impart their specific biological and functional properties (Marcelo & Rizvi, 2008) initiated further changes in the approach to whey processing.

This chapter will focus on updating the role of membranes in whey processing.

9.3 Problems of traditional whey processing

Over the last three decades utilisation of whey for manufacturing of protein ingredients has become an integral part of the dairy industry worldwide. Yet whey processing did not provide the expected economic or functional benefits over other milk or food proteins (Schmidt *et al*., 1984). In the early 1900s, whey processing was initiated by processes involving hot roller driers, heat concentration followed by extrusion, or steam heating followed by drum drying. Later the introduction of spray drying for whey processing resulted in the availability of a gritty product with poor solubility and limited use. The later half of the 20th century witnessed the introduction of membranes in processing of whey (Tunick, 2008). In all these processes, the major obstacles that hindered the commercial viability of whey processing included:

- changes due to denaturation and interactions of whey proteins with other constituents during processing
- high variability in composition of processed whey
- poor functional properties of the final product
- variations in product composition and quality
- high capital and operating costs.

Traditionally, cheese whey was used to produce spray-dried WPC (35–80 g 100 g^{-1} protein) or WPI (80–95 g 100 g^{-1} protein). Processing steps involved in cheesemaking and whey processing altered the native conformation of proteins, which affected their functionality and thereby, their utility. The aspects of whey processing that are critical to the development of the final product (such as WPC) that need to be identified and their effects on its functionality understood, irrespective of the method of whey processing, are (Melachouris, 1984):

- variability in whey composition
- pre-treatment of whey
- characterisation of final product
- applications of the final product.

Membrane separation technology has been accepted as a logical choice for processing milk and milk products and has been extensively adopted by the dairy industry for more than three decades. During this period the developments in membrane technology have influenced dairy processing. At the same time, challenges of dairy and food processing have contributed the most to advances in application research for membrane processing (Henning *et al.*, 2006).

9.4 Membranes in whey processing

Separation of liquid into two streams, namely permeate and retentate, is achieved by passing the liquid through a semi-permeable membrane. This fundamental principle of membrane filtration has now grown into a wide field of membrane science and technology, having a wide range of application from water treatment to food processing.

Whey processing was one the first applications of membranes in the dairy industry (Rektor & Vatai, 2004). Advances in membrane science have adapted this technology to suit the changing demands of whey processing. Now, it is not just about membrane processing of whey for products with better functionality, it is also about creating specific raw materials with the enhanced possibility of being processed by membranes to produce products or ingredients with specific functionality and utility. In this context, the concept of virgin whey proteins, liquid virgin WPI and microfiltered whey (permeate obtained by microfiltered milk) have evolved in recent years, all of which help maintain the native conformation of whey proteins and therefore their biological functionality (Marcelo & Rizvi, 2008; Akbache *et al.*, 2009).

Cost and energy-efficient membrane technologies have been applied in whey processing to:

- concentrate whey proteins and lactose
- separate individual whey constituents
- isolate specific products of whey protein hydrolysis
- remove minerals and fat before further processing.

Currently, four membrane technologies are in use for whey processing: (a) microfiltration (MF), (b) ultrafiltration (UF) including diafiltration, (c) nanofiltration (NF) and (d) reverse osmosis (RO) (Cuartas-Uribe *et al.*, 2009). With the realisation that separation of whey into its well-defined fractions having specific functionality would enable the optimal use of whey and also result in development of newer product formulations, newer membrane processes, such as electrodialysis (ED), ion exchange (IE) membrane chromatography and affinity membrane processes, are also being explored. This highlights the versatility acquired by membrane processes over the years and their ever-increasing applications in whey processing (Fig. 9.2). However, membrane performance, fouling, cleanability and specificity of separation continue to be issues that spur ongoing research in plant design and membrane development.

The major membrane technologies and their application to whey processing follow.

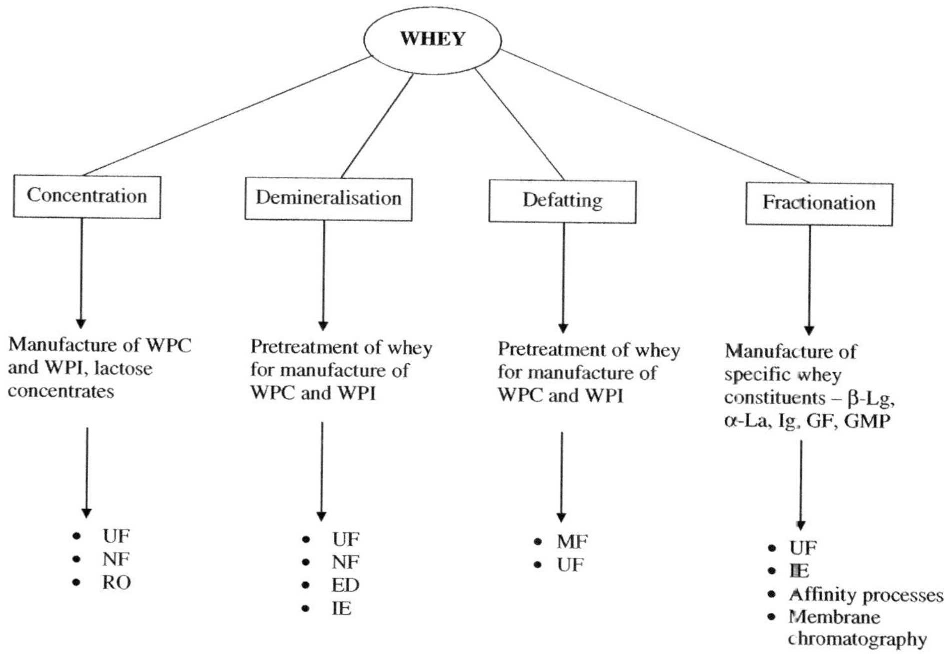

Fig. 9.2 Role of membranes in various processes of whey. WPC, whey protein concentrate; WPI, whey protein isolate; β-Lg, β-lactoglobulin; α-La, α-lactalbumin; Ig, immunoglobulin; GF, growth factor; GMP, glycomacropeptide; ED, electrodialysis; IE, ion exchange; MF, microfiltration; NF, nanofiltration; UF, ultrafiltration; RO, reverse osmosis.

9.4.1 *Microfiltration*

MF is a low-pressure driven membrane filtration process. The membranes used in this process have an open structure that allows all the dissolved components to pass through while retaining the undissolved components. Ceramic and spiral wound MF membranes are available for use in whey processing.

This process is widely used as a pre-treatment for improving the bacterial quality of whey and for removing fat. The presence of fat in whey limits the protein levels attainable in WPC and WPI, more specifically in WPI where protein levels of 80–95 g 100 g^{-1} are required. Tangential MF of whey has been used for producing WPI having a protein content of 80–85 g 100 g^{-1} (Baumy *et al.*, 1990; Gésan-Guiziou *et al.*, 1993). Morin *et al.* (2006) used ceramic MF membranes for concentrating the milk fat globule membrane (MFGM) from whey buttermilk (by 50 g 100 g^{-1}) obtained by churning whey creams compared with doubling of its content in regular buttermilk obtained by churning cream. They reported that MF of whey buttermilk was a better option that gave MFGM of higher purity. MF has also been used to obtain caseinomacropeptide-free WPC from sweet whey (Tolkach & Kulozik, 2005). The use of MF in combination with some physico-chemical modifications has been reportedly applied for separating β-Lg from α-La (Pouliot, 2008). Although whey is a valuable source for isolation of the GF for use in therapeutic and infant food preparations (Pouliot & Gauthier, 2006), use of microfiltered whey is reported to be a better option for obtaining a product enriched with GF. It was concluded that microfiltered whey better preserved the bioactivity of the minor compounds such as, GF than cheese whey (Akbache *et al*., 2009).

9.4.2 *Ultrafiltration*

UF is a medium-pressure driven membrane process. Membranes used in this process possess an intermediary open structure that permits most dissolved components and some undissolved components to pass through while larger components are rejected by the membrane. UF was one of the first and most common membrane processes used successfully in the whey industry. UF of acid and sweet cheese whey is commonly practised (Musale & Kulkarni, 1998). As a consequence, substantially more research has been carried out in this process. It is primarily used

- to fractionate and concentrate whey into WPC and lactose-rich permeate, which can be further processed to recover the lactose
- to pre-concentrate the lactose-rich permeate obtained from WPC production for lactose production
- to decalcify the permeate obtained from WPC production, rendering the permeate suitable for concentration by other processes, such as RO
- as a secondary treatment to MF of whey; UF of the microfiltered whey increased the concentration of transforming GF (TGF-β2) by a factor of 13 (Akbache *et al*., 2009).

The use of UF membranes to fractionate individual whey proteins close in molecular mass, such as glycomacropeptide (GMP), and β-Lg has been limited. However, in recent years the introduction of charged UF membranes has changed this concept. Bhushan & Etzel (2009) successfully separated GMP and β-Lg from cheese whey using positively charged membranes that increased the selectivity of the separation by over 600%. Similarly, Metsämuuronen & Nyström (2009) used UF membranes of different hydrophobicity and pore size to produce α-La-enriched permeate from diluted whey solutions. Hydrophilic 30 kg mol^{-1} regenerated cellulose membrane exhibited highest selectivity and minimum fouling. Temperature and pH controls further improved the separation of α-La from β-Lg and, although permeation of α-La decreased with time and transmembrane pressure, the selectivity of the membrane increased because of narrowing of pores that resulted in better retention of β-Lg.

The most common UF modules in use are tubular, plate and frame, spiral-wound and hollow (large) fibre modules. With regards to UF processing of whey, the spiral-wound and hollow fibre units appear to be more economical than plate and tubular units in terms of recirculation pump energy, but have the disadvantage of relatively low fluxes. Hence, selection of membrane module for any particular plant would depend not only on processing costs, but also on membrane replacement cost, cleanability of the membrane module and the composition and functionality of the final product. Very little research has been directed towards studying the operability of UF plants for processing whey (Barba *et al*., 2001; Guadix *et al*., 2004; Atra *et al*., 2005).

The loss of membrane productivity is a continual problem reported in UF processes. Flux and rejection are important characteristics that relate to membrane productivity. Fouling is the single largest cause for decrease in membrane flux during operation. Flux decline has been identified to occur in three definite stages: (a) initial rapid drop in flux due to concentration polarisation; (b) second stage of continuous decline in flux due to deposit formation (fouling); and (c) quasi-steady-state period where the flux settles to a steady-state value.

Of these, the first stage of flux decline is generally reversible, and usually takes place in less than a minute. Fouling on the other hand takes place over the length of the processing time period. Fouling within the membrane pores can change the size, size distribution and density of the pores, which can reduce membrane flux (Nigam *et al*., 2008). In whey processing, complex protein–protein interactions, protein–calcium interactions, protein–lipid interaction and calcium phosphate precipitation contribute to fouling of membranes (Schmidt *et al*., 1984). Of these, precipitation of calcium phosphate is mainly responsible for loss in membrane productivity.

Membrane flux in whey processing may be optimised by manipulating: (a) transmembrane pressure; (b) recirculation rate; and (c) temperature. All these manipulations affect energy consumption and therefore the cost of the process.

Loss of membrane performance due to fouling by whey proteins and salts can be controlled by

- back flushing at regular intervals (Breslau & Kilcullen, 1977)
- careful selection of the type of membrane module used, which in turn influences ease of cleaning, need for pre-treatment and membrane replacement (Cheryan & Kuo, 1984)
- pre-treatment to whey, such as decalcification of whey by clarification (Musale & Kulkarni, 1998), adjusting pH to values around 7 followed by heat treatment to 70°C for 2 min before passing through the membrane (Fachin & Viotto, 2005)
- mid-run washing of membranes (Yee *et al*., 2007)
- monitoring the total solids and protein concentration of the final product to determine processing variables that will influence the product flow rate and composition (Yee *et al*., 2007)
- variability in whey composition is another factor that can have an impact on the performance of UF whey processing plants and the concentration of protein in the final product.

9.4.3 *Diafiltration*

Diafiltration (DF) is a technique that uses UF membranes to concentrate whey proteins or their fractions. The process selectively utilises permeable (porous) membrane filters to separate the components of solutions and suspensions based on their molecular size. The technique is adapted in three modes (Schwartz, 2011).

- Continuous DF (also referred to as constant volume diafiltration, CVDF) that involves washing out the low molecular weight species in the sample by adding water or a new buffer at the same rate as filtrate is being generated. As a result, the sample volume and product concentration do not change during the DF process.

- Discontinuous DF by sequential dilution involves first diluting the sample with water or replacement buffer to a predetermined volume. The diluted sample is then concentrated back to its original volume by UF.
- Discontinuous DF by volume reduction wherein the sample is first concentrated to a predetermined volume, and then diluted back to its original volume with water or replacement buffer.

The inclusion of DF in whey processing has been studied over the last decade, and will be included later under Section 9.4.6. In general, DF has been used to increase the purity and concentration of fractionated whey constituents, such as α-La (Cheang & Zydney, 2003) and β-Lg (Bhattacharjee *et al*., 2006). Recently, Pan *et al*. (2011) used NF membrane for demineralising whey at pH 4.6 using the DF process, and were able to achieve 72% demineralisation without any loss of lactose.

9.4.4 *Nanofiltration and reverse osmosis*

NF is a medium- to high-pressure driven membrane filtration process. It can be regarded as a type of RO where the membrane has a slightly more open structure to allow most monovalent ions to pass through while rejecting most of the divalent and multivalent ions. RO is a high-pressure driven membrane process using very dense membranes that allow only water to pass through. Both NF and RO are mainly used to demineralise or concentrate whey or ultrafiltered permeates. However, NF is considered to be a better cost-effective alternative to RO in concentrating/demineralising whey because of the higher flux that can be achieved with NF (Cuartas-Uribe *et al*., 2007). Hence, the interest has shifted towards NF than RO, particularly for producing demineralised whey powders suitable for production of baby foods. NF is more applied as a downstream process, such as to concentrate the lactose-rich permeate from UF to obtain concentrated lactose with a higher degree of purity. Some such integrated application studies will be described below.

As with other membrane processes, fouling is the major problem with NF application. Whey proteins, such as α-La and β-Lg as well as fat and calcium phosphate ions, can adsorb onto the membranes and clog the pores. As a consequence, the cost of whey processing increases because of the added cleaning and membrane replacement costs. To prevent fouling, use of hydrophilic membranes and passing the whey through UF before NF are considered useful options (Cuartas-Uribe *et al*., 2007). Similarly, when using RO, minerals causing fouling influence the type for membrane that can be used for whey processing (Muller & Harper, 1979).

9.4.5 *Electrodialysis and other related processes*

Over the past two decades the significance of individual whey constituents as functional ingredients has introduced functionalised membranes to whey processing. Selectivity and resistance to fouling are key requirements for fractionation of whey proteins. Novel applications of ED using ion exchange (IE) membranes and bipolar membrane electroacidification (BMEA) have been developed mainly for efficient demineralisation of whey for special uses, such as infant formulas, delipidation of whey for improving their functionality and for separating individual whey constituents having similar physico-chemical properties, such as α-La and Ig.

ED is the separation process that involves transport of the charged species across one or more semi-permeable membrane under the influence of a direct current. This process mainly targets demineralisation of whey. ED processes used for demineralisation of whey can achieve a 60–75% reduction in mineral content in continuous process and 90–95% by recirculating the whey in ED cells (Bazinet, 2005).

The development of IE membranes allowed the use of ED for whey protein fractionation on the principles of chromatography. In this technique, the selectivity of IE chromatography is combined with the productivity of membrane processes. An IE membrane is made from a base material that carries ionisable groups covalently attached to membrane surfaces. These membranes function on the basis of the reversible interaction between target protein and charged functional groups. The various categories of IE membranes are shown in Fig. 9.3.

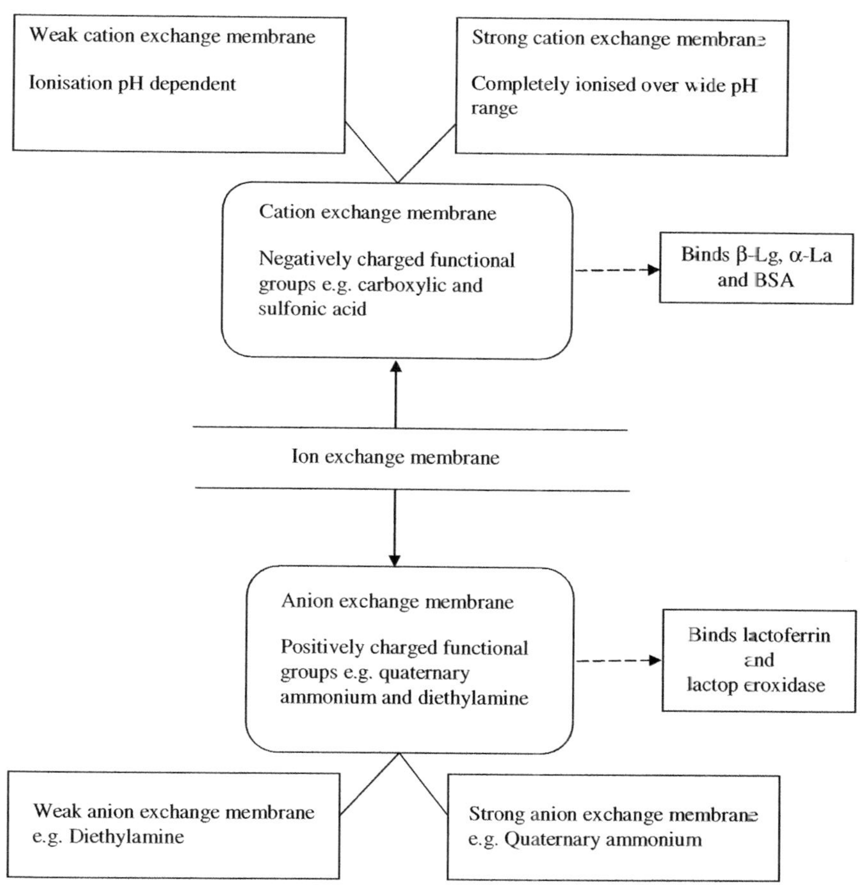

Fig. 9.3 Categories of ion exchange membranes. β-Lg, β-lactoglobulin; α-La, α-lactalbumin; BSA, bovine serum albumin.

Ion exchange mixed matrix membranes (MMMs) are prepared by incorporating an ion exchange resin into a membrane polymer solution at a specified ion exchanger resin loading prior to membrane casting. This procedure eliminates the need for chemical modification to induce a charged group in the membrane (Saufi & Fee, 2011). Some important applications of IE membrane chromatography technique are given in Table 9.2. The major advantages of IE membrane chromatography over the traditional column chromatography are:

- rapid rate of association between target proteins and functional group
- reduction in processing times
- ease of scale-up and operation
- maintaining native structure and functionality of individual proteins.

BMEA is another application of ED that is being attempted for processing whey. This technique couples the effect of acidification and demineralisation by combining the ability of bipolar membranes to dissociate water molecules at their interface to cause acidification and that of cationic exchange membranes to demineralise the whey. It is claimed that this process permits high protein recoveries up to 53% with >95% purity (Bazinet *et al*., 2004). Shee *et al*. (2007) have suggested the use of BMEA as the first step of WPC treatment to reduce the lipid content by 73% in the acidification step followed by demineralisation to produce a WPC with low-ash content. The major advantages of BMEA are: (a) greater precision in process control; (b) *in situ* generation and reuse of dangerous acids and bases for the process; (c) ease of transfer to industrial scale; and (d) high protein recovery yield.

Affinity membranes with immobilised triazinic dyes have also been developed for protein downstream processing. Hollow fibres grafted with glycidyl methacrylates/diemthyl acrylamide copolymer and attached with Red HE-3B dye have been synthesised for a one-step purification method for isolating lactoferin from whey. However, these membranes have not received much acceptance mainly due to their low protein-binding capacity (Wolman *et al*., 2007).

The electrochemical processes also face the problem of membrane fouling by proteins and minerals. The separation of concentrate loops in two different loops and use of pulsed electrical fields are reported to minimise fouling, although the processing times increase (Casademont *et al*., 2009).

Membrane processes based on ED have not yet entered commercial whey processing in a big way mainly because of the low binding capacity of the membranes and their limited operational life (Pouliot, 2008). Some suggestions for enhancing the separation of the specific whey proteins using IE membranes are:

- maintaining pH of pre-treated whey to 2.8 and that of loading whey to pH 5.0 enhances separation of β-Lg from α-La (Bhattacharjee *et al*., 2006)
- manipulating loading volume, pH and elution gradient for better separation of individual proteins (Goodall *et al*., 2008)
- selection of suitable buffer to maintain native protein structures (Kim *et al*., 2003)
- controlling buffer pH and ionic strength for improved selectivity and selection of membranes with suitable charge (Saksena & Zydney, 1994; Cheang & Zydney, 2003).

Table 9.2 Some applications of ion exchange (IE) membranes in whey processing.

Membrane characteristics	Type of whey	Pre-treatments before IE	Constituents separated using IE membrane	Product characteristics	Reference
Cation exchanger (patented process)	Acid whey	ED	β-Lg α-La Lactose	–	Stack *et al*. (1995)
Sartobind® cation exchange membrane	Sweet whey	Cross-flow filtration	LF	95% purity	Ulber *et al*. (2001)
Cation exchange chromatography	Whey		Growth factors	–	Smithers (2004)
Strong anion exchange membrane Quaternary ammonium ion	Casein whey	Centrifugation MF Two-stage DF	β-Lg	88% purity	Bhattacharjee *et al*. (2006)
Weak anion exchange membrane – diethylamine Strong anion exchange membrane – quaternary ammonium group	Rennet whey	Centrifugation	β-Lg α-La BSA	–	Goodall *et al*. (2008)
Anion mixed matrix membrane Lewatit®MP500 anion exchange resin Quaternary amine	Casein whey	Filtration through 0.45 μm membrane pH adjusted to 6.0	β-Lg		Saufi & Fee (2009)
Cation mixed matrix membrane; SP Sepharose™cation exchange resin	Casein whey	Filtration through 0.45 μm membrane pH adjusted to 6.0	LF	91% purity	Saufi & Fee (2011)

β-Lg, β-lactoglobulin; α-La + α-lactalbumin; BSA, bovine serum albumins; LF, lactoferin; ED, electrodialysis; MF, microfiltration; DF, diafiltration.

Table 9.3 Integrated membrane processes for whey processing.

Membrane processes integrated	Outputs	Reference
UF-NF	Bioactive peptides	Butylina *et al.*, 2006
UF-DF	WPC with 99% protein	Henning *et al.*, 2006
UF-DF	Liquid virgin WPI, unique physicochemical properties	Marcelo and Rizvi, 2008
NF-DF (continuous variable volume)	Demineralized WPC, less expensive than ED/IE	Román *et al.*, 2009
MF-UF-IE-RO	Pure lactose (99.8%), overall recovery 74%	De Souza *et al.*, 2010

DF, diafiltration; IE, ion exchange; MF, microfiltration; NF, nanofiltration; RO, reverse osmosis; UF, ultrafiltartion.

9.4.6 *Integrated processes*

Although each membrane process has defined capabilities to process whey, an integration of these processes is often suggested as a means of completely extracting all beneficial components of whey in a single process leaving behind a permeate that can easily be disposed off without any major processing. Moreover, for most ED processes, other membrane processes are usually integrated as stages of pre-treatment of the whey. Some of these integrated processes studied are compiled in Table 9.3.

9.5 Conclusions

Membrane science and technology is the answer to the problem of whey disposal in economic and technological terms. Whey is suited to be processed by various membrane technologies and the number of successful applications for whey processing is increasing. Newer techniques are being developed with a view to enhance the quality and functionality of whey products. Separation of whey constituents, in particular, is the current area of interest and research, and is also broadening the scope of whey utilisation. However, there is still great need to scientifically prove the health benefits of whey constituents in particular, which will go a long way in the development of a suitable market for whey ingredients. The price of whey products and ingredients, their functionality, compositional consistency and the availability of alternative proteins will decide the market for these products. All this in turn will govern how this area of whey processing develops in the coming years. Furthermore, development of newer membrane types tailor-made for specific fractionation and of membrane modules and plant design will also shape the future course of whey processing. Although membrane science has constantly grown to cater for the specific needs of whey processing, there is still a need for developing membranes with higher specificity, productivity and resistance to fouling, and these will continue to be the main areas of research in the coming years. Moreover, the environmental concerns and rising cost of energy and fuels will deem necessary the design of cost-efficient, energy-efficient and sustainable membrane processes as well as development of protocols that will ensure complete utilisation of all the streams produced by membrane processing of whey.

References

Akbache, A., Lamiot, É., Moroni, O., Turgeon, S., Gauthier, S.F. & Pouliot, Y. (2009) Use of membrane processing to concentrate TGF-β2 and IGF-I from bovine milk and whey. *Journal of Membrane Science*, **326**, 435–440.

Atra, R., Vatai, G., Bekassy-Molnar, E. & Balint, A. (2005) Investigation of ultra-and nanofiltation for utilization of whey protein and lactose. *Journal of Food Engineering*, **67**, 325–332.

Barba, D., Beolchini, F., Cifoni, D. & Veglio, F. (2001) Whey protein concentrate production in a pilot scale two-stage diafiltration process. *Separation Science Technology*, **36**, 587–603.

Baumy, J.J., Gestin, L., Fauquant, J., Boyaval, E. & Maubois, J.-L. (1990) Technologies de purification des phospholipids du lactosérum. *Process*, **1047**, 29–33.

Bazinet, L. (2005) Electrodialytic phenomena and their applications in the dairy industry: A review. *Critical Reviews in Food Science and Nutrition*, **45**, 307–326.

Bazinet, L., Ippersiel, D. & Mahdavi, B. (2004) Fractionation of whey proteins by bipolar membrane electroacidification. *Innovative Food Science and Emerging Technologies*, **5**, 17–25.

Bhattacharjee, S., Bhattacharjee, C. & Datta, S. (2006) Studies on the fractionation of β-lactoglobulin from casein whey using ultrafiltration and ion-exchange membrane chromatography. *Journal of Membrane Science*, **275**, 141–150.

Bhushan, S. & Etzel, M.R. (2009) Charged ultrafiltration membranes increase the selectivity of whey protein separations. *Journal of Food Science*, **74**, E131–E139.

Brans, G., Schröen, C.G.P.H., van ser Sman, R.G.M. & Boom, R.M. (2004) Membrane fractionation of milk: state of the art and challenges. *Journal of Membrane Science*, **243**, 263–272.

Breslau, B.R. & Kilcullen, B.M. (1977) Hollow fiber ultrafiltration of cottage cheese whey: Performance study. *Journal of Dairy Science*, **60**, 1379–1386.

Butylina, S., Luque, S. & Nyström, M. (2006) Fractionation of whey-derived peptides using a combination of ultrafiltration and nanofiltration. *Journal of Membrane Science*, **280**, 418–426.

Casademont, C., Sistat, P., Ruiz, B., Pourcelly, G. & Bazinet, L. (2009) Electrodialysis of model salt solution containing whey proteins: Enhancement by pulsed electric field and modified cell configuration. *Journal of Membrane Science*, **328**, 238–245.

Cheang, B. & Zydney, A.L. (2003) Separation of α-Lactalbumin and β-Lactoglobulin using membrane ultrafiltration. *Biotechnology and Bioengineering*, **83**, 201–209.

Cheryan, M. & Kuo, K.P. (1984) Hollow fibres and spiral wound modules for ultrafiltration of whey: Energy consumption and performance. *Journal of Dairy Science*, **67**, 1406–1413.

Cuartas-Uribe, B., Alcaina-Miranda, M. I., Sorano-Costa, E. & Bes-Piá, A. (2007) Comparison of the behaviour of two nanofiltration membranes for sweet whey demineralization. *Journal of Dairy Science*, **90**, 1094–1101.

Cuartas-Uribe, B., Alcian-Miranda, M.I., Sorano-Costa, E., Mendoza-Roca, J.A., Iborra-Clar, M.I. & Lora-Garca, J. (2009) A study of separation of lactose from whey ultrafiltration permeate using nanofiltration. *Desalination*, **241**, 244–255

Fachin, L. & Viotto, W.H. (2005) Effect of pH and heat treatment of cheese whey on solubility and emulsifying properties of whey protein concentrate produced by ultrafiltration. *International Dairy Journal*, **15**, 325–332.

Gésan-Guiziou, G., Daufin, G., Merin, U., Labbé, J.-P. & Quémerais, A. (1993) Fouling during constant flux cross-flow microfiltration of pretreated whey-Influence of transmembrane pressure gradient. *Journal of Membrane Science*, **80**, 131–145.

Goodall, S., Grandison, A.S., Jauregi, P.J. & Price J. (2008) Selective separation of the major whey proteins using ion exchange membranes. *Journal of Dairy Science*, **91**, 1–10.

Guadix, A., Sorensen, E., Papageorgiou, L.G. & Guadix, E.M. (2004) Optimal design and operation of continuous ultrafiltration plants. *Journal of Membrane Science*, **235**, 131–138.

Henning, D.R., Baer, R.J., Hassan, A.N. & Dave, R. (2006) Major advances in concentrated and dry milk products, cheese and milk fat-based spreads. *Journal of Dairy Science*, **89**, 1179–1188.

Jelen, P. (2003) Whey processing: Utilization and products. *Encyclopedia of Dairy Sciences* (eds H. Roginski, J.W. Fuquay and P.F. Fox), Vol. **4**, pp. 2739–2745, Academic Press, New York.

Kim, J.I., Choi, D.Y. & Row, K.H. (2003) Separation of whey proteins by anion-exchange membranes. *Korean Journal of Chemical Engineering*, **20**, 538–541.

Marcelo, P.A. & Rizvi, S.S.H. (2008) Physicochemical properties of liquid virgin whey protein isolate. *International Dairy Journal*, **18**, 436–246.

Melachouris, N. (1984) Critical aspects in development of whey protein concentrate. *Journal of Dairy Science*, **67**, 2693–2700.

Metsämuuronen, S. & Nyström, M. (2009) Enrichment of α-lactalbumin from diluted whey with polymeric ultrafiltration membranes. *Journal of Membrane Science*, **337**, 248–256.

Morin, P., Pouliot, Y. & Jiménez-Flores, R. (2006) A comparative study of the fractionation of regular buttermilk and whey buttermilk by microfiltration. *Journal of Food Engineering*, **77**, 521–528.

Muller, L.L. & Harper, W.J. (1979) Effect of membrane processing of pretreatments of whey. *Journal of Agriculture and Food Chemistry*, **27**, 662–664.

Musale, D.A. & Kulkarni, S.S. (1998) Effect of whey composition on ultrafiltration performance. *Journal of Agriculture and Food Chemistry*, **46**, 4717–4722.

Nigam, M.O., Bansal, B. & Chen, X.D. (2008) Fouling and cleaning of whey protein concentrate fouled in ultrafiltration membranes. *Desalination*, **218**, 313–322.

Pan, K., Song, Q., Wang, L. & Cao, B. (2011) A study of demineralization of whey by nanofiltration membrane. *Desalination*, **267**, 217–221.

Pouliot, Y. (2008) Membrane processes in dairy technology – From a simple idea to worldwide panacea. *International Dairy Journal*, **18**, 735–740.

Pouliot Y. & Gauthier, S.F. (2006) Milk growth factors as health products: Some technological aspects. *International Dairy Journal*, **16**, 1415–1420.

Rektor, A. & Vatai, G. (2004) Membrane filtration of mozzarella whey. *Desalination*, **162**, 279–286.

Román, A., Wang, J., Csanádi, J., Hodúr, C. & Vatai, G. (2009) Partial demineralization and concentration of acid whey by nanofiltration combined with diafiltration. *Desalination*, **241**, 288–295.

Saksena, S. & Zydney, A.L. (1994) Effect of solution pH and ionic strength on the separation of albumin from immunoglobulins (IgG) by selective filtration. *Biotechnology and Bioengineering*, **43**, 960–968.

Saufi, S.M. & Fee, C.J. (2009) Fractionation of β-lactoglobulin from whey by mixed matrix membrane ion exchange chromatography. *Biotechnology and Bioengineering*, **103**, 138–147.

Saufi, S.M. & Fee, C.J. (2011) Recovery of lactoferrin from whey using cross-flow cation exchange mixed matrix membrane chromatography. *Separation and Purification Technology*, **77**, 68–75.

Schmidt, R.H., Packard, V.S. & Morris, H.A. (1984) Effect of processing on whey protein functionality. *Journal of Dairy Science*, **67**, 2723–2733.

Schwartz, L. (2011) Diafiltration: A Fast, Efficient Method for Desalting, or Buffer Exchange of Biological Samples. Available from http://www.pall.com/main/Laboratory/Literature-Library.

Shee, F.L.T., Angers, P. & Bazinet, L. (2007) Delipidation of a whey protein concentrate by electroacidification with bipolar membranes. *Journal of Agriculture and Food Chemistry*, **55**, 3985–3989.

Smithers, G.W. (2004) Isolation of growth factors from whey and their application in food and biotechnology industries – a brief review. *Advances in Fractionation and Separation Processes for Novel Dairy Application*, Document No. **389**, pp. 16–19, International Dairy Federation, Brussels, Belgium.

de Souza, R.R., Bergamasco, R., da Costa, S.C., Feng, X., Faria, S.H.B. & Gimenes, M.L. (2010) Recovery and purification of lactose from whey. *Chemical Engineering and Processing*, **49**, 1137–1143.

Stack, F.M., Hennessy, M., Mulvihill, D. & O'Kennedy, B.T. (1995) Process for fractionation of whey constituents. World Patent Application No. 9534216-C1.

Tolkach, A. & Kulozik, U. (2005) Fractionation of whey proteins and caseinomacropeptide by means of enzymatic crosslinking and membrane separation techniques. *Journal of Food Engineering*, **67**, 13–20.

Tunick, M.H. (2008) Whey protein production and utilization : A brief history. *Whey Processing, Functionality and Health Benefits* (eds C.I. Onwulata and P.J. Huth), pp. 1–14. Wiley-Blackwell Publishing, Ames, IA.

Ulber, R., Plate, K., Weiss, T., Demmer, W., Buchholz, H. & Scheper, T. (2001) Downstream processing of bovine lactoferrin from sweet whey. *Acta Biotechnologica*, **21**, 27–34.

Wolman, F.J., Maglio, D.G., Grasselli, M. & Cascone, O. (2007) One-step lactoferrin purification from bovine whey and colostrums by affinity membrane chromatography. *Journal of Membrane Science*, **288**, 132–138.

Yee, K.W.K., Wiley, D.E. & Bao, J. (2007) Whey protein concentrate production by continuous ultrafiltration: Operability under constant operating conditions. *Journal of Membrane Science*, **290**, 125–137.

10 Concentrated Milk and Powders

G. Gésan-Guiziou

10.1 Introduction

As in whey processing, membrane operations are widely used for the manufacture of milk and milk protein concentrates (MPCs). The milk concentrates can be marketed either in liquid or in drying form but, in recent years, the production of liquid form has tended to decline whereas the production of powder has increased.

Initially, concentrated milks (full-fat and skimmed) with considerably reduced water content were considered as liquid milk preserves. They were used as a shelf-stable product, particularly in tropical regions, where milk production is not widely available, and were considered as cheap substitutes for fresh milk. Concentrated milks were mostly packaged in small metal cans, the content of which often being diluted with water before consumption to resemble plain milk. Currently, alternative products are used more often, such as whole milk powder or recombined milk. Some other forms of concentrated milks products have been developed that are, for instance, used as a pour-over accompaniment for canned fruit and other desserts. However, the consumption of concentrated milks, and in particular the consumption of sweetened condensed milk, has greatly declined in recent years.

Simultaneously, world production figures of dry dairy products have increased steadily over the past few decades because of the following main advantages of the powders (Schuck, 2011): (a) retention of high quality and no special storage conditions required; (b) reduction of mass and volume compared with fluid products; (c) adjustment of balance between milk supply and consumption; (d) supply of an irreplaceable food component in hot climate regions and valuable food reserve for emergencies; and (e) suitability for use in various tailor-made food products.

Regardless of the liquid or dried form of the concentrates, membrane operations are largely applied in the concentration of milk and milk proteins. Among the membrane operations widely applied during the manufacture of these products are ultrafiltration (UF), reverse osmosis (RO) to a less extent and microfiltration (MF).

10.2 Concentrated milks and powders

10.2.1 *Background*

The manufacturing stages of concentrated milks and powders are well established and, in general, the technologies have not seen significant changes in the last 25 year

Membrane Processing: Dairy and Beverage Applications, First Edition. A. Y. Tamime.

(Henning *et al*., 2006). However, concentrated milk products are defined as the liquid food obtained by partial removal of water from milk by evaporation and sometimes such processes may be combined with RO to remove water. The preservation of the concentrated milks can be achieved by sterilisation, and the product is known as 'evaporated' or 'condensed' milk. The approximate composition of some types of evaporated milk available on the market is shown in Table 10.1. In general, the fat and solids-not-fat (SNF) contents of evaporated milk are 7–9 and 18–22 g 100 g^{-1}, respectively. The preservation of evaporated milk can also be achieved by creating conditions restricting the growth of micro-organisms. This is generally achieved by addition of a large quantity of sugar, and the product is known as 'sweetened condensed milk'. Dulce de Leche (milk jam in Europe, milk caramel spread in the USA), which is a specific type of sweetened condensed milk that is very popular in some South American countries (Argentina, Uruguay, Brazil, Mexico), can be classified as a form of sweetened condensed milk prepared by boiling whole milk with added sucrose until 70 g 100 g^{-1} total solids is reached (Oliveira *et al*., 2009). The objective of this type of preservation is to decrease the water activity (a_w) of the products under the threshold of micro-organism development.

In the manufacture of milk powders, the concentration of milk is usually performed prior to drying, classically by evaporation and more rarely by membrane operations. Concentrating milk prior to drying is actually useful because it has a positive effect on milk powder quality: milk powder produced from concentrated milk consists of larger powder particles containing less occluded air, and therefore results in better storage stability (Schuck, 2011). For milk powders, the preservation of the products is ensured by dehydration, which is most frequently performed using roller drying or spray drying. The dehydration is carried out in order to decrease water activity down to the preservation criteria: $0.18 < a_w < 0.22$ for milk powders.

10.2.2 *Production of concentrated whole milk and powder*

Membrane processing of milk is rarely employed during the manufacture of whole milk, concentrated milk or powders. Fig. 10.1 outlines the manufacturing processes of sterilised concentrated milk and milk powder. The flow chart of milk concentrates and powder

Table 10.1 Statutory specifications of the chemical compositions (g 100 g^{-1}) and concentration factor of evaporated milk and sweetened condensed milk produced in different countries.

	Fat content	Milk solids-not fat	Concentration factor
Evaporated milk (American standard)	7.8	18.1	2.10
Evaporated milk (British standard)	9	22	2.60
Low-fat evaporated milk	4	20	2.25
Evaporated skimmed milk	0.1	22	2.35
Sweetened condensed milk (American standard)	8	20	4.60
Sweetened condensed milk (British standard)	9	22	5.00

Source: Adapted from Walstra *et al*. (2006).

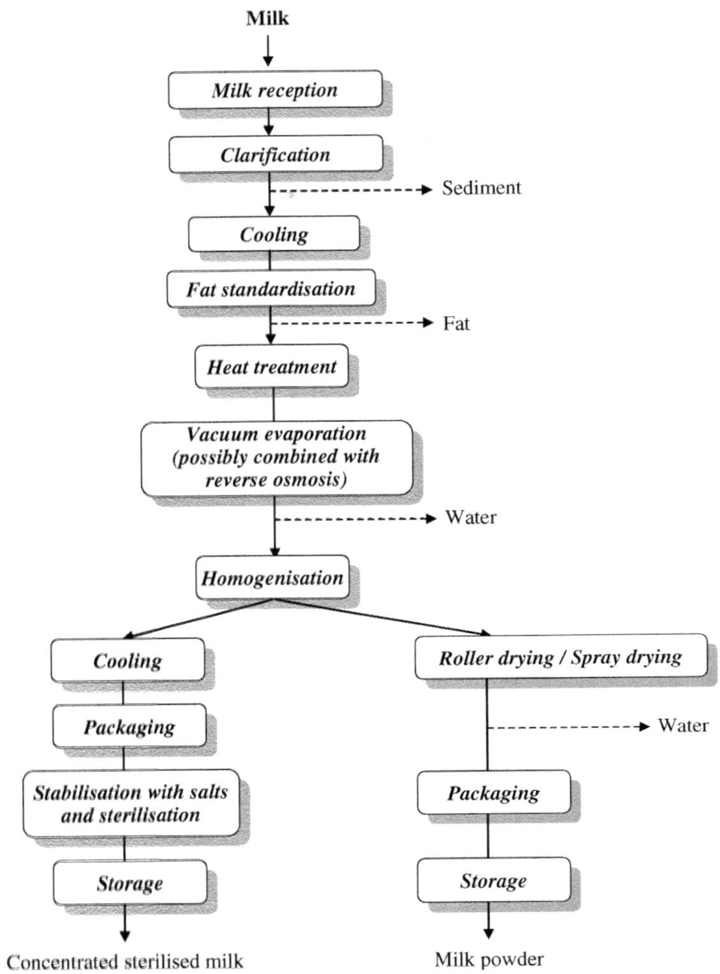

Fig. 10.1 Flow diagram showing the manufacturing stages of concentrated milk and milk powder.

production consists of milk reception, clarification, cooling, standardisation, heat treatment, concentration mainly by evaporation and homogenisation, followed by either drying for powder or sterilisation for concentrated milk. Raw full-fat milk used for powder production must be of high chemical, sensory and bacteriological quality and follow the standards approved by the government in many countries. At reception, milk is clarified, usually by centrifugal separators, cooled to 4°C in plate heat exchangers (PHEs) and stored at the same temperature. After standardisation, which is used to adjust the ratio of milk fat to total solids as required in the final product, milk is heated using tubular or PHEs. This heating serves to destroy pathogenic micro-organisms, inactivate enzymes (especially lipase) and inhibit oxidative changes (i.e. by activating the SH groups in β-lactoglobulin (β-Lg), which results in an antioxidative effect). Heating (i.e. temperature–time relationship) is determined by the specificity of the final product. For evaporated milk, a long heat treatment

(10–25 min) at a temperature below 100°C is often applied to enhance the heat stability of the milk, but currently ultra-high-temperature processing (UHT) is preferred. For sweetened condensed milk, UHT heating is also commonly applied (Walstra *et al*., 2006). However, for milk powder, heat treatment is commonly performed at 85–95°C for 15–30 s because the shelf-life of the products is extended by the decrease in the water activity by high dehydration level.

After the heat treatment stage, the milk is usually concentrated by evaporation using multiple-effect vacuum evaporators with mechanical and thermal steam recompression, where energy consumption is about 10–30 times lower than in spray drying (Schuck, 2011). In some plants, RO of milk is also used as a system to complement existing milk evaporators and increase the capacity by pre-concentration, as is more extensively done for whey processing. Indeed, an RO filtration system ahead of the evaporation stage is generally a less expensive solution to increasing the throughput of the dryers. RO is rarely applied to pre-concentrate the milk because the flux rate is small, especially with the spiral-wound membrane configuration that is generally used for such an application; however, the maximum concentration attainable is low. Owing to low heat stability, the concentration level of evaporated milk cannot be by more than about 2.6 times, which corresponds to a level of about 22 g 100 g^{-1} SNF (Table 10.1). The milk is, however, often concentrated to a higher level and is then restandardised to the correct dry-matter content during a further stabilisation step: for powder production, 30–35 and 45–50 g 100 g^{-1} total solids content in the evaporated milk to be dried using the roller or spray dryers, respectively.

After concentrating the milk solids, homogenisation is applied, which serves to prevent creaming and coalescence of the fat content in the concentrated milk. Homogenisation also reduces the 'free-fat' content in whole-fat milk powder, resulting in an increase in powder solubility and decrease in oxidation phenomena (Schuck, 2011).

10.2.3 *Production of concentrated skimmed milk and powder*

Membrane technologies are more often used in the manufacture of concentrated skimmed milk and powders than in the manufacture of similar products from whole milk. Because world production figures of skimmed milk powder are increasing every year more than whole milk powder production, membrane applications are expected to slightly grow in that sector. From 2007 to 2009, the production of whole milk powder has slightly decreased from 3.9 to 3.8 million tonnes (−3%), and simultaneously the production of skimmed milk powder has grown moderately from 3.3 up to 3.6 million of tonnes (+9%) (the production figures were obtained from the compilation of data issued by 50 selected countries, representing at least 90% of world production for butter, cow's milk cheese and milk powders, and 70% for liquid milk) (CNIEL, 2010).

The procedure for the manufacture of skimmed concentrated milk and powders differs in several features from similar products made from whole milk. First, standardisation of the fat content leads to very low fat levels in the skimmed milk (0.05–0.10 g 100 g^{-1}). The homogenisation stage is omitted, and the skimmed milk is concentrated to a higher solids content; however, concentration of skimmed milk using RO can be achieved to a maximum total solids of 30 g 100 g^{-1}. Moreover, the regime of skimmed milk heat treatment may be more intense than whole milk, and may be adapted according to the intended applications of the powder being produced; these aspects are known as 'heat specification'. The classification of skimmed milk powder according to this criterion is shown in Table 10.2. Low-heat milk powder, which is used in the production of cheese, baby foods, etc., is usually obtained by simple pasteurisation (i.e. 72°C for 15 s). High-heat milk powder requires additional heat treatment, such as 85–88°C for 15–30 min (Schuck, 2011). This intense heat treatment is generally applied in the production of skimmed milk powders used in the bakery industry and chocolate industry, and other foods where a high degree of protein denaturation (i.e. low whey protein nitrogen index (WPNI) – Table 10.2) is desired. Medium- and high-heat milk powders are used in cheese manufacture, but they lead to most of the problems encountered in cheesemaking. The cumulative effect of heat treatments applied for purposes of microbiological control and thermal efficiency during concentration and drying considerably affect the cheesemaking capabilities of the resulting powders. In that context, membrane technologies and more particularly MF offer two alternatives to improve the cheesemaking potential of skimmed milk concentrates and powders.

Microfiltration: 0.1 μm membrane

Twenty years ago a patented process using MF was proposed to lower the extent of formation of β-Lg/κ-casein complex during heat treatment, resulting in a better aggregation of the renneted casein micelles (Quiblier *et al*., 1992). During heat treatment, β-Lg actually forms complexes with the κ-casein surrounding the casein micelles. This association results in a lower accessibility of the coagulating enzyme (rennet) and in a reduction of the rennet coagulability of milk. MF of skimmed milk using a 0.1 μm mean pore diameter membrane was then used to separate in one single operation the milk into a retentate enriched specifically with casein micelles (~150 nm) and a permeate containing whey proteins (2–10 nm) and in particular β-Lg. The modified whey protein-depleted milk (e.g. MF retentate) was low or medium heat treated and possessed cheesemaking properties higher than that of low-heat milk powders. The process is illustrated in Fig. 10.2, and consists of several successive steps:

- partial or total removal of whey proteins and β-Lg from the skimmed milk using MF 0.1 μm diameter pore membrane

Table 10.2 Classification of skimmed milk powders according to heat treatment application during production.

Class	Whey protein nitrogen index (WPNI)[a]
Ultra-low heat	≥9.0
Low heat	≥6.0 to <9.0
Medium heat	≥4.5 to <6.0
Medium–high heat	≥1.5 to <4.5
High heat	≤1.5

Source: Adapted from Schuck (2011). Reproduced with permission of Elsevier.
[a]A low WPNI corresponds to a high degree of protein denaturation.

- UF of the microfiltrate using a 10–20 kg mol^{-1} molecular weight cut-off membrane
- blending of the microfiltered retentate with the ultrafiltrated skimmed milk in order to reinstate the original skimmed milk composition, except for whey protein depletion
- vacuum evaporation and spray drying (low or medium heat treatment) of the blend.

Mozzarella produced by recombination of the heat-treated new whey protein-depleted skimmed milk powder has the same composition and exhibits similar cheesemaking ability to that of raw milk. But when the whey protein-depleted skimmed milk was used, the cheesemaking yield increased up to 7.3 ± 1.8 % compared with the control cheese (Garem *et al*., 2000). The recovery of total solids, fat and total nitrogen contents showed similar trends.

Today, the MF 0.1 μm application is mainly conducted using ceramic membranes with either the uniform transmembrane pressure (UTP) system, developed by Alfa-Laval (currently known as Tetra Pak) (Bactocatch® system) (Sandblöm, 1974; Holm *et al*., 1986), or membranes with linear hydraulic resistance gradient (Garcera & Toujas, 1998; Skrzypek & Burger, 2010), as is also used for the removal of bacteria (Gésan-Guiziou, 2010). Separation is classically performed at 50–55°C, at high crossflow velocity (approximately 7 m s^{-1}),

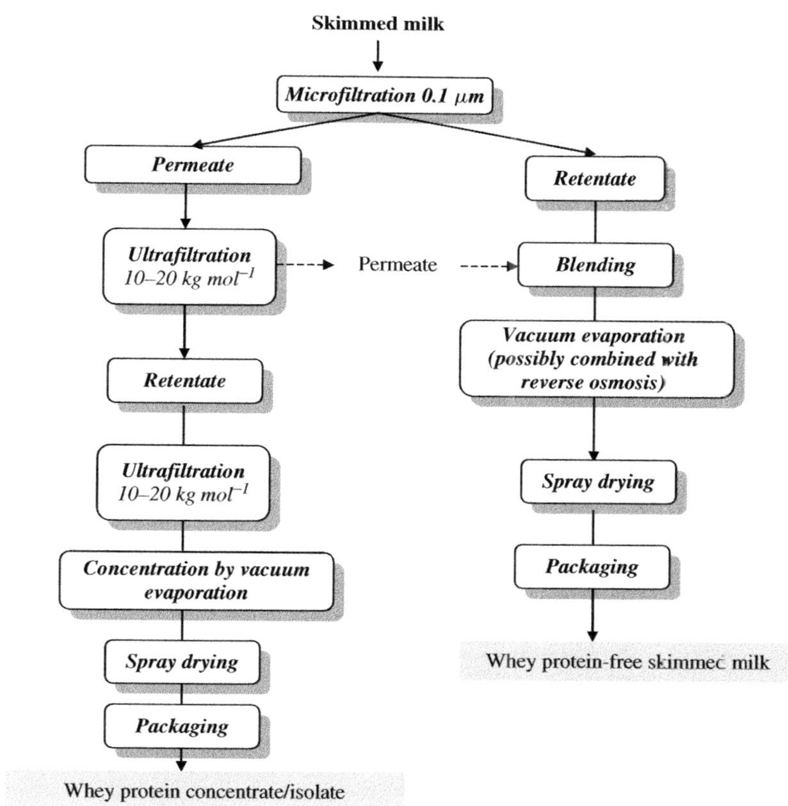

Fig. 10.2 Flow chart for the production of whey protein-free skimmed milk powder.

low transmembrane pressure (~50 kPa) and with a volume reduction ratio (VRR) in the range of 2–4. At VRR = 3, the transmission of whey proteins varies from 65% to 80% depending on the milk heat treatment (pasteurisation, thermisation or raw shimmed milk). Permeation flux is ~75–80 L h^{-1} m^{-2} for 10 h. The control of the membrane flux by controlling fouling is very important in this operation, and the operating conditions should be chosen according to the critical stability criterion in order to limit the accumulation of casein micelles at the membrane surface and perform long runs with very moderate fouling and high selectivity (Gésan-Guiziou *et al*. 1999; Jimenez-Lopez *et al*. 2008).

For the past few years and due to the high running costs and investments imposed by tubular ceramic equipment, the industry has started to operate MF with spiral-wound polymer membranes (Zulewska *et al*., 2009; Beckman *et al*., 2010). Contrary to ceramic tubular membranes, these membranes cannot be used with the UTP system, which leads to significant heterogeneity of transport along the membrane. Performance of the system is quite different from the ceramic ones: filtration is performed at low temperature (<10°C) to reduce the growth of bacteria, crossflow velocity is low (~0.5 m s^{-1}) and transmembrane pressure is relatively high (1.0–1.5 × 10^5 Pa). In these conditions, the permeation flux is reported to be very low (i.e. <10 L h^{-1} m^{-2}) and the protein transmission ranges from 20% to 30% at VRR = 3.

In this process (Fig. 10.2), the concentration of whey proteins is carried out by UF, which is mainly constructed of multi-stage spiral-wound systems with polyethersulphone membranes. It is currently operated at temperature either around 50°C or 10°C. The flux rate at 50°C is around twice as high as the flux rate at 10°C. Owing to the rapid decrease in membrane prices, a low temperature process (10–12°C) is now favoured because it limits bacterial growth in the filtration plants (i.e. retains high microbiological quality of the end product), and limits membrane fouling due to the precipitation of calcium phosphate.

Microfiltration: 1.4 μm membrane

MF using a 1.4 μm mean pore diameter also makes it possible to propose an alternative to heat treatment of raw milk and offer another opportunity to improve the properties of concentrated milks and powders. As well as the use for the drinking milk sector (Gésan-Guiziou, 2010), MF pre-treatment of skimmed milk using a 1.4 μm mean pore membrane can actually be expanded to all skimmed milk used for the production of milk powder. In such a process, MF treatment of raw skimmed milk is carried out prior to vacuum evaporation and spray drying, and leads to a high-quality milk powder (Fig. 10.3). MF is classically and specifically applied to skimmed milk applications, because the size ranges of fat globules and bacteria overlap. No heat treatment is required to obtain an ultra-low-heat powder with a maximum bacterial count of 3 × 10^3 colony forming units (cfu) g^{-1} (Table 10.2). Thus, such powder has the same renneting time after reconstitution as the original raw milk, and can be used as a reference powder for either industrial or scientific purposes.

This separation is classically performed with a multi-channel ceramic membrane with a pore diameter of 1.4 μm in order to retain bacteria and let the proteins (both casein micelles and whey proteins) pass through the membrane (see Chapter 6, Section 6.4). The temperature of the separation ranges from 50°C to 55°C. In order to achieve good separation, high

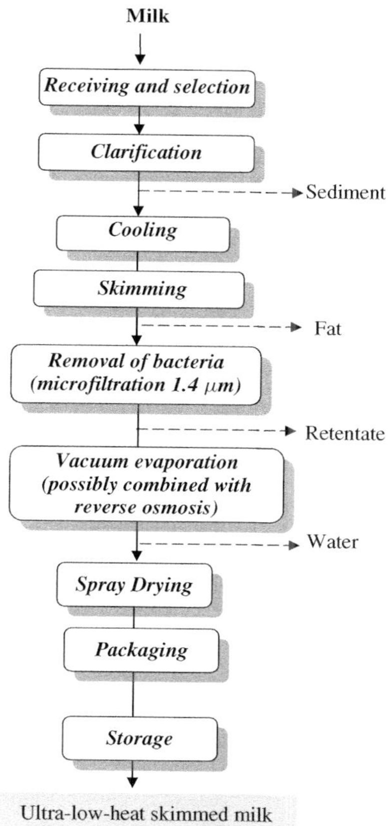

Fig. 10.3 Manufacture of ultra-low-heat skimmed milk powder.

crossflow velocity in combination with subcritical conditions and uniform transmembrane pressure are required. Thus in order to overcome membrane fouling phenomena, most MF plants in the dairies operate using either the hydraulic concept of uniform transmembrane pressure (UTP) or a specific ceramic membrane with linear hydraulic resistance.

10.2.4 *Applications of reverse osmosis concentrated milks*

RO is widely used for processing whey, but rarely applied to the concentration of milk; the flux is rather low, the maximum concentration attainable is low, and this technology is not as attractive to the manufacturers of cheese and yoghurt as UF.

Traditionally, cheese makers used to fortify the solids content of the base milk by adding milk powder to improve the body characteristic of the product and increase the yield but, at present, RO milk is used to replace the addition of powder. The latter approach was proposed for the manufacture of Cheddar and cottage cheeses (Agbevavi *et al*., 1983; Barbano & Bynum, 1984; Bynum & Barbano, 1985; Schmidt *et al*., 1986) and several advantages were claimed. The use of RO retentate increases the solids content of the milk base, enhances the rate of equipment utilisation and greater plant throughput based on existing production lines. For Cheddar cheesemaking, RO retentate was proposed for pre-concentrating the whole milk to 20–25 g 100 g^{-1} solids, which leads to a similar gross chemical composition of the resulting cheese compared with that of cheese from unconcentrated milk. The amount of starter and rennet required are reduced by 50% and 60%, respectively, and with a 20% milk volume reduction (i.e. removal of water) by RO, a 2–3% increase in cheese yield can be expected. For cottage cheese, 8% of the skimmed milk volume reduction leads to a 5% increase in cheese yield. Such yield increases result from the entrapment of concentrated whey proteins within the casein network.

However, RO also results in several drawbacks. In the manufacture of Cheddar cheese, fat losses in the whey are shown to increase with increasing solids concentration due to a partial homogenisation effect during concentration of the milk. A sudden release of pressure during RO also induces lipolysis in the milk and cheese, and a concentration of 1.8 has then been shown to be optimal (Barbano *et al*., 1983). Excess minerals and lactose in the curd considerably modify cheese manufacture. Minerals are concentrated in the cheese milk and, therefore, the same consequences for cheese quality are observed as described for UF retentates; for example, increase in the buffer property capacity of the milk, differences in acidification rate of the starter culture, and lipolysis and proteolysis kinetics in cheese during the maturation period (Mistry & Maubois, 2004). The higher retention of residual lactose in the cheese may lead to a resumption of lactic acid production by the starter organisms after several days in the ripening room. This is almost detrimental to organoleptic qualities of the product. To minimise such drawbacks, several plants in the United States of America (USA) have already used thermal evaporators to pre-concentrate cheese milk (Sandfort, 1983; Honer, 1984), but they remove the excess lactose and minerals in the curd by washing with water. The benefits of the overall process must be examined closely.

RO of skimmed milk is also used in the ice-cream industry for increasing the SNF content and removal of water. This allows an ice-cream manufacturer to source whole milk for cream harvesting and then make concentrated skimmed milk for in-house use. The ability to have a direct source of cream and concentrated skimmed milk are ingredient cost savings compared with buying value-added cream and concentrated skimmed milk from the marketplace. An additional advantage is that RO filtration is a cold concentration technique. The ice-cream maker has a better dairy flavour from RO milk than from evaporated milk that is produced by adding heat to induce a phase change and evaporate the water from the milk. Evaporated milk is often associated with a cooked flavour, whereby RO skimmed milk imparts a fresh milk flavour in the ice-cream.

For the production of yoghurt and fermented products, such as Koumiss, RO has been used for the enrichment of the SNF in the milk base (Tamime & Robinson, 2007). For organic yoghurt manufacturers who cannot often find a supply of organic powder for increasing the milk solids, RO can be advantageous because it makes it possible to take fresh organic skimmed milk and concentrate the solids content to the desired level for yoghurt making. Because RO concentrates the total solids of milk, the operation can be compared with evaporation. Using RO of skimmed milk allows for less heat damage or denaturation of the whey proteins found in evaporated milk. Products obtained with RO are generally considered as better in terms of texture and flavour than those made from milk with added powder. Such results probably originated from a drastic reduction of the

Maillard reaction initiated in milk powders and the absence of insoluble particles that are more or less present even in high-quality powders.

So far, RO has not been used extensively in the production of fermented milk products, but the number of production plants and the competitiveness and reliability of RO modules is increasing. Today, UF is used in many fermented milk plants around the world, and this technology is often preferred to RO filtration because of the reduced level of lactose in the milk base. In UF, selective concentration takes place, i.e. the fat and proteins are concentrated whereas the other solids contents of the milk remain almost unchanged. The main advantage of UF is that the protein content can be substantially increased without significant changes to the lactose content. The protein level is normally concentrated to 3.8–4.7 g 100 g^{-1}. UF results in a better yoghurt gel strength than those obtained with the addition of skimmed milk powders or concentrating the milk base by vacuum evaporation. Compared with cheesemaking, the increase in protein level is not accompanied in the manufacture of yoghurt by an improvement in yield, because some lactose and other soluble components present in the milk are lost in the permeate.

10.2.5 *Dulce de Leche*

As mentioned earlier, Dulce de Leche (milk jam, milk caramel) is a form of sweetened condensed milk that is very popular in some South American countries, such as Argentina, Uruguay, Brazil and Mexico. There are two types of Dulce de Leche product: first, for household use, which is consumed as a spread or a dessert and, second, for confectionery use, which has a higher viscosity. The industrial manufacturing stages of these products are similar to that of sweetened condensed milk, where the milk solids are concentrated by heating and evaporation. In the particular case of Dulce de Leche, sucrose is added to the milk prior to the concentration until 70% total solids are reached.

The most technological problem in Dulce de Leche production is the sandy texture of the product, which, as a consequence, has a negative impact in reducing product acceptability. This phenomenon is mainly attributed to the high concentration of lactose in the product leading to crystallization; the lactose crystals can be up to 1500 μm in size. Much effort has been made to resolve this problem by breaking down lactose using bacteria or enzymes, by seeding with lactose micro-crystals or by using UF in order to reduce the processing time and lowering the final lactose content of the milk (Oliveira *et al*., 2009) Using the last process to treat the milk, an ultrafiltered concentrated product with a reduced concentration of lactose can be obtained. Thus, the UF process may be adapted to produce a low-lactose type of Dulce de Leche and, by reducing the heat processing time, the production costs may be lowered. The major advantage of the UF process is that it yields higher protein and lower lactose with excellent nutritional and functional properties (Lee & White, 1991).

In the past, Christiansen *et al*. (1987) and Edelsten *et al*. (1987) investigated the use of UF processing for the production of Dulce de Leche in which sandiness was prevented, but no detailed technical information was given. More recently, Oliveira *et al*. (2009) studied the impact of UF (pre-concentration of the milk base to 55 g 100^{-1} total solids) of the milk prior the manufacture of Dulce de Leche. A higher VRR would have required a modification of the formulation due to the loss of chemical components (lactose, minerals and some non-protein nitrogen) in the permeate during the UF process. Ultrafiltered milk was shown to be suitable for Dulce de Leche manufacture, but more research work is still required to confirm the pilot-scale studies are suitable for industrial-scale operation(s). UF retentates offer several advantages, and are briefly described as follows:

- it decreases energy consumption of the process by decreasing the time taken by evaporation
- it prevents the formation of large lactose crystals which cause sandiness and, therefore, improves consumer acceptability in the absence of a grainy texture
- it improves also the organoleptic properties of the product even after being in storage for some months and possesses a firmer texture than the control.

10.3 Milk protein concentrates

Contrary to concentrated milk and powders, MPCs are mainly produced using membrane technology and, in particular, using UF, and they owe their origin to the technological development of UF in the dairy industry. As their names suggest, MPCs contain both casein and whey proteins at a ratio that exists in raw milk. UF increases the milk protein content above the typical protein level in dry matter content of skimmed milk powder (~35 g 100 g^{-1}) by concentrating all the milk proteins (serum proteins and casein micelles) and fat (if present), and simultaneously allowing the lactose and minerals to pass through the membrane. MPCs could then be loosely described as protein-enriched forms of skimmed milk. Since there are no specific standards of identity in any country for MPCs, they are available in a wide range of protein levels from 42 to 85 g 100 g^{-1}, and the lactose level falls as the protein level increases in the product (Table 10.3). In Table 10.3 hyphenated double-digit figure accompanying the MPC denotes the protein content of the product, expressed as a percentage of protein in dry matter. MPCs with protein contents of $\geq$90 g 100 g^{-1} are frequently described as milk protein isolates.

10.3.1 *Manufacture of milk protein concentrates*

MPCs, which are usually processed and marketed in spray-dried form, are produced from skimmed milk by using UF and/or diafiltration, evaporation and drying technology. For the MPC production, the use of high-quality raw milk (low total and spore counts) is essential

Table 10.3 Typical chemical composition (g 100 g^{-1}) of milk protein concentrates (MPCs).

Product	Protein	Fat	Lactose	Minerals	Moisture
MPC-42	42	1.0	46.0	7.5	3.5
MPC-70	70	1.4	16.2	8.2	4.2
MPC-75	75	1.5	10.9	7.6	5.0
MPC-80	80	1.8	4.1	7.4	3.9
MPC-85	85	1.6	1.0	7.1	4.9

Source: Adapted from Kelly (2011). Reproduced with permission of Elsevier.

because only relatively modest levels of heat treatment are used, particularly during the early stages of the process (milk pre-heating, cream separation, milk pasteurisation) and that residual bacteria and spores are retained by UF and concentrated in the retentate.

The manufacture of MPCs typically includes UF of skimmed milk with or without the use of diafiltration. The diafiltration makes it possible to reduce the level of lactose in the retentate, and then to achieve higher protein content than 65 g 100 g^{-1} in the final powder. The resulting retentate, depending on the desired protein level, is then subjected to pre-heat treatment. Since the viscosity of the retentate increases as proteins are selectively enriched, the capacity of UF plants to maximise solids concentration is affected by a rapid decline in permeate flux across the UF membranes. Additional thermal evaporation is then necessary in order to increase retentate solids further so that the resulting spray-dried powder has the desired physical characteristics, such as bulk density. High concentration is carried out by evaporation in a one- or two-stage falling-film tubular evaporator, and spray drying, preferably in a tall-form spray dryer, followed by 'after-drying' and cooling in an external fluid bed drier (Kelly, 2011).

Regardless of the applications and the level of protein concentration, the membranes used for the production of MPCs are normally of spiral-wound configurations and, to a lesser extent, tubular ceramic membranes; both with a molecular weight cut-off of 10–50 kg mol^{-1}, and operated at a transmembrane pressure of 200–400 kPa with a permeation flux of 30–120 L h^{-1} m^{-2}. Tubular membranes and plate and frame modules may be used, especially for the last loops in UF plants when high viscous concentrates are treated. UF processing at an industrial level may be undertaken at either ~50°C or ~10°C, but the lower temperature is now generally preferred. At high temperature filtration, permeation fluxes are higher, but this temperature favours both mineral precipitation of calcium phosphate in the membrane pores and bacterial growth in the final retentate. At low-temperature (9–12°C) filtration, no or little bacterial growth is observed for process duration of 10–12 h. Moreover, with organic spiral-wound membranes, no calcium phosphate precipitation in the membrane is observed, and fouling is mainly due to the adsorption and accumulation of proteins at the membrane surface, which could greatly simplify cleaning procedures of the membrane plants.

Concentrates with a high content of casein fractions have also been developed using MF with a 0.1 μm mean pore diameter membrane. The MF membrane separates casein micelles from whey proteins. The casein content of such powder is ~90 g 100 g^{-1}, and forms almost 96 g 100 g^{-1} of total proteins (Saboya & Maubois, 2000). This powder is mainly used in cheesemaking, and is known to increase the cheese yield. As we have already discussed, such powders also provide flexibility in usage regardless of the extent of heat treatment because the κ-casein/β-Lg complex does not occur because of the removal of the β-Lg during MF.

10.3.2 *Applications of milk protein concentrates*

The establishment of the unique functional and nutritional properties of MPCs has expanded the markets for high-protein dairy ingredients, which were traditionally dominated by casein and caseinates. For nearly 30 years, MPCs have been increasingly used as milk protein ingredients in recombined dairy products, cheeses and health-related products.

The variability in the functionality (solubility, viscosity, foaming and emulsification properties, etc.) of MPCs on the market is wide. This is attributed both to the various compositions and changes in the mineral-to-protein ratio of the available MPC ingredients, and to the processing conditions used for their manufacture. Indeed, concentration and drying considerably affect the quality of milk powders to varying degrees, mainly because of the exposure of highly concentrated constituents to the prevailing thermal conditions. Recent spray-drying approaches have evolved to minimise some negative effects through the use of multi-stage dryers and reduction of residence time in the primary drying zone. Some applications of the MPCs will be briefly discussed.

Cheesemaking

The dried or liquid forms of MPCs can be used to supplement the milk constituents for cheesemaking. Since the beginning of the 1980s, numerous cheese varieties have been made from medium-concentrated retentates obtained with UF in what is largely known as the MMV process after the investigators Maubois, Mocquot and Vassal who originally developed it in 1969 (Maubois, 1981). The MMV process using UF technology is completely different from the traditional cheesemaking process (Fig. 10.4). The general idea behind the MMV process is to concentrate all the milk proteins (casein and whey proteins) and simultaneously to remove the excess water, some of the mineral ions and lactose by an initial UF step carried out at the neutral pH of the milk, prior to coagulation (induced by addition of enzymes (rennet) and micro-organisms). Thus, UF results in the reduction or elimination of the need to separate the whey from the curd (Fig. 10.4).

In the 1980s, the driving force behind the development of membrane filtration was based on cheese production where the desired main objective was to obtain a continuous process

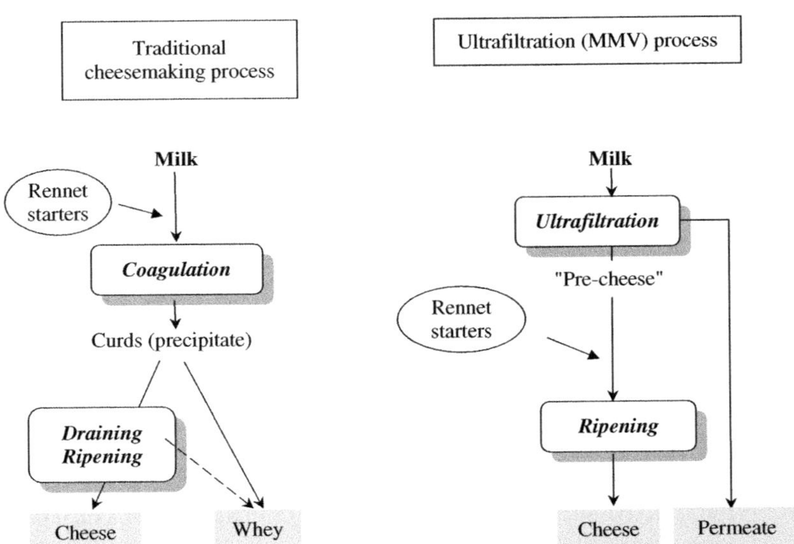

Fig. 10.4 Schematic representations of traditional ways of making cheese and the Maubois, Mocquot and Vassal (MMV) process (with concentration to the final protein composition of the cheese).

with increased yield compared with traditional cheesemaking processes. The advantages of the MMV process were numerous (increase in yield of about 10–30% compared with the traditional process; reduction of enzyme usage; savings in capital investments and energy costs, etc.), but the increase in whey proteins and mineral contents in the concentrated milk gave rise to textural defects accompanied by bitter flavours during the maturation period. As a consequence, the cheese manufacturing industry took a conservative view and continued to rely on traditional processes.

UF milk concentrates became firmly established in the market in the form of dried MPCs with varying protein contents. Increasingly, these MPCs are finding their way back into cheesemaking for the purpose of protein standardisation of cheese milk. The dried and liquid products can actually be used to supplement milk constituents for cheesemaking, where such an approach is the main application of MPCs. For example, in experiments with Gouda cheese, it was shown that fortification of milk with 1 g 100 g^{-1} MPC increased cheese yield (Mistry & Pulgar, 1996). This was attributed to reduced losses of casein and better retention of whey proteins.

In the meantime, enthusiastic research was developed to find solutions to the problematic early applications of UF in cheesemaking, and MPCs became part of an innovative processes for the manufacture of ultrafiltered cheeses. MPCs introduced new possibilities in cheesemaking, as demonstrated by numerous researchers claiming innovations and the many patents filed (Bhaskar *et al*., 2001; Moran *et al*., 2001). Moran *et al* (2001), for instance, developed a continuous method for the manufacture of cheese in which acidified milk is ultrafiltered and diafiltered to a concentration factor of four- to seven-fold, followed by evaporation up to 70 g 100 g^{-1} total solids to give reduced calcium in the pre-cheese. Processed cheese is made from this pre-cheese in the traditional manner using flavouring agents and emulsifying salts. No cheese base was required but, instead, high protein concentrates were used as the base material along with flavouring agents.

MPCs may also be used to modify cheese functionality by adapting the MPC processing conditions. For example, partial depletion of β-casein using cold MF of the milk results in less firm cheese and higher meltability and flow.

Yoghurt and related fermented milk products

MPCs (UF, VRR ~1.5) are also used in the manufacture of yoghurt and related fermented products in order to increase the protein levels without addition of milk powders (Tamime & Robinson, 2007). UF results in a better yoghurt gel strength than those obtained with addition of milk powders or concentration by vacuum evaporation. MPCs may be used to replace concentrated skimmed milk, skimmed milk powder and whey protein concentrate in low-fat set-type yoghurt milks with comparable viscosity and syneresis index (Kelly, 2011). UF is also preferred to RO because of the reduced level of lactose in the milk base.

Food emulsions

The native aggregated structure of casein micelles, which is maintained by UF and, therefore found in MPCs, results in a more stable emulsion to creaming than those made with caseinate. It is believed that the protein aggregates in MPCs may be too large to cause significant depletion flocculation, or alternatively that they trigger a high degree of depletion flocculation and thus form very viscous emulsions which reduce creaming.

Miscellaneous products

Various other applications of MPCs are being developed. The high protein, low lactose ratio makes MPCs suitable for protein-fortified beverages and low-carbohydrate foods. The whey protein is usually still in the native form, but can easily be denatured prior to processing if required. Replacing the protein content of ice-cream mix with MPC (up to 50%) results in a: (a) higher viscosity of the mix base; (b) larger amount of fat destabilisation; (c) narrower ice melting curves; and (d) better shape retention. MPC addition may also be used to improve the keeping quality of coffee cream and whipping cream.

10.4 Conclusion and future trends

Membrane technologies have, during more than 30 years, opened new avenues for improving traditional cheesemaking procedures and developing new ingredients. The manufacture of concentrated milk protein and powders represents one of the first applications for membranes in the dairy sector. Although membrane technologies are rarely applied today for concentrating milk, both in liquid or in dried form, MF using either a 0.1 μm or a 1.4 μm mean pore diameter membrane is expected to grow. The best understanding of the limiting phenomena occurring during MF operation conditions make this technology more and more attractive. The MF operation may give rise to new generations of milk powders, especially suitable for cheesemaking in countries where there is a shortage of milk (MF 0.1 μm), or with ultra-low-heat treatment (MF 1.4 μm) and offers another opportunity to improve the properties of concentrated milk and powders.

In contrast, UF have been largely used for the production of MPCs. Because of difficulties encountered 30 years ago, for example in making cheeses with satisfactory organoleptic qualities by membrane operations, an impressive amount of research and knowledge has been acquired in various fields of dairy science, such as physico-chemistry, microbiology, process engineering). The future of the use of membrane technologies is very promising in that domain. Many new cheese varieties may be prepared/developed taking into account the properties of UF retentates and microbial/enzymatic activities of starter cultures and rennet. Numerous ideas for technologists have also originated from filtration research, but the future development of these new applications will depend on the added value associated with new nutritional and/or functional properties of the concentrates.

References

Agbevavi, T., Rouleau, D. & Mayer, R. (1983) Production and quality of Cheddar cheese manufactured from whole milk concentrated by reverse osmosis, *Journal of Food Science*, **48**, 642–643.

Barbano, D.M. & Bynum, D.G. (1984) Whole milk reverse osmosis retentates for Cheddar cheese manufacture: cheese composition and yield. *Journal of Dairy Science*, **67**, 2839–2849.

Barbano, D.M., Bynum, D.G. & Senyk, G.F. (1983) Influence of reverse osmosis on milk lipolysis. *Journal of Dairy Science*, **66**, 2447–2451.

Bhaskar, G.V., Singh, H. & Blazey, N.D. (2001) Milk protein products and processes, World Patent Application WO 01/41578 A1.

Beckman, S.L., Zulewska, J. Newbold, M. & Barbano, D.M. (2010) Production efficiency of micellar casein concentrate using polymeric spiral-wound microfiltration membranes. *Journal of Dairy Science*, **93**, 4506–4517.

Bynum, D.G. & Barbano, D.M. (1985) Whole milk reverse osmosis retentates for Cheddar cheese manufacture: chemical changes during aging. *Journal of Dairy Science*, **66**, 415–421.

Christiansen, P.S.D., Edelstel, J.R. Kristiansen, J. & Siegaard, P. (1987) Production of low-lactose dulce de leche. *Dairy Science Abstract*, **49**, 482.

CNIEL (2010) *L'économie Laitière en Chiffres*, CNIEL, Paris.

Edelsten, D., Christiansen, P.S., Meersohn, M & Jensen, P.F. (1987) Production of low-lactose dulce de leche from UF milk. *Dairy Science Abstract*, **49**, 398.

Garcera D. & Toujas E. (1998) Macroporous support with permeability gradient and it manufacturing process, European Patent Application EP 0870534A1 FR 9704359 1–11.

Garem, A., Schuck, P. & Maubois, J.-L. (2000) Cheesemaking properties of a new dairy-based powder made by a combination of microfiltration and ultrafiltration. *Lait*, **80**, 25–32.

Gésan-Guiziou, G. (2010) Removal of bacteria, spores and somatic cells by centrifugation and microfiltration techniques. *Improving the Safety and Quality of Milk* (ed. M.W. Griffiths), Vol. 1, pp. 349–372, Woodhead Publishing, Cambridge.

Gésan-Guiziou, G., Boyaval E., & Daufin G. (1999) Critical stability conditions in crossflow microfiltration of skimmed milk: transition to irreversible deposition. *Journal of Membrane Science*, **158**, 211–222.

Henning, D.R., Baer, R.J., Hassan A.N. & Dave R. (2006) Major advances in concentrated and dry milk products, cheese, and milk fat-based spreads. *Journal of Dairy Science*, **89**, 1179–1188.

Holm S., Malmberg R. & Svensson K. (1986) Method and plant for producing milk with a low bacteria content, World Patent Application WO 86/01687.

Honer, C. (1984) Preconcentrating milk for cheesemaking. *Dairy Records*, **85**, 68–70.

Jimenez-Lopez, A.J.E., Leconte N., Dehainault O., Geneste C., Fromont, L. & Gésan-Guiziou, G. (2008) Role of milk constituents on critical conditions and deposit structure in skimmilk microfiltration (0.1 μm). *Separation Purification and Technoogy*, **61**, 33–43.

Kelly, P. (2011) Milk protein products. *Encyclopedia of Dairy Science* (eds. J.W. Fuquay, P.F. Fox and P.L.H. McSweeney), Vol. 3, 2nd edn., pp. 848–854, Academic Press, San Diego.

Lee, R. & White, C.H. (1991) Effect of UF retentates and whey proteins concentrates on ice cream quality during storage. *Journal of Dairy Science*, **74**, 1170–1180.

Maubois, J.-L. (1981) Perspectives d'utilisation des techniques à membranes dans les industries agro-alimentaires. Académie d'Agriculture de France, *Procès-verbal* 1980, **26 nov**, 1451–1461.

Mistry, V. V. & Maubois, J.-L. (2004) Application of membrane separation technology to cheese production. *Cheese Chemistry, Physics and Microbiology* (eds. P.F. Fox, P.L.H. McSweeney, T.M. Cogan and T.P. Guinee), Vol. 1, pp. 261–285, Elsevier, London.

Mistry, V. V. & Pulgar, J.B. (1996) use of high milk protein powder in the manufacture of Gouda cheese. *International Dairy Journal*, **6**, 205–216.

Moran, J.W., Trecker, G.W. & Monckton, S.P. (2001) Continuous manufacture of process cheese, United States of America Patent Application No. 6183805.

Oliveira, M.N., Penna, A.L.B. & Garcia Nevarez H. (2009) Production of evaporated milk, sweetened condensed milk and 'Dulce de Leche'. *Dairy Powders and Concentrated Products* (ed. A.Y. Tamime), pp. 149–179, Wiley-Blackwell, Oxford.

Quiblier, J. P., Ferron-Baumy, C., Garric, G., & Maubois, J. L. (1992) Procédé de traitement des laits permettant au moins de conserver leur aptitude fromagère, European Patent Application EP 0 542 583 B2.

Saboya, L. V. & Maubois, J. L. (2000) Current developments of microfiltration technology in the dairy industry. *Lait*, **80**, 541–553.

Sandblöm R.M. (Alfa-Laval) (1974) Filtering process, Swedish Patent Application No. 7 416 257.

Sandfort, P. (1983) Preconcentrated cheese-milk by evaporation. *Dairy Records*, **84**, 16–17.

Schmidt, D., Fedrick, I.A. & Donovan, H.M. (1986) Quality and yield of Cheddar cheese manufactured from reconstituted reverse osmosis milk concentrates. *New Zealand Journal of Dairy Science and Technology*, **21**, 125–131.

Schuck, P. (2011) Dehydrated dairy products. Milk powder: types and manufacture. *Encyclopedia of Dairy Science* (eds. J.W. Fuquay, P.F. Fox & P.L.H. McSweeney), Vol. 2, 2nd edn, pp. 108–116, Academic Press, San Diego.

Skrzypek, M. & Burger, M. (2010) Isoflux® ceramic membranes – Practical experiences in dairy industry. *Desalination* **250**:1095–1100.

Tamime, A.Y. & Robinson, R.K. (2007) *Tamime and Robinson's Yoghurt: Science and Technology*, 3rd edition pp. 13–161, Woodhead Publishing, Cambridge.

Walstra, P., Wouters, J.T.M. & Geurts, T.J. (2006) Concentrates milks. *Dairy Science and Technology*, pp. 497–512, CRC Press (Taylor & Francis group), Boca Raton, FL.

Zulewska, J., M. Newbold, and D. M. Barbano (2009) Efficiency of serum protein removal from skim milk with ceramic and polymeric membranes at 50°C. *Journal of Dairy Science*, **92**,1361–1377.

11 Further Applications of Membrane Filtration in Dairy Processing

J.A. O'Mahony and J.J. Tuohy

11.1 Introduction

The dairy industry can be described as an 'early adopter' of membrane filtration technology, and today the production processes of whey protein or milk protein concentrate ingredients by ultrafiltration (UF), fat and bacteria removal from whey or skimmed milk by microfiltration (MF) and pre-concentration of skimmed milk or whey by reverse osmosis (RO) or nanofiltration (NF) are often regarded as mature technologies in the industry (Pouliot, 2008). Of course, there remain many operational and efficiency challenges with these unit operations, particularly in relation to membrane fouling, cleaning and process efficiency parameters, such as yield, purity and energy consumption (Brans *et al.*, 2004). A further challenge is finding useful and profitable outlets for the by-product streams from these processes, such as UF and NF permeates, particularly those with high ash contents, or the fat-containing MF retentate from the whey clarification step in the production of whey protein isolates.

Previous chapters have covered in detail the main membrane unit operations and applications in dairy processing. This chapter will, however, concentrate on the opportunities for other applications of membrane filtration in the dairy industry with the possibility of developing new high-value ingredients targeted for specific end uses in food, nutraceuticals, cosmoceuticals and biopharmaceuticals. Applications to be considered are based on the major milk constituents, i.e. proteins, fat, carbohydrates and salts.

The isolation and purification of native (so-called micellar or phospho-) casein is of interest because of the unique physico-chemical properties of casein micelles, and native casein would be ideal for standardising the casein level of milk in cheese or yoghurt manufacture (Govindasamy-Lucey *et al.*, 2007). The separation of the casein and non-casein protein groups into fractions enriched in specific individual protein components (e.g. β-casein, α-lactabumin (α-La) or immunoglobulin-Ig), which could be incorporated into products for target populations, such as infants, the elderly or the immunocompromised population groups, is of considerable commercial interest. Membrane filtration technology has applications for the purification of hydrolysed casein or whey proteins and fractionation into enriched peptides and amino acid products (Pouliot, 2008). These purified products containing, in specific peptides of physiological interest have promising applications in sports nutrition, and in foods for those susceptible to protein allergies, those with specific dietary requirements (e.g. phenylketonuria) and potentially in bioactive therapy applications

Membrane Processing: Dairy and Beverage Applications, First Edition. A. Y. Tamime.

(Fitzgerald & Meisel, 2003). The isolation and purification of osteopontin and human growth factors could open up opportunities in infant nutritional products and possibly in the field of therapeutics.

Membrane applications in the dairy industry are not limited to protein materials. Native milk fat globule fractions have been shown to influence the texture and rheological properties of butter, low-fat cheese and acid milk protein, gels and here membrane technology offers a natural solution to texture (and possibly flavour) modification in a range of dairy products. Milk fat globule membrane (MFGM) materials including phospholipids and sphingolipids have desirable emulsifying and biological properties as food ingredients (Corredig & Dalgleish, 1997), and the supplementation of milk and fresh dairy products with long-chain unsaturated fatty acids is at the commercialisation stage.

Cow's milk is a potential source of naturally occurring oligosaccharide ingredients for inclusion in infant milk formulae (Kunz *et al*., 2000). Membrane technology is core to the purification and concentration of these carbohydrates which are present at very low concentrations accompanied by high levels of lactose in cow's milk. Consumer demand for lactose-reduced and lactose-free milk and milk products has grown significantly in recent years alongside the growth of milk consumption in regions where lactose intolerance is prevalent in the population. In-package enzymatic lactose hydrolysis, whereby lactase can be added post pasteurisation or post ultra-high temperature (UHT) treatment using filter sterilisation is a good example of the innovative use of membrane technology for new product development in a high-volume processing environment. The recovery, concentration and reuse of lactic acid is another potential high-volume application for membrane technology in dairy processing.

Whereas milk itself is an excellent source of bio-available mineral salts important in human nutrition, direct acidification in the production of acid casein and concentration steps in other processes results in whey streams with high ash contents, which limit their applications as food ingredients (Jelen, 2009). Membrane processes, such as NF and electrodialysis (ED), offer the industry the means to selectively demineralise whey and whey ingredients.

11.2 Fractionation of milk proteins using membranes

11.2.1 *Separation of casein and whey proteins in milk*

Cow's milk has an average protein content of $3.5\,g\,100\,mL^{-1}$ (Fox & McSweeney, 1998). The two major protein fractions in bovine milk are the caseins and whey proteins, which account for approximately 80 and $20\,g\,100\,g^{-1}$ of total protein, respectively (Walstra & Jenness, 1984). Many strategies have been used for separation of casein from whey protein in milk, such as rennet coagulation (Mulvihill & Ennis, 2003), acid coagulation (Mulvihill & Ennis, 2003), precipitation of casein with ethanol (Hewedi *et al*., 1985), precipitation of casein with an anionic polysaccharide (Molochnikov & Zetteir, 1994), ED to convert the casein to a metastable state under which it can be separated by centrifugation (Morel, 1991; Noel, 1992) and precipitation of casein using carbon dioxide under pressure (Tomasula *et al*., 1995; 1997). Commercially, mainly for economic reasons, casein is normally separated from whey proteins by acid or rennet coagulation (Mulvihill & Fox, 1994; Mulvihill & Ennis, 2003).

Over the past 15 years, membrane technology (principally MF) has also been investigated for the separation of casein from whey protein in milk. Numerous studies have been conducted on the production of native micellar casein (also termed phosphocaseinate or micellar casein concentrate) using inorganic (almost invariably ceramic) and polymeric (generally spiral-wound polyvinylidene) MF membranes (Fauquant *et al.*, 1988; Pierre *et al.*, 1992; Schuck *et al.*, 1994; Rosenberg, 1995; Singh *et al.*, 1995; Pouliot *et al.*, 1996; Dinkov *et al.*, 1997; Maubois, 1997, 2002; Brandsma & Rizvi, 1999; Kelly *et al.*, 2000; Rizvi & Brandsma, 2002; Papadatos *et al.*, 2003; Nelson & Barbano, 2005; Lawrence *et al.*, 2008; Marcelo & Rizvi, 2008; Beckman *et al.*, 2010). Examples of some common membrane constructions can be seen in Fig. 11.1.

MF membranes, because of high permeation fluxes, are traditionally prone to rapid fouling (Horton, 1997). The application of MF membrane separation technology to the removal of whey proteins from casein in skimmed milk only became possible with the advent (by Alfa Laval – currently known as Tetra Pak) of uniform transmembrane pressure (UTMP)

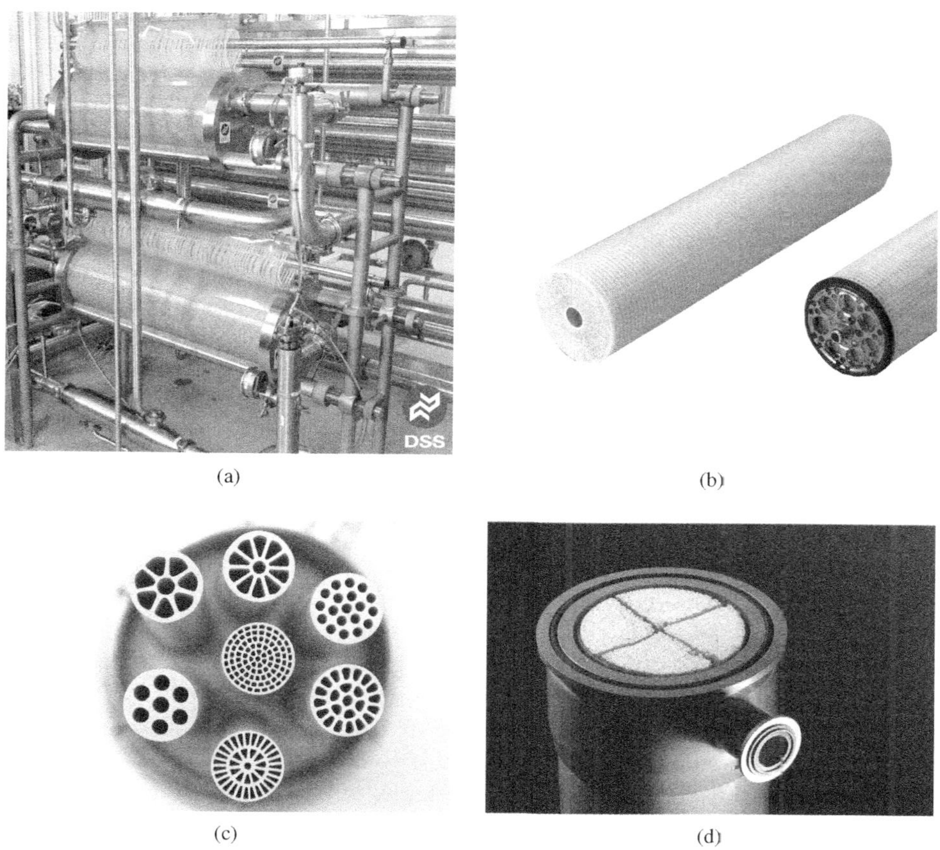

Fig. 11.1 Examples of some common membrane constructions used in membrane processing. (a) Plate and frame housing for flat sheet ultrafiltration/nanofiltration (UF/NF) membranes (Reproduced courtesy of DSS Silkeborg AS). (b) Spiral-wound organic UF/NF membranes (Reproduced courtesy of General Electric Company). (c) Tubular ceramic MF membranes (Reproduced courtesy of TAMI Industries SAS). (d) Hollow-fibre organic UF elements within housing (Reproduced courtesy of Microdyn-Nadir GmbH).

(Baruah *et al*., 2006). This development in membrane separation technology, involving the implementation of high-velocity, cocurrent (with the feed/retentate) circulation of microfiltered permeate inside the membrane module, greatly reduced fouling during MF (Maubois, 1997; Saboya & Maubois, 2000; Al-Akoum *et al*., 2002). The early research on MF of skimmed milk was almost invariably conducted using ceramic MF membranes (Fauquant *et al*., 1988; Pierre *et al*., 1992; Schuck *et al*., 1994; Pouliot *et al*., 1996; Maubois, 2002; Rizvi & Brandsma, 2002). Ceramic membranes have good chemical and thermal stability, can withstand pH in the range 0.5–13.5 and temperatures greater than 100°C (Baruah *et al*., 2006). For efficient and most economical removal of whey protein from casein, ceramic membrane-based MF plants generally need to be operated at a temperature in the range 45–55°C. Processing of milk at temperatures in this range can cause problems with growth of thermophilic bacteria and associated biofouling problems; fouling is caused by many factors, such as calcium phosphate formation and deposition, enzyme activity resulting in reduced levels of intact casein and/or whey proteins, protein aggregation/gelation (if the in-process pH drop is severe; see Rizvi & Brandsma, 2002) and shortening of membrane life (Smith, 2004). Indeed, UF processing of cheese whey and permeate from MF of milk, at 50°C, for the production of whey protein concentrate (WPC) and whey protein isolate (WPI), respectively, has been associated with structural alteration of β-lactoglobulin (β-Lg) and deleterious effects on functional properties of the product (Britten & Pouliot, 1996).

Compared with ceramic membranes, polymeric membranes used for separation of casein and whey proteins in milk generally have reduced chemical stability to pH and cleaning agents (Karasu *et al*., 2010). Polymeric membranes also have broader pore size distributions, and are relatively hydrophobic (facilitating protein adhesion) compared with ceramic membranes (Baruah *et al*., 2006). However, the low capital and running costs for polymeric membranes compared with ceramic membranes (Lawrence *et al*., 2008; Beckman *et al*., 2010) have led to recent investigation of the suitability and cost implications of using (predominantly spiral wound) polymeric membranes to achieve separation of whey protein from casein in milk (O'Mahony *et al*., 2007; Lawrence *et al*., 2008; Beckman *et al*., 2010; Karasu *et al*., 2010). In a recent survey, it was reported that all of the polymeric membranes offered by different membrane supply companies were made from polyvinylidene fluoride (PVDF) (Pearce, 2007).

Recent work conducted to compare the performance of spiral-wound polymeric (0.3 μm pore size) and ceramic (0.1 μm pore size) membranes showed that the polymeric membrane was capable of removing just 70.3% of total serum protein, compared with >95% removal with the ceramic membrane on MF of pasteurised skimmed milk at 50°C to a volume concentration factor of 3, using a three-stage process with two diafiltration (DF) stages using RO water (Beckman *et al*., 2010). Similar work was conducted by Karasu *et al*. (2010), and they concluded that polymeric (0.3 and 0.5 μm pore size) and ceramic (0.2 μm pore size) membranes provide comparable permeate flux performance during MF of skimmed milk when operated at comparable levels of transmembrane pressure (TMP) (30–175 kPa), shear stress (cross-flow velocity) and temperature (10 or 50°C). While this work was conducted at laboratory-scale, Karasu *et al*. (2010) acknowledged that the inability to generate UTMP in polymeric membrane systems is a limitation in the scale-up of such polymeric MF systems. One approach, which is practised in the implementation of such polymeric membrane-based processes at industrial-scale, is to use shorter membrane modules to minimise pressure drop across the membrane. In the past 15 years, cross-flow, spiral-wound, polymeric membranes now account for >70% of the total number of membrane system installations in the dairy industry (Bird, 1996), with the majority of these membrane systems currently used for UF, but with the proportion of MF systems likely set to increase over the coming years.

The permeate fraction from MF of milk, unlike regular cheese whey, is effectively sterile, is free from curd fines, lipid material (which may cause development of off-flavours and impairment of foaming (Morr & Ha, 1991; Pearce *et al*., 1992), added colour (e.g. annatto, if used), glycomacropeptide, lactic acid, residual coagulant activity and starter culture (Britten & Pouliot, 1996; Kelly *et al*., 2000; Nelson & Barbano, 2005; O'Mahony *et al*., 2007; Lawrence *et al*., 2008). Milk microfiltered permeate also has a lower ash content than cheese whey and has a pH (~6.70–6.80) similar to that of skimmed milk. Owing to the absence of such 'contaminating' materials and a reduction in the number of thermal processes involved (e.g. pasteurisation of cheese whey to inactivate residual coagulant and starter cultures), whey proteins extracted directly from skimmed milk using MF are commonly referred to as being in the native, virgin or unadulterated state ('ideal whey' or 'milk serum protein'), may be able to command a premium price on the market and are reported to have superior physico-chemical functionality compared to traditional WPC/WPI ingredients (Bacher & Konigsfeldt, 2000; Maubois, 2002; Rizvi & Brandsma, 2002; Nelson & Barbano, 2005; Heino *et al*., 2007; O'Mahony *et al*., 2007; Marcelo & Rizvi, 2008). One of the major problems of food product formulators experience with whey protein products is inconsistency with respect to functionality (Fox, 2001; Rizvi & Brandsma, 2002), with many such inconsistencies caused by several of the factors listed above (e.g. presence of residual lipids and denatured whey proteins).

The casein-enriched retentate fraction from MF of skimmed milk has many interesting applications, such as: (a) direct use in protein fortification and standardisation of milk intended for cheesemaking (Papadatos *et al*., 2003; Nelson & Barbano, 2005; Govindasamy-Lucey *et al*., 2007); (b) further concentration and spray drying in the manufacture of an ingredient for use in premium nutritional and clinical applications; and (c) a platform starting material for further ingredient development, e.g. isolation of glycomacropeptide from whey fraction of rennet-treated micellar casein solutions.

11.2.2 *Fractionation of individual casein proteins*

Caseins/caseinates have excellent functional properties, such as foaming and emulsification, and are one of the principal food ingredients globally (Mulvihill & Ennis, 2003). β-Casein is the most surface-active (i.e. superior foaming/emulsification properties) of all the milk proteins (Mulvihill, 1991). The molecule is also the source of numerous peptides with biological/physiological activity (e.g. peptides that have anti-hypertensive activity, promote mineral absorption or have β-casomorphin-like activity) (Maubois & Leonil, 1989; FitzGerald & Meisel, 2003). Hence, there has been considerable research invested in developing strategies for the fractionation and purification of β-casein from various sources of casein (e.g. milk, rennet casein, sodium caseinate and calcium caseinate) (Allen *et al*., 1985; Famelart *et al*., 1989; Murphy & Fox, 1991; Woychik, 1992; Famelart & Surel, 1994; Le Berre & Daufin, 1994; Le Magnen & Maugas, 1995; Ward & Bastian, 1996;

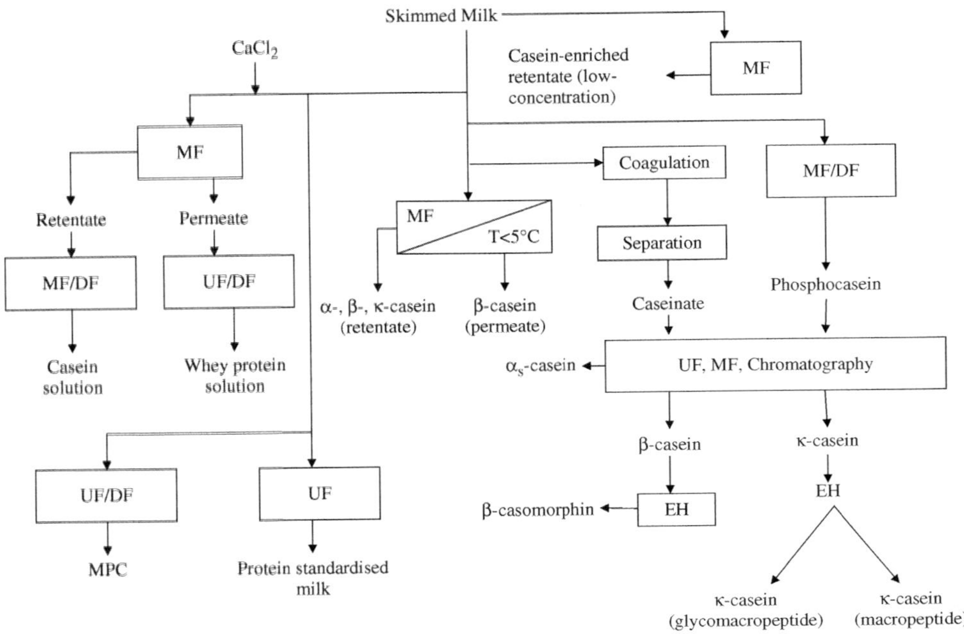

Fig. 11.2 Schematic representation of possible schemes for the fractionation of skimmed milk proteins based on application of microfiltration (MF), ultrafiltration (UF), diafiltration (DF). $CaCl_2$, calcium chloride; T, temperature; EH, enzyme hydrolysis; MPC, milk protein concentrate (Adapted from Rosenberg, 1995).

Huppertz *et al.*, 2006; O'Mahony *et al.*, 2007). Several of these fractionation/purification processes employed UF and/or MF membrane technology. A schematic diagram summarising the range of fractionation options for the major proteins in skimmed milk by Rosenberg (1995) is shown in Fig. 11.2.

Murphy & Fox (1991) developed a laboratory-scale process for the fractionation of sodium caseinate solutions (1 g 100 g^{-1} protein), at pH 7.0 and 4°C, into a β-casein-enriched permeate and an α_s-casein- and κ-casein-rich retentate using tangential flow, flat-sheet, UF membranes with molecular weight cut-off (MWCO) of 300 000 Da. Yields of β-casein (expressed as a proportion of total β-casein content of the feed material) ranged from 5.2 to 18.8 g 100 g^{-1}. Optimisation work conducted as part of this study also demonstrated increasing yield of permeable protein with increasing MWCO/pore size of the filters used; the yield of permeable protein on filtration of sodium caseintate solutions (0.1 g 100 g^{-1}) at pH 7.0 and 4°C was 3, 18 and 20 g 100 g^{-1} of total protein using 100 000 or 300 000 Da MWCO UF membranes and 0.1 μm pore size MF membranes, respectively.

Woychik *et al.* (1992) reported on filtration of pasteurised skimmed milk with a laboratory-scale, tangential flow UF system (Minitan Acrylic Ultrafiltration System from Millipore) using 0.1 and 0.2 μm pore size polymeric (PVDF) membranes to effect enrichment of casein micelles based on size (and consequently individual casein protein profile). The permeate fractions were concentrated 10-fold using 100 000 Da MWCO UF membranes and freeze dried before being analysed by transmission electron microscopy to confirm differences in casein micelle size between retentate and permeate streams.

Woychik (1992), using the laboratory-scale apparatus described above, performed cold (4°C) UF/MF of skimmed milk (held for at least 4 h at $\leq$4°C) using 100 000 Da MWCO, 0.1 μm or 0.2 μm pore size membranes to generate a permeate material enriched in whey proteins and β-casein as a platform in the manufacture of protein ingredients with protein profile potentially suitable for infant nutritional product formulation. Permeate fractions had a crude protein level of 0.41 g 100 g^{-1} (of which 26.6 g 100 g^{-1} was β-casein) on MF of skimmed milk, at 4°C, using PVDF membranes with a MWCO of 100 000 Da. The level of β-Lg in the permeate fraction was subsequently decreased by pH adjustment and salting out with the addition of sodium chloride .

O'Mahony *et al.* (2007) used low-temperature (2–6°C) MF (volume concentration factor of 3, with DF using skimmed milk UF permeate) of skimmed milk, using polymeric spiral-wound membranes (0.45 μm pore size) to simultaneously deplete skimmed milk of whey proteins and β-casein. The resultant β-casein-depleted milk protein concentrate showed altered rennet coagulation properties (O'Mahony *et al.*, 2009), and resulted in enhanced heat-induced functionality (especially meltability) in direct acid cheese made from there (O'Mahony *et al.*, 2008). Further enrichment of the β-casein protein fraction from the cold MF permeate stream was achieved by concentration (using the same polymeric MF membranes as used for removal of whey proteins and β-casein from skim milk) at 26°C, of hydrophobically induced, thermo-reversible β-casein micelles in the retentate fraction.

Several of the methods developed over the years focused on development of a β-casein-enriched supernatant or permeate fraction from different feed materials, such as sodium caseinate (Famelart *et al.*, 1989; Murphy & Fox, 1991) and calcium caseinate (Allen *et al.*, 1985; Ward & Bastian, 1996) involved thermocalcic aggregation and precipitation for recovery of β-casein. Issues with the purity of β-casein, residual rennet activity in β-casein product, poor solubility and consequently poor functionality of the product, low yield and the burden of processing and finding potential uses for the β-casein-depleted material have largely prevented these methods from being applied commercially.

Skimmed milk can be electro-acidified either by water electrolysis with monopolar membranes or by electrodialysis using bipolar membranes (EDBM). However, the electrical efficiency of EDBM is higher than water electrolysis with monopolar membranes. A bipolar membrane is a laminate comprising an anion-exchange layer (AEL), a cation-exchange layer (CEL) and an adhesive interfacial layer to which they are chemically bonded (Lutin & Bar, 2007). The membrane is porous to water so when it is placed in an electric field, with the AEL facing the anode, water molecules in the interfacial layer dissociate into hydrogen cations and hydroxyl anions in order for the current to be carried across the membrane. Hydroxyl anions pass through the AEL towards the anode and hydrogen cations migrate towards the cathode. Fig. 11.3 shows a two-compartment bipolar membrane cell. Bipolar membranes can be configured into membrane stacks to generate an acidified stream and a diluted base stream. Thus, in a stack of membranes they can be configured for the purpose of acidifying or making a liquid alkaline. The electrical conductivity of skimmed milk is relatively low, due to a lack of mobile monovalent ions, but Bazinet *et al.* (2000) demonstrated that the electrical energy consumption by the EDPM process could be reduced by up to 45% by supplementing skimmed milk with potassium chloride. The whey resulting from acid casein produced using EDBM has a lower mineral

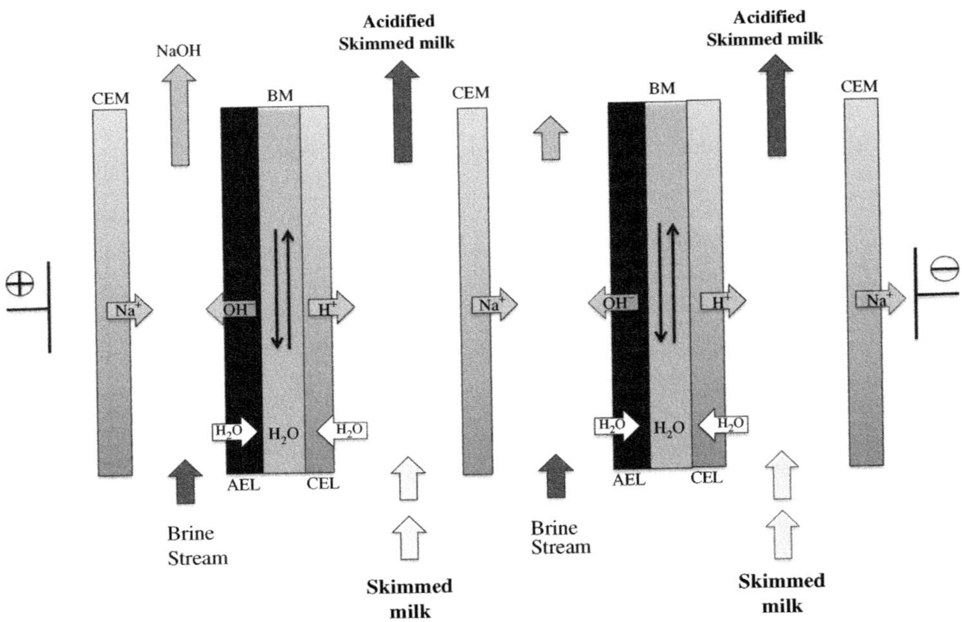

Fig. 11.3 Schematic representation of a two-compartment bipolar membrane cell. The bipolar membrane (BM) is flanked on either side by cation-exchange membranes (CEM) to form two compartments between each pair of BM; an acidic one between the cathode exchange layer (CEL) of the first BM and the cation-exchange membrane and a basic one between the cation-exchange membrane and the anion-exchange layer (AEL) of the next BM . Under the driving force of the electrical field, water within a bipolar membrane interface layer dissociates into hydroxyl (OH^-) and hydronium (H^+) ions which migrate to the respective surfaces of the membrane and combine with the cations and anions of the salt to form the respective acids and bases (Reproduced by permission of Pourcelly & Bazinet (2009) and of *Critical Reviews in Food Science and Nutrition*).

content than that produced by acid precipitation and resembles rennet casein whey, which is more suitable for further processing. Furthermore, the cost and health and safety issues associated with the transport, storage and use of concentrated acids are avoided. The main technical limitations of the EDBM process for the production of casein is protein fouling, with curd build up in the spacers between membrane cells through which the skimmed milk is recirculated (Pourcelly & Bazinet, 2009). Mineral fouling of the cationic membrane, due to the formation of calcium carbonate, could also affect performance and shorten the lifetime of these membranes, but can be prevented by controlling the pH of the baseified stream compartment during processing.

11.2.3 *Fractionation of individual whey proteins*

Bazinet *et al*. (2004) used electroacidification using EDBM to precipitate β-Lg in high purity from reconstituted WPI, and found that up to $53\,g\,100\,g^{-1}$ of the total protein in a $10\,g\,100\,g^{-1}$ solution of the isolate could be recovered in the precipitate compared with the $17–33\,g\,100\,g^{-1}$ range using chemical acidification reported by Amundson *et al*. (1982)

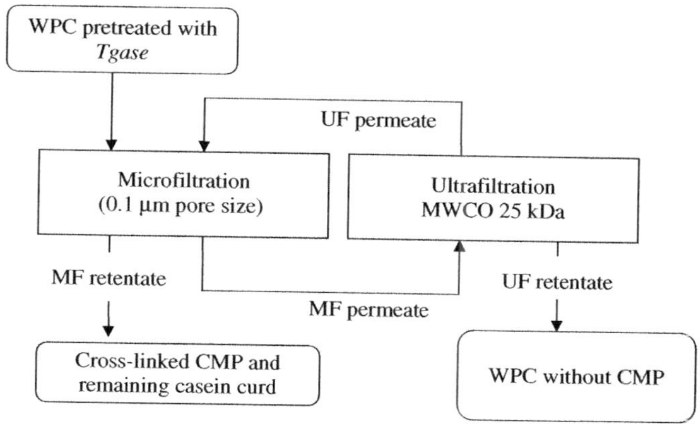

Fig. 11.4 Schematic representation of a fractionation scheme for the isolation and concentration of casein macropeptide (CMP) and the production of CMP-free whey protein concentrate from sweet whey. MWCO, molecular weight cut off; *Tgase*, transglutaminase (Reproduced by permission of Tolkach & Kulozik, 2005, and of the *Journal of Food Engineering*).

and Slack *et al*. (1986). This fraction contained $98\,g\,100\,g^{-1}$ protein and $97\,g\,100\,g^{-1}$ β-Lg. However, protein recovery diminished as the concentration of the WPI solution was increased beyond $10\,g\,100\,g^{-1}$ due to a decrease in the conductivity of the solution.

Tolkach & Kulozik (2005) used cross-flow MF with ceramic membranes to remove casein macropeptide (CMP), which had previously been cross-linked using the enzyme transglutaminase (Fig. 11.4). The degree of CMP cross-linking achieved was $92\,g\,100\,g^{-1}$ of the original concentration, and the particle size increase resulting from the aggregation enabled the CMP to be separated from the native whey proteins with 0.1 μm pore size membranes operating at 0.04 MPa TMP and 55°C. Approximately, $5–10\,g\,100\,g^{-1}$ of total protein in the cheese whey was retained in the retentate, comprising the cross-linked CMP and other casein fines.

Garcin & Paviot (1991) patented a method for recovery of 'serocolostrum', a whey fraction containing IgG, from skimmed colostrum using MF and DF with a saline solution. Fat globules, casein micelles, somatic cells and blood cells were retained in the microfiltered retentate and the clear permeate contained the whey fraction with elevated IgG levels. Piot *et al*. (2004) using 0.1 μm pore size ceramic membranes operating at a TMP of 0.04–0.05 MPa at <40°C and DF with a saline solution recovered up to $95\,g\,100\,g^{-1}$ of the IgG in the skimmed colostrum (which in turn contained approximately $90\,g\,100\,g^{-1}$ of the IgG in the colostrum itself) in the MF permeate.

11.2.4 *Fractionation of milk protein hydrolysates*

Nutrition companies, such as Mead Johnson Inc. and Nestle SA have pioneered the development of milk protein hydrolysate ingredients for inclusion in hypoallergenic infant formulae and dietetic foods, and dairy ingredient manufacturers have responded to the growing market opportunities for these high-value speciality ingredients. Protein hydrolysates used in these applications are generally prepared using purified pancreatic enzymes, such as trypsin, chymotrypsin and pancreatin, and typically contain 35–75 g 100 g^{-1} total nitrogenous material (i.e. total nitrogen converted to protein using a conversion factor of 6.38). Experience from the use of UF for the concentration of proteins in milk led to the use of UF for the purification of crude hydrolysates, which can contain residual unhydrolysed protein, proprietary proteolytic enzyme(s), macropeptides, various smaller peptides and amino acids along with any salts or lactose present in the substrate protein source.

In principle, any unhydrolysed protein, the proteolytic enzymes used and macropeptides are rejected by UF membranes with MWCO in the range 5–30 kDa, and the mixture of low molecular weight peptides and amino acids pass through the membrane into the permeate along with any salts or lactose present. Thus, the UF treatment has a purification effect only, and if DF is needed to optimise the yield of peptides some dilution, relative to concentration in the crude hydrolysate, may occur. The process of UF of crude hydrolysates can be configured as a batch, continuous or integrated (membrane enzyme reactor) process.

Severe membrane fouling, manifested in a decline in flux, was a feature of early processes for the purification of casein hydrolysates using UF membranes (Brulé *et al*., 1980). Also because of the bitter flavour commonly associated with casein hydrolysates (due to the production of peptides with hydrophobic terminal amino acids) and the superior biological value and ready availability at lower cost of whey protein, it was the preferred substrate in the case of cow's milk (Olofsson *et al*., 1981; Maubois *et al*., 1984; Turgeon & Gauthier, 1990).

Olofsson *et al*. (1981) introduced a UF step in their process for the preparation of a purified whey protein hydrolysate using organic or inorganic membranes with MWCO in the range 1–10 kDa. Flux rates in the range 60–120 L m^{-2} h^{-1} were reported for the purification (with or without DF) and concentration (volume concentration factor of 3) of hydrolysed α-La solutions (9–13 g 100 g^{-1} total solids (TS)).

Maubois *et al*. (1984) described an enzyme reactor fitted with hollow-fibre membranes. A range of commercially available membranes with nominal MWCO of 2, 5, 10 and 50 kDa were used. Total hydrolysis of the whey protein was achieved resulting in peptides of less than 2 kDa with 70–90 g 100 g^{-1} of peptides having less than 10 amino acids and 50 g 100 g^{-1} of the peptides comprising 2–5 amino acids. No difference in peptide profile was found between 5 and 10 kDa MWCO membranes. Information on yields of purified hydrolysates achieved by the early pioneers is scant, but by the following decade 10–50 kDa MWCO UF membranes had become the norm for the purification of crude hydrolysates prepared from whey protein (Turgeon & Gauthier, 1990; Wijers *et al*., 1998; Pouliot *et al*., 1999), sodium caseinate (Pouliot *et al*., 1993; Gourley *et al*., 1995, 1998) and milk protein (Pouliot *et al*., 1995). The next advance in membrane processing of milk protein hydrolysates was the technology to prepare peptide-enriched fractions with low amino acid contents. A two-stage membrane filtration process (Fig. 11.5) was used by Turgeon & Gauthier (1990) to prepare a purified peptide mixture from WPC. In the first stage, a purified hydrolysate was prepared from the crude hydrolysate using a 30 kDa UF membrane. In the second stage, the purified hydolysate was further fractionated into an amino acid stream (permeate) using a 1 kDa membrane with the retentate containing a purified mixture of peptides smaller than 5 kDa suitable for nutritional and cosmetic ingredient applications.

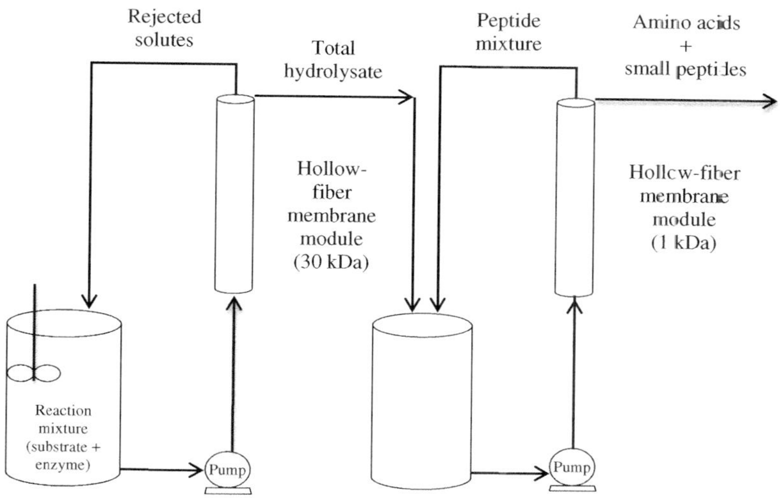

Fig. 11.5 Schematic representation of a membrane bioreactor for the hydrolysis of whey protein concentrate and two step purification of the hydrolysate. kDa, kilo Daltons (Reproduced by permission of Turgeon & Gauthier, 1990, and of the *Journal of Food Science*).

Research at laboratory-scale with model systems comprising select amino acid and short peptide mixtures (Garem *et al*., 1996; Kimura & Tamano, 1986; Tsuru *et al*., 1994) indicated that NF membranes with a MWCO of 500 Da or below were unsuitable for the separation of amino acids from peptides because of high steric rejection for all the amino acids. The efficacy of charged UF membranes with MWCO in the range 1–10 kDa was highly dependent on membrane type, charge, TMP, operating pH and to a lesser extent on the ionic strength.

A significant decline in permeate flux was found by Pouliot & Gauthier (1990) and Pouliot *et al*. (1993) when they fractionated a purified hydrolysate that had been prepared from sodium caseinate using polysulphone membranes. Earlier, Brulé *et al*. (1980) had, in addition to fouling, observed variation in the selectivity of membranes under the influence of physico-chemical operating conditions, including the calcium (Ca^{2+}) sequestering effect on phosphoserine residues resulting in aggregation and rejection by UF membranes. The flux rate decline subsequently found by the same authors was modified by pH adjustment, and the molecular mass profile was also modified. Brulé *et al*. (1980) concluded that peptide–membrane interactions were responsible for the flux rate and specific rejective properties of polysulphone for charged and hydroxylated amino acids. In follow-up work (Gourley *et al*., 1994) found that a high degree of orientation of water molecules at the surface of polysulphone membranes led to an unstable hydration layer, which promoted the adsorption of casein peptides and speculated that adsorption might also involve hydrophobic interactions between polysulphone and peptides.

Pouliot *et al*. (1995) purified a total milk protein hydrolysate prepared from reconstituted low-heat skimmed milk powder in a two-stage UF process, using 30 kDa and 1 kDa MWCO membranes, respectively. The same authors found a greater proportion of short peptides (<2 kDa) in low-heat skimmed milk hydrolysate than hydrolysed sodium caseinate prepared according to the same process. More significantly, the peptide-enriched fraction from the low-heat skimmed milk hydrolysate contained 43 g 100 g^{-1} lactose compared with 0.3 g 100 g^{-1} when sodium caseinate was used as the substrate.

Although a process for the preparation of caseinophosphopeptides using UF membranes was patented as early as 1980 (Brulé *et al*., 1980), further advances in biological chemistry and nutrition over the following decades and the identification of potentially bioactive peptides within milk proteins led to a demand for membrane technology to prepare various crude and purified individual peptide fractions. Thus, the application of pressure-driven membrane processes, such as UF and NF, to the fractionation of peptides became a focus of membrane technology research efforts.

Gourley *et al*. (1998) recovered and identified the casein peptides adsorbed on polysulphone under static conditions (Table 11.1). The same authors confirmed that the absorption of peptides on to hydrophobic polysulphone membranes generally increased with increasing hydrophobicity of the peptide, but found that hydrophobicity alone was not the determining factor and that size and charge of the molecule were also key factors. Peptides were classified into three groups based on their partition coefficient between the hydrolysate solution and the adsorbed layer on the surface of the membrane: first, hydrophobic peptides with a high affinity for the membrane ($a > 1.25$), and second, peptides with no specific affinity ($0.75 < a < 1.25$) and third, peptides with little or no affinity for the membrane ($a < 0.75$), which were further classified as acidic, neutral and basic peptides. Six different flat sheet, thin-film, composite NF membranes (five negatively charged) having MWCO in the range 300–2500 Da were compared by Wijers *et al*. (1998) for the simultaneous desalting and fractionation of purified hydrolysates prepared from WPC and WPI. The best fractionation of peptides was achieved with the two highest MWCO membranes (2000–2500 Da).

Pouliot *et al*. (1999) used a 10 kDa MWCO UF membrane to purify a crude tryptic hydrolysate prepared from WPI, and compared five different UF/NF membranes including cellulose acetate and polyamide for the fractionation of peptide mixtures. The best separation consistent with high flux was obtained using a cellulose acetate membrane (SG13) with a MWCO in the range 1–5 kDa. An improvement in flux, nitrogen transmission and a decreased tendency to foul the membrane was found when the pH of the feed was increased from pH 5 to pH 9, and the same authors observed that the proportion of negatively charged peptides relative to neutral and positively charged peptides in the permeate decreased. Pouliot *et al*. (1999) also demonstrated that the membrane permeability could be increased by increasing the ionic strength of the feed with added sodium chloride, but at the cost of a greater tendency to fouling.

At this point in time it can be concluded that purified peptide mixtures containing only trace amounts of amino acids can be obtained using charged UF/NF membranes, but the efficient recovery and enrichment of peptides of less than 1 kDa and particularly neutral or positively charged ones remains the challenge facing membrane technologists. Although the capability of using NF to fractionate specific peptides in model systems has been demonstrated, attempts to separate heterogeneous mixtures of peptides, such as is found in casein and whey protein hydrolysates prepared using pancreatic enzymes, on the basis of charge by manipulating hydrodynamic and/or physico-chemical conditions have met with limited success thus far.

Table 11.1 Peptide sequence identification of 46 peaks isolated from a casein hydrolysate by HPLC (high pressure liquid chromatography).

Peak number	Peptide sequence	Molecular mass (Dalton)	Hydrophobicity (kcal mol^{-1})	Isoelectric point	Net charge at pH 8.0
1	α_{s1} 80–83	525	4.5	9.4	+1
2	β 164–169	649	3.9	9.4	+1
3	α_{s1} 4–7	494	7.1	9.4	+1
4	α_{s1} 125–132	910	5.2	7.0	0
5	β 100–105	646	6.9	6.2	0
6	α_{s1} 120–124	615	5.2	9.4	+1
7	α_{s1} 84–90	831	5.1	4.1	−2
8	α_{s1} 37–42	689	5.6	6.2	0
9	α_{s1} 35–42	947	7.1	6.2	0
10	κ 1–10	1253	6.3	4.1	−2
11	α_{s1} 80–90	1338	9.5	5.4	−1
12	β 177–183	830	11.3	9.0	+1
13	κ 17–21	643	6.8	6.3	0
14	β 33–48	2063	7.0	4.1	−5
15	β 170–176	781	12.5	9.4	+1
16	α_{s2} 200–203	491	10.1	5.8	0
17	β 49–52	513	8.2	7.0	0
18	α_{s2} 189–193	632	9.2	9.4	+1
19	β 108–113	748	10.7	6.2	0
20	α_{s2} 138–149	1467	10.2	4.1	−2
21	α_{s2} 115–125	1196	14.7	10.2	+1
22	κ 98–105	971	10.3	7.0	0
23	α_{s1} 81–91	1369	13.1	6.2	0
24	α_{s1} 106–119	1662	15.5	4.1	−2
25	κ 25–30	796	14.2	5.8	0
26	κ 35–43	1140	16.4	5.8	0
27	α_{s1} 104–119	1662	15.5	4.1	−1
28	α_{s2} 92–98	958	13.4	5.8	0
29	β 184–190	821	10.3	4.1	−1
30	β 191–202	1384	19.4	6.2	0
31	α_{s2} 174–181	980	15.2	9.0	+1
32	β 1–25	3124	21.1	1.7	−5
33	α_{s1} 194–199	748	10.2	5.8	0

(continued)

Table 11.1 *(Continued)*

Peak number	Peptide sequence	Molecular mass (Dalton)	Hydrophobicity (kcal mol^{-1})	Isoelectric point	Net charge at pH 8.0
34	α_{s1} 152–164	1547	17.9	4.1	−1
35	α_{s1} 166–193	3028	31.9	4.1	−3
36	α_{s1} 152–165	1710	20.7	4.1	−1
37	α_{s2} 99–113	1771	23.7	6.3	0
38	α_{s1} 165–193	3191	34.8	4.1	−3
39	β 203–209	742	15.5	5.8	0
40	α_{s1} 91–100	1268	16.1	6.2	0
41	α_{s1} 23–34	1385	21.5	6.2	0
42	α_{s1} 133–151	2318	28.5	4.1	−2
43	β 69–97	3115	43.2	6.2	0
44	α_{s2} 92–113	2711	37.1	6.3	0
45	β 49–97	5360	73.0	7.0	0
46	β 114–143	3469	40.5	4.1	−4

Source: After Gourley *et al*. (1998).

11.2.5 *Enrichment of osteopontin from milk/whey*

Osteopontin is a highly phosphorylated, calcium-binding glycoprotein found in the extracellular matrix of several tissues and organs, such as bones, immune system, mammary gland and kidneys. It is a sialic acid-rich protein with several cell adhesive domains (Liaw *et al*., 1995). Osteopontin was first shown to be present in milk in 1987 by Azuma & Yamauchi (1987). Cow's milk osteopontin has been shown to contain 28 phosphorylation sites and three glycosylation sites (Sorensen *et al*., 1995). The expression of osteopontin in mammary gland epithelia is enhanced during pregnancy and lactation with a considerable amount of the protein being present in milk (Senger *et al*., 1989). Although osteopontin is a multifunctional protein implicated as having an important role to play in several biological processes, such as bone maintenance and repair, inhibition of ectopic calcification (Jono *et al*., 2000), inhibition of cellular adhesion (Liaw *et al*., 1995) and multiple immune functions (Giachelli & Steitz, 2000), the exact biological roles(s) of osteopontin are still somewhat unclear.

It is known that there are major differences in the concentrations of osteopontin in human milk (138 mg L^{-1}), cow's milk (18 mg L^{-1}) and cow's milk-based infant formulae (9 mg L^{-1}) (Schack *et al*., 2009). Such differences in composition have generated interest in the development of processing schemes for isolation and enrichment of osteopontin from milk and whey at industrial-scale. Several chromatography-based approaches, including for example anion- and cation-exchange chromatography, gel filtration chromatography, reversed-phase and hydrophobic chromatography, have been developed over the past 15–20 years for purification of osteopontin from milk (Sorensen & Petersen, 1993;

Bayless *et al.*, 1997; Sorensen *et al.*, 2003). However, these analytical approaches are not suitable for scale-up in the industrial-scale production of osteopontin. To our knowledge, there is only one scale-able membrane technology-based approach developed for the isolation of osteopontin from milk or whey (i.e. that of Sorensen *et al.*, 2007). This process employs MF membranes to recover osteopontin which has been selectively aggregated by pH adjustment or addition of insoluble calcium salt. In one such process put forward by Sorensen *et al.* (2007), acid whey (pH 4.55, containing approximately 0.002 g 100 g^{-1} osteopontin) was concentrated at 50°C by polymeric UF membranes (MWCO of 10 kDa) by a volume concentration factor of 10. Osteopontin was concentrated in the retentate fraction, and then it was heat treated at 68°C for 15 s, followed by cooling to 50°C, before adjusting pH to 7.0 with sodium hydroxide. After equilibrating for 2 h, the retentate was then microfiltered at 50°C using 1.4 μm pore size ceramic membranes and diafiltered with hot demineralised water adjusted to pH 7.0 to a target permeate conductivity of <100 μS. The osteopontin protein, which aggregates at neutral–basic pH, was concentrated in the retentate fraction while the other soluble components of the whey were removed in the permeate stream. The osteopontin-enriched retentate fraction was spray dried to yield a powder with 96.4 g 100 g^{-1} dry matter, 13.6 g 100 g^{-1} protein and 2.3 g 100 g^{-1} osteopontin. Several other variants of this basic process are also detailed by Sorensen *et al.* (2007), with ceramic MF membranes (pore size 0.8–1.4 μm) being employed in each option as a means of enriching selectively aggregated osteopontin protein from acid whey.

11.2.6 *Production of microparticulated whey protein*

Microparticulation refers to the controlled production of protein aggregates (i.e. microparticles) in whey protein solutions by means of heat and shear energy (Steventon *et al.*, 1994). The size of the microparticles varies, typically in the range 0.1–3.0 μm, and is controlled by varying protein concentration, temperature, shear energy input and hold time. It is a very convenient means of modifying the functional properties of whey proteins with such products (e.g. Simplesse® and LeanCreme™) finding applications as fat replacers and texture modifiers in a wide range of formulated food systems. Microparticulation can be achieved by dynamic heating and shearing (i.e. using scraped surface heat exchanger, plate heat exchanger or shear agglomerator technology) or by microfluidisation (Steventon *et al.*, 1994; Iordache & Jelen, 2003; Tobin *et al.*, 2010). While various types of whey (with or without pre-treatment, such as fermentation, cream addition etc.) may be used in the manufacture of microparticulated ingredients, common to essentially all ingredient offerings is the fact that microparticulation is conducted on pre-concentrated whey protein solutions (i.e. 28–80 g 100 g^{-1} protein). UF, using 10–20 kDa MWCO membranes, is commonly used in the pre-concentration of liquid whey for microparticulation. Different protein concentration microparticulated products are used for different application, e.g. microparticulated WPC-60 is most suitable for cheesemaking (typical addition levels of 2–8 g 100 g^{-1} in cheese milk prior to pasteurisation) while microparticulated WPC-35 is most suitable for use in fermented milk products (e.g. drinking yoghurt) and neutral flavoured (e.g. chocolate and protein fortified) milk-based beverages. Microparticulation is relatively new technology with several of the dairy processing technology vendors offering such equipment, e.g. MICRO FORMULA™ from GEA Filtration (Skanderborg, Denmark), LeanCreme™

from APV Systems – Invensys (Silkeborg, Denmark; currently known as SPX Flow Technology) and Tetra Therm MicroPart™ from Tetra Pak (High Wycombe, UK).

11.2.7 *Isolation and enrichment of growth factors from milk/whey*

Growth factors are synthesised in the mammary gland and are present in milk, colostrum and whey. Examples of such growth factors include insulin-like growth factor-I, insulin-like growth factor-II, transforming growth factor-β1, transforming growth factor-β2 and fibroblast growth factor (Grosvenor *et al.*, 1993). Although it is still unclear what the exact biological activities of growth factors are, they are generally believed to be involved in cell proliferation and differentiation. There has been much interest in the recovery and enrichment of such growth factors from milk and whey over the last 10–15 years due to their potential for application in infant nutrition and health care products (e.g. for skin disorders, gastrointestinal diseases and bone health).

Enrichment of growth factors from milk is not without difficulty due to the very low concentrations (5.0–1000 ng mL^{-1}) of such components in milk and the wide range of molecular weights and iso-electric points (approximate P_I range of 4.0–10.0) of the individual growth factors (see Pouliot & Gauthier (2006) for detailed review). Most of the enrichment schemes proposed involve the use of whey as starting material due to the pre-removal of casein micelles by acid or rennet. In some of the methods published using cheese whey, MF (0.1 or 0.8 μm pore size membranes) is used for clarification of whey before selective adsorption of growth factors using cation-exchange chromatography. These methods also detail the use of UF and DF in concentration of such alkaline chromatography eluates, followed by sterile membrane MF prior to freeze drying of growth factor-enriched material (Francis *et al.*, 1995; Rogers *et al.*, 1996; Ballard *et al.*, 1999). Another method details the use of MF membrane technology (0.1 μm pore size) for the recovery and concentration of transforming growth factor-β2 which had been precipitated from native whey by combined heat and acid (Maubois *et al.*, 2003).

11.3 Fractionation of milk fat using membranes

11.3.1 *Isolation and enrichment of native milk fat globules*

The fat fraction of whole cow's milk is predominantly present as spherical globules, ranging from 0.1 to 20 μm in diameter, with an average diameter of 3.0–4.0 μm (Walstra, 1994). Greater than 90 g 100 g^{-1} of the total mass of milk fat globules is accounted for by triacylglycerols, with the remainder comprised mainly of mono- and diacylglycerols, sterols, sterol esters, phospholipids, glycolipids, enzymes and proteins (Keenan & Dylewski, 1994; Mather, 2000). The majority of these minor components orientate at the surface of fat globules in a rather specific manner (Danthine *et al.*, 2000) forming a membrane of 10–50 nm in thickness (Freudenstein *et al.*, 1979). This membrane is termed the MFGM and acts to confer stability on the fat globules (Anderson & Cawston, 1975; McPherson & Kitchen, 1983).

A proprietary milk fat fractionation process, based on MF technology, was developed at the Institut National de la Recerché Agronomique (INRA – the English equivalent is the National Institute for Agricultural Research) in France ~15 years ago to separate native milk fat globules from milk according to size (Goudédranche *et al*., 1998). The process developed used tubular inorganic alumina multichannel membranes in tangential configuration with a surface area of 0.24–0.72 m^2, operated at 50°C under UTMP conditions (Goudédranche *et al*., 1998; Michalski *et al*., 2006). The unit was tested using different pore size membranes, ranging in diameter from 2 to 12 μm, with the process subsequently optimised for industrial use with a pore size of 5 μm (Membralox P1940, Pall Exekia, Tarbes, France). The process was effective at separating raw, whole cow's milk, with a mean fat globule diameter of 4.2 μm, into a retentate fraction containing large fat globules with mean diameter of 5.0–7.5 μm and a permeate fraction with small fat globules with mean diameter of 0.9–3.3 μm. The MF system was designed (e.g. minimal dead volume, residence time and number of 90° bends) to minimise mechanical damage to native fat globules – as confirmed by measurement of the zeta-potential of fat globules.

The physical, chemical and technological properties of milk fat globules are influenced by their size. Small fat globules have a greater surface area and quantity of MFGM material per unit mass of fat than large fat globules (Kernohan *et al*., 1971; Timmen & Patton, 1988). Chemical characterisation of fractions of native milk fat globules prepared by MF, confirmed earlier findings that the fatty acid profile of fat globules is slightly influenced by fat globule size (Fauquant *et al*., 2005, Michalski *et al*., 2005). In addition, Michalski *et al*. (2004) demonstrated that crystallisation temperature increases logarithmically with size of native fat globules. These differences in composition and physicochemical properties of native fat globules have important implications for the manufacture and quality of dairy products, such as butter and cheese.

St-Gelais *et al*. (1997) used the proprietary milk fat fractionation process described above for the manufacture of low-fat Cheddar cheese from milks enriched in small or large fat globules. Cheese made from milk containing primarily large fat globules (combined with modified mineral profile) was scored substantially higher for texture, flavour and colour than cheese made from milk containing primarily small fat globules. Michalski *et al*. (2003) reported that Camembert cheese made from milk with small fat globules (diameter 3.30 μm) had a higher moisture content, accelerated proteolysis and a less firm texture than cheese made from milk with large fat globules (diameter 6.03 μm). Emmental cheese produced with milk containing native small fat globules (average diameter 3.31 μm) was reported to have improved sensory, textural and stretching characteristics compared with that made from control milk containing unmodified fat globules (average diameter 4.50 μm) (Michalski *et al*., 2007).

It has also been shown that the size of native milk fat globules influences the rheological properties of milk protein gels. In rennet milk gels, containing $1.50\,g\,100\,g^{-1}$ fat, the storage modulus decreased from 70 to 35 Pa on increasing the size of native fat globules from 3.0 to 5.5 μm, whereas the opposite trend was observed for acid gels (Michalski *et al*., 2002). Large deformation rheological testing of rennet milk gels, showing that shear stress increases with native fat globule size, suggests that a larger number of smaller fat globules facilitate structural fracture and collapse by providing numerous weak points (Michalski *et al*., 2002, 2006). Butter made from cream with predominantly large fat globules has lower cholesterol content, is more yellow in colour, and it also has a softer texture (Goudédranche *et al*., 2000). Preliminary data has also suggested potential improvement of texture and stability of whipped dairy products using small fat globules fraction compared with unmodified cream (Michalski *et al*., 2006).

11.3.2 *Isolation and enrichment of milk fat globule membrane*

The MFGM surrounding fat globules consists of approximately $60\,g\,100\,g^{-1}$ protein and $40\,g\,100\,g^{-1}$ lipid (Fox & McSweeney, 1998). Xanthine oxidase and butyrophilin are two of the major proteins of the MFGM, whereas the lipid fraction is mainly comprises cholesterol, phospholipids and sphingolipids (Mather, 2000). Because of this compositional profile, the MFGM fraction of milk has many interesting physico-chemical and biological functionalities. Phospholipids have good interfacial activity, with MFGM ingredients shown to have good emulsifying properties (Corredig & Dalglesigh, 1997). Interest in the nutritional/biological role of MFGM material arises partly from the fact that sphingolipids reduce the absorption of cholesterol (Eckhardt *et al*., 2002), protect against bacterial infections (Pfeuffer & Schrezenmeir, 2002), and have an inhibitory role in colon cancer (Schmelz, 2004). The MFGM material is released into the serum fraction (e.g. buttermilk or butter serum) of milk/cream on phase inversion of the fat globules as occurs during churning of cream during the manufacture of butter. Buttermilk has similar gross chemical composition to skimmed milk except that it also contains the MFGM material. Butter serum is the aqueous phase of butter, and is a by-product of anhydrous milk fat (AMF) production. Butter serum has gross chemical composition similar to that of buttermilk except that it has a higher fat content ($\leq 4.5\,g\,100\,g^{-1}$) (McPherson & Kitchen, 1983) and, consequently, MFGM material content. Both buttermilk and butter serum have been used as substrates in the enrichment of MFGM material in the manufacture of new MFGM-enriched ingredients.

MF membrane technology has been used over the last 20 years or so for fractionation of buttermilk/butter serum in the enrichment of MFGM material (Surel & Famelart, 1995; Astaire *et al*., 2003; Corredig *et al*., 2003; Morin *et al*., 2004, 2007; Rombaut *et al*., 2006, 2007). There are several challenges incurred in the application of MF for enrichment of MFGM material from buttermilk. MFGM particles in buttermilk are quite variable in size ranging from 0.1 μm to several micrometres (Morin *et al*., 2007), resulting in poor retention using membranes with pore size greater than 0.1 μm; however, the MFGM particles and casein micelles in buttermilk are comparable in size, making physical separation with membranes of pore size 0.1–1.0 μm very difficult (Rombaut *et al*., 2006). Several studies have been conducted over the last 10 years focused on developing and evaluating fractionation approaches, which address and overcome these challenges in retaining as much MFGM material as possible while removing as much non-MFGM buttermilk solids as possible. Such studies are based on four main approaches:

- dissociation of casein micelles (e.g. using sodium citrate) in buttermilk prior to MF (Corredig *et al*., 2003)
- rennet-induced coagulation of caseins in buttermilk to yield a casein-free whey material from which to recover MFGM material (Sachdeva & Buchheim, 1997)

- use of washed (with skimmed milk UF permeate) cream buttermilk for recovery of MFGM material using MF (Morin *et al*., 2007)
- use of starting materials naturally low in casein protein content. such as buttermilk obtained from churning of whey cream in the manufacture of whey butter (Morin *et al*., 2006; Costa *et al*., 2010), butter serum (Rombaut *et al*., 2006) and acid whey (Rombaut *et al*., 2007).

Polymeric and ceramic MF membranes have been used for enrichment of MFGM material from buttermilk/butter serum. In the study of Corredig *et al*. (2003), a pilot-scale filtration system (NIRO filtration, Hudson, WI, United States of America – USA) was used with stackable polymeric UF/MF membranes made of PVDF (250 and 500 kDa MWCO) or polyethersulphone (PES) (0.1 μm pore size) for fractionation of reconstituted buttermilk powder. The plant was operated in a batch-wise fashion with batch volumes of 8–10 L of buttermilk. Filtration was generally carried out at 50°C, with multiple batch-wise DF using deionised water, carried out after initial concentration. In the study of Morin *et al*. (2007), the Tetra Alcross Pilot M1 (Tetra Pak Filtration Systems, Lund, Sweden) was used for fractionation of fresh normal and washed cream buttermilk. The plant was fitted with a 0.5 μm pore size ceramic Membralox P35–37 membrane (Pall Corporation, Ontario, Canada) with a filtration surface area of 0.35 m^2. The plant was configured with permeate cocurrent recirculation to provide a low and uniform TMP. The plant was operated in a batch-wise fashion with batches of 20 kg of buttermilk. Filtration was generally carried out at 50°C, using a 50 kPa TMP and a pressure drop of 140 kPa to a volume concentration factor of 4. The retentate was DF (6-dilution) in a batch-wise fashion at 50°C using deionised water after initial concentration. Other approaches to the fractionation of buttermilk, such as the use of supercritical fluid extraction of phospholipids (Costa *et al*., 2010) and thermocalcic aggregation of MFGM fragments (Rombaut & Dewettinck, 2007), are also likely to rely on the use of membrane filtration technology (UF and MF) for pre-concentration or pre-fractionation and/or downstream processing of enriched fractions of buttermilk or its derivatives.

11.3.3 *Removal of phospholipids from liquid whey*

WPC ingredients can contain 5–15 g 100 g^{-1} lipid depending on whey protein content. Much of this whey lipid material is derived from the MFGM and causes multiple issues with respect to processing and ultimate quality of such ingredients. Examples of such issues include contribution to fouling of UF membranes during processing, turbidity and lack of clarity of liquid whey and reconstituted product, and it also causes reduced shelf-life and impaired product quality due to oxidation and impaired physico-chemical functionality (e.g. foaming). Strategies have been developed and employed commercially over the years to remove fat from high-protein content WPC and isolate WPI (see Chapter 9). Chitosan is a natural polycationic polymer derived from chitin, which has been shown to selectively aggregate and precipitate MFGM-derived phospholipids from cheese whey (Hwang & Damodaran, 1995; Damodaran, 1995). MF with membranes of 2–3 μm pore size may be used to remove the chitosan-phospholipid complex from chitosan-treated (0.01–0.02 g 100 g^{-1} chitosan at pH 4.5) whey (Hwang & Damodaran, 1995). Cheese whey contains 2–8 g L^{-1} lipids, of which approximately one-third are polar phospholipids.

For the production of high-protein WPI and WPC powders with optimum functional properties, it is necessary to have a lipid removal step in the process. The most effective method for lipid removal is thermocalcic precipitation of phospholipoproteins followed by MF, which enables the production of high-protein whey powders containing less than $1\,g\,100\,g^{-1}$ fat (Maubois *et al.*, 1987; Rombaut & Dewettinck, 2007).

Whey electroacidification by EDBM has been shown to produce phospholipoprotein precipitates containing up to $32\,g\,100\,g^{-1}$ lipid (Lin *et al.*, 2007). By decreasing the ionic strength of the concentrate and by dilution prior to EDBM, Bazinet *et al.* (2008) were able to remove $73\,g\,100\,g^{-1}$ of the original lipid in a $55\,g\,100\,g^{-1}$ protein ultrafiltered concentrate. It was claimed that this phospholipid-rich fraction would be suitable for cosmoceutical and nutraceutical applications.

11.3.4 *Filter sterilisation of polyunsaturated fatty acids*

In-line dead-end liquid filter sterilisation is commonly used in the beverage industry, most notably in the case of high-clarity products such as premium ciders and lager beers immediately before filling; however, the main applications in the dairy industry are in aseptic processing for the filtration of product-contact air and steam used for direct-injection or infusion heating. Another significant application in aseptic processing is for the sterile addition of heat labile ingredients in-line post-heat treatment and immediately before the sterile tank or aseptic packaging machine. An illustration of the Tetra Aldose™ sterile filtration unit is shown in Fig. 11.6. Applications include the addition of heat labile vitamins, unsaturated fatty acids, such as omega-3 (or oils rich in such fatty acids), flavourings and colours.

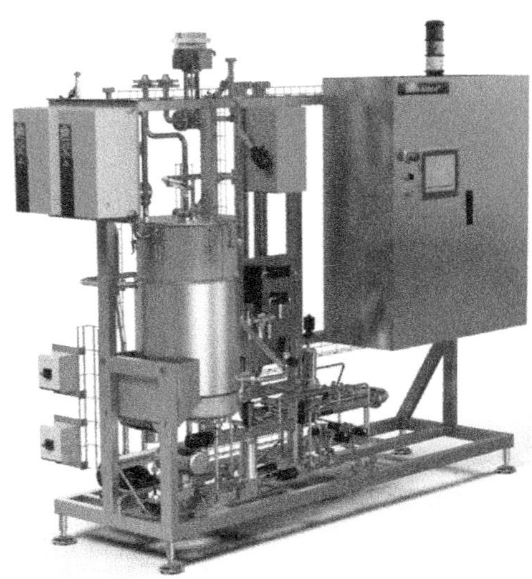

Fig. 11.6 Tetra Aldose™ plant for the sterile filtration of heat sensitive products and aseptic in-line addition post-heat treatment (Reproduced courtesy of Tetra Pak AB).

In-line dead-end sterilising membranes are vulnerable to suspended solids and, to avoid blockage of the pores, require pre-filters to reduce the particle load on the sterilising filter and optimise performance of the filtration system (Olsson, 2008). Pre-filters come in a range of different materials of construction, configurations, surface area, pore sizes and pore size distribution, and the choice of pre-filter for a particular application is often dictated by the filtration characteristics of the of the material to be filtered. In some cases, the best solution is to use a coarse and fine pre-filter in series to maximise particle retention and optimise the flow rate. Fouling of sterilising filters can occur both on the membrane surface and within the pores by a variety of methods – pore constriction, intermediate blocking, pore blockage and cake formation. In addition, in-line sterile filtration is unable to handle conventional oil-in-water emulsions (or water in oil) due to the droplet size formed, but laboratory-scale filtration trials with nano-emulsions carried out by Olsson (2008) with 0.2 μm PVDF filters indicated the feasibility of this approach for dosing heat labile oils subject to process temperature control.

11.4 Fractionation of milk carbohydrates using membranes

11.4.1 *Isolation and purification of bovine milk oligosaccharides*

The development of oligosaccharide-enriched ingredients from cow's milk is of considerable interest due to the demonstrated bioactive functionality of these compounds in humans, particularly in infants (Kunz *et al*., 2000) and potential opportunities for incorporation of such ingredients into infant formulae. This is despite the fact that the concentration of naturally occurring oligosaccharides in cow's milk is very low relative to human milk, and the main group found in human milk, the fucosylated oligosaccharides, has not been detected in cow's milk (Mehra & Kelly, 2006). The oligosaccharides common to bovine and human milk are- 3′- and 6′-sialylactose and 3′- and 6′-galactosyl-lactose. Oligosaccharides, which are defined as carbohydrates containing 3–10 monosaccharides, covalently bonded through glycosidic bonds, accompany lactose in bovine milk (present at 40–50 g L^{-1}) and the various commercially available milk and whey fractions. Research on methods for the enrichment of milk oligosaccharides has been reviewed by Mehra & Kelly (2006) and includes membrane filtration.

A laboratory-scale method of recovering oligosaccharides from milk, using a UF and a NF stage, was published by Sarney *et al*. (2000) with a yield of more than 50 g 100 g^{-1} claimed. In order to separate the oligosaccharides from lactose, an enzymatic lactose hydrolysis step was included, followed by a UF step to remove the lactase prior to NF. Martinez-Ferez *et al*. (2006) used a small-scale two-stage filtration process to isolate oligosaccharides from skimmed goat's milk. The first stage involved continuous DF to four diavolumes through a 50 kDa MWCO ceramic UF membrane to isolate the oligosaccharide fraction, along with lactose, salts and low molecular weight peptides, from the major milk proteins. In the second stage, the permeate from stage 1 was diafiltered to three diavolumes through a 1 kDa MWCO NF membrane to purify the oligosaccharide fraction. Oligosaccharide recoveries of 98 g 100 g^{-1} and 80 g 100 g^{-1} were reported for stages 1 and 2, respectively, whereas only 5 g 100 g^{-1} of the original lactose was retained in the virtually salt-free retentate. Flux rates ranged from 28 L m^{-2} h^{-1} (initial) to 19 L m^{-2} h^{-1}

(final) during the NF stage. Martinez-Ferez *et al*. (2006) attributed a greater number of peaks found in an oligosaccharide chromatographic profile of the NF retentate to the partial degradation of some oligosaccharide structures as a consequence of membrane processing parameters such as shear stress.

A method for the isolation and enrichment of sialyllactose (SL) from whey permeate, involving UF and NF, which yielded products containing up to 25 g 100 g^{-1} SL was outlined in a patent application by Holst *et al*. (2007). The applicants cited the use of low temperature UF (10°C) at a pressure of 0.5–0.7 MPa using GE Osmonics membranes (GH 8040 C1566/GH3840-30D) followed by batch DF of the retentate until the refractive index of the permeate dropped to <0.2 or <0.1 brix. The diafiltered retentate was then concentrated by RO and spray dried. Using their patented plate and frame membrane cassette system, Kopf & Kopf (2005) described sequential fractionation procedures for the isolation and purification of SL from skimmed milk or whey (Fig. 11.7) and for the preparation of a WPI, or α-La fraction, enriched in SL suitable for use as ingredients in infant milk formulae.

SL was isolated from the whey proteins using a low surface charge membrane material, such as regenerated cellulose (MWCO of 5–10 kDa), and then separating the SL from lactose in the permeate stream using a high surface charge PES membrane (MWCO of 1–5 kDa) and DF to further reduce lactose and mineral levels. An SL concentration factor of 1000:1 and a yield of 65% was claimed for the process.

A WPI enriched in SL was prepared using a high surface charge PES membrane (MWCO of 1–5 kDa). An α-La fraction enriched in SL was prepared by first fractionating whey into an α-La-enriched fraction plus a fraction containing the other whey proteins

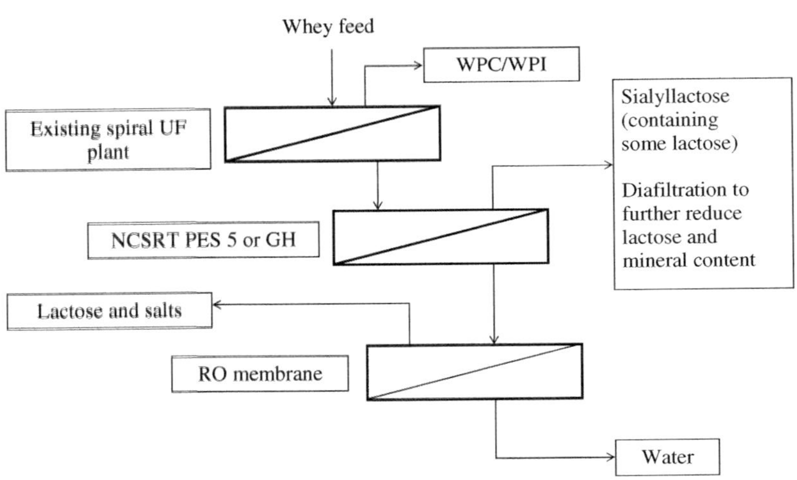

Fig. 11.7 Schematic representation of a fractionation scheme for the recovery of sialyllactose from whey using sequential membrane filtration. Conventional spiral wound ultrafiltration plant followed by a NCSRT plate and frame plant containing either PES 5 (PTI Advanced Filtration Inc., Oxnard Ca., USA) or GH (Osmonics, Minnetonka, MI, USA) membranes and configured according to the patent filed by Kopf & Kopf (2005). WPI, whey protein isolates; WPC, whey protein concentrates; NCSRT, North Carolina Science Research Triangle Inc.; PES, polyethersulphone; GH, not relevant – Osmonics Co. serial #; RO, reverse osmosis (Reproduced courtesy of SmartFlow Technologies Inc.).

(β-Lg, bovine serum albumin (BSA) and immunoglobulins) using a regenerated cellulose membrane with a MWCO of 20–100 kDa. Then in a second step separating out the lactose from the α-La-enriched fraction using a PES membrane to yield a retentate enriched in both α-La and SL.

11.4.2 *Filter sterilisation of lactase*

Although in-line dead-end liquid filter sterilisation applications include the addition of heat labile vitamins, unsaturated fatty acids, such as omega-3 (see Section 11.2.4), flavourings and colours, the most widely used application is the dosing of the enzyme lactase in the production of lactose-free or lactose-reduced milk and milk products using the Tetra Aldose system (Anonymous, 2011). With this approach, the lactose hydrolysis takes place in the package, between packaging and the point of sale or consumption.

Sterile liquid filtration is defined as a process that produces a sterile filtrate when challenged with more than 10^7 colony forming units (cfu) of the organism *Brevinudius deminuta* cm^{-2} (Anonymous 2004). For liquid sterile filtration processes a 0.22 μm membrane is used and suppliers include Millipore and Pall Corporation. The two most widely used membrane materials are PES and PVDF. Both have low protein binding affinity, but PES is reported to be unsuitable for solutions with high concentrations of polyethylene glycol (PEG), and PVDF is unsuitable for highly alkaline solutions (Olsson, 2008). Also the pressure drop across PES is less than across PVDF due to the asymmetric pore structure of PES membranes.

Commercially available lactase preparations consist of a concentrated solution of the enzyme in a water and glycerol base, and come in a range of activities, degrees of purity and concentrations. Haze formation due to lactase denaturation is the main fouling component and, since the haze particles are smaller than the pore size, fouling in this instance is probably due to absorption (Olsson, 2008). Dilution of concentrated lactase down to working strength promotes haze formation particularly if the water is slightly acidic or the temperature exceeds 8°C, and Tetra Pak recommend that diluted lactase should be processed within 12 h (Anonymous 2011). Olsson (2008) concluded from full-scale trial experience with the Tetra Aldose sterile filtration system that it is difficult to choose a pre-filter suitable for lactase preparations based solely on the rated pore size and theoretical considerations and recommend performing laboratory-scale trials before going to full-scale runs.

11.4.3 *Lactic acid removal and purification*

The production of lactic acid by fermentation is a core technology in the dairy industry and a key step in the manufacture of cheese, yoghurt, cultured butter, lactic casein and various fermented milk products. While the production of ethanol from whey permeate has been in commercial practice for almost 40 years, the production of purified lactic acid by the dairy industry has not received much attention despite its wide range of applications in the food industry. A two-step ED process for the isolation and purification of lactic acid produced by fermentation of whey permeate has been described by Norddahl (2001) with

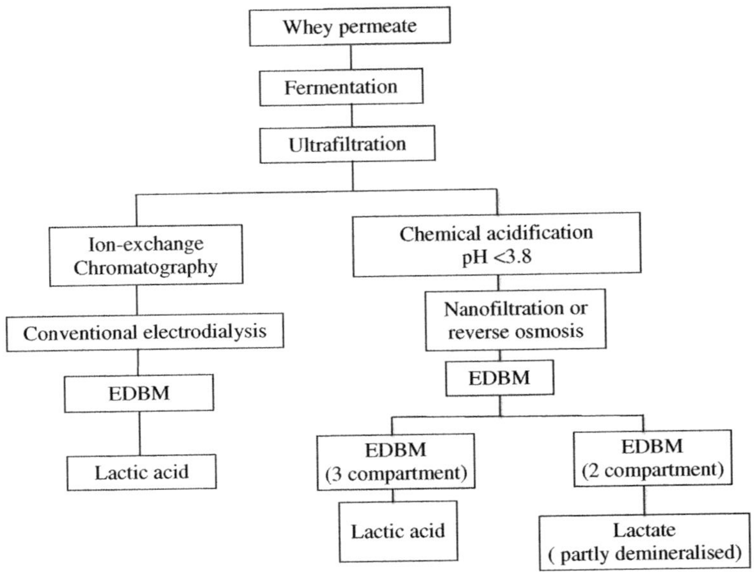

Fig. 11.8 Schematic representation of three fractionation processes for the recovery of lactic acid/ demineralised lactate from the ultrafiltration permeate of fermented whey EDBM, electrodialysis with bipolar membranes (Adapted from Pourcelly and Bazinet (2009), and reproduced by permission of *Critical Reviews in Food Science and Nutrition*).

an overall recovery rate 85–90% lactic acid (based on the sugar content of the permeate). The isolation process involved UF, ion exchange and a two-stage ED treatment. The first ED step involved concentration of the lactate using conventional ED membranes. In the second step, using EDBM, lactate was converted to lactic acid and the corresponding salt hydroxide, which were recovered as two separate streams. An improved process was the subject of a patent application published by Norddahl *et al*. (2004) for which a recovery of 90–95% was claimed. In this process, the ultrafiltered permeate was acidified below pH 3.9, nanofiltered and the permeate subjected to EDBM. A summary of fractionation processes for the isolation, purification and concentration of lactic acid is shown in Fig. 11.8 (Pourcelly and Bazinet 2009).

11.5 Fractionation of milk salts using membranes

11.5.1 *Demineralisation using membranes*

Cow's milk contains ~7–8 g L^{-1} inorganic mineral matter (ash) (Lucey & Horne, 2009), whereas acid whey powder can contain up to 15 g 100 g^{-1} ash (FAO, 2010). The high ash levels and particularly the levels of sodium and chloride in acid whey powders is a limiting factor as regards their use as an ingredient in infant milk formulae and other ingredient applications. Thus, demineralisation processes afford the opportunity to processors to add value to whey. For very high demineralisation levels (>90 g 100 g^{-1}) ion exchange is the most economical method for whey demineralisation, but membrane-based demineralisation approaches are interesting alternatives where only partial demineralisation is required.

Partial and selective demineralisation of skimmed milk, whey, ultrafiltered permeate and various milk by-product streams can be achieved using NF. As the prefix suggests, NF refers to size exclusion in the nanometre range (10^{-9} m), and its principal application is for the separation of ions with molecular weight less than 1 kD. Water and molecules with a molecular weight of <200 Da can permeate organic NF membranes, whereas the permeation of electrolytes and other low MW components, such as non-protein nitrogen or organic acids, is selective and based on charge. According to the Donnan equilibrium principle, an equal electrochemical potential is required on each side of a membrane (Donnan, 1995). Accordingly, NF membranes have a selectivity based on the magnitude of charge such that monovalent ions (e.g. sodium, potassium and chlorine, Na^+, K^+ and Cl^-, respectively) pass through the membrane, whereas divalent ions, such as Ca^{2+} and magnesium (Mg^{2+}) and polyvalent ions, are rejected. The operating pressures used in NF processes, which generally range from 2 to 4 MPa, are significantly higher than for UF (<1 MPa), but considerably lower than those used for RO where pressures up to 12 MPa can be used.

The main application of NF in the dairy industry is the concentration and partial demineralisation of acid casein whey, which typically contains 50% more ash than sweet whey. In terms of overall extent of demineralisation, up to 35% demineralisation can be achieved with NF when concentrating to a volume concentration ratio of 3.5–4.0, although higher extents of demineralisation can be achieved by DF with deionised water (Kelly & Kelly, 1995). Above 22–24 g 100 g^{-1} TS, the viscosity of nanofiltered whey retentates increases exponentially, and precipitation of calcium phosphate can also occur. Based on membrane flux, the most economical concentration target for whey NF is 18 g 100 g^{-1} TS, which corresponds to an overall demineralisation rate of ~28 g 100 g^{-1}. Most commercial NF membranes are thin-film composite membranes that have an active layer formed from an aromatic cross-linked polyamide, and have a negative charge at neutral pH (Petersen, 1993). A typical example of constituent losses in nanofiltered permeate when sweet whey is concentrated to 18.5 g 100 g^{-1} TS is shown in Table 11.2. Some commercial polyamide NF membranes were compared by Rasanen *et al.* (2002) for the NF of sweet whey. They found that the NF membranes evaluated could be classified into two groups: 'tight' and 'looser'. The rate of demineralisation was higher for the looser membranes (58 g 100 g^{-1} versus 53 g 100 g^{-1} at a volume concentration ratio of 5) but, in the case of the tighter membranes, the loss of lactose was only 0.1 g 100 g^{-1} compared with 1.0 g 100 g^{-1} for the looser membranes. According to Kelly & Kelly (1995), NF is the most economic process in terms of operating costs for the partial demineralisation of whey for up to 32 g 100 g^{-1}, and ideally suited to the demineralisation of acid whey down to the ash levels of sweet whey, while at the same time concentrating to 20 g 100 g^{-1} TS.

11.5.2 *Demineralisation using electrodialysis*

ED, although generally classified as an electrochemical process, can equally be regarded as a membrane process in which the driving force is electrical charge rather than pressure. In

Table 11.2 Typical loss of constituents in the permeate when raw sweet whey ($6.5\,g\,100\,g^{-1}$ total solids) is concentrated to $18.5\,g\,100\,g^{-1}$ by nanofiltration.

Constituent	Raw whey ($g\ 100\,g^{-1}$)	Concentrated whey ($g\ 100\,g^{-1}$)	Permeate ($g\ 100\,g^{-1}$)	Lost in permeate (%)
Total solids	6.5	18.5	0.37	
Fat	0.7	0.8	0	0
Total protein	12.8	13.5	0	0
NPN	0.5	0.36	4.32	31
Lactose	74.4	77.0	26.2	1.3
Organic acids	3.4	3.37	4.9	5
Total Ash	7.7	5.3	68.8	28
Minerals				
Ca	0.42	0.41	0.7	6
Mg	0.12	0.12	0.21	6
Na	0.61	0.38	6.5	40
K	2.36	1.48	25.2	40
S	0.16	0.15	0.55	12
P	0.62	0.57	2.05	12
Cl	1.44	0.68	21.1	54

NPN, non-protein nitrogen.
Source: Reproduced with permission from Eurodia Industries.

an ED plant, ion-permeable membranes (anionic and cationic) are arranged alternatively in stacks with a spacer between them to create membrane cells. A dilute salt solution (brine) is circulated on either side of each cell; thus, when the solution to be demineralised (whey or skimmed milk) flows through the cell, the anions and cations migrate towards the oppositely charged electrode through the respective membranes into the brine solution. Furthermore, the concentration of ions in the product falls as the concentration in the flushing brine solution rises. As the process continues, the flushing solution is diluted to maintain a concentration gradient and, as the ionic concentration of the product falls, its conductivity decreases and the voltage needs to be increased to compensate for the progressive reduction in the rate of ionic migration. An industrial ED installation with multiple membrane stacks is shown in Fig. 11.9. ED lends itself to batch operation whereby significant demineralisation can be achieved through continuous product recirculation. Thus, ED is best suited to applications where only partial demineralisation is required.

Ion-permeable ED membranes are comprised of an ion-exchange resin supported on a styrene polymer cross-linked with divinyl benzene, which provides mechanical strength and flexibility, and range from 50 to 200 mm in thickness. Cation-permeable membranes are produced by sulphonation with 98% sulphuric acid resulting in a negatively charged sulphonate group ($-SO_3$) chemically attached to the polymer chains. Oppositely charged ions to the fixed charge (counter ions), such as Na^+, are freely exchanged with the flushing brine solution at these sites. The concentration of counter ions on cation-permeable

Fig. 11.9 Electrodialysis membrane stacks (with protective covers removed to show membranes) in an industrial installation for whey demineralisation (Reproduced courtesy of General Electric Company).

membranes is relatively high, and these counter ions carry most of the electrical current though the membrane. In a cation-exchange membrane, the fixed negative charges attached to the polymer chains reject ions of the same charge (co-ions), for example anions such as Cl^-, due to electrostatic repulsion (Donnan effect). Therefore, the concentration of anions on the membrane is relatively low, and anions carry relatively little electric current through a cation-permeable membrane. Conversely, anion-permeable membranes are produced by post treating cross-linked polystyrene resulting in fixed positively charged quaternary ammonium groups (e.g. $-NCH_3^+$), and are selective to the transport of negative ions. Ion-exchange membranes based on perfluorocarbon, polypropylene and polyvinyl chloride plastisol are also commercially available. ED has been used in the dairy industry for whey demineralisation for several decades and, historically, was the preferred demineralisation process of infant formula manufacturers (Houldsworth, 1980). In the case of whey, 50–70% demineralisation is commonly the target of industrial-scale ED processes, although it is technically feasible to achieve over 90% demineralisation. For efficient operation of an ED plant, pre-concentration of whey to 18–28 g 100 g^{-1} TS is necessary to increase the conductivity of the ED feed, and NF is the preferred method of concentration as, in addition to the concentration factor, some 'free' removal of monovalent ions takes place during NF.

11.6 Conclusions and future trends

The role of membranes in the development of new and valuable ingredients, such as micellar casein, individual protein-enriched fractions derived from casein or whey protein, osteopontin, peptides, oligosaccharides, phospholipids, milk fat globule membrane materials and growth factors, has been discussed in this chapter. It is clear from the work reported in this chapter that there are many minor biologically interesting components of milk for which membrane filtration-based approaches have been developed and validated at laboratory- and pilot-plant-scale with little commercial uptake to date. The low commercial uptake of such ingredients is almost invariably a reflection of scale-up and/or non-technical constraints, such as limited market size, underdeveloped market and high ingredient costs, resulting from supply and demand imbalances. Reduction of these constraints would open up commercial implementation of several such membrane-based processes for the isolation and enrichment of various components from milk.

In the short term, in modifying or expanding existing facilities and in the installation of new membrane filtration equipment in the dairy sector, enhanced emphasis will be placed on energy and environmental criteria, such as energy efficiency, water usage and chemical usage of such installations.

In the longer term, the successful adoption and commercialisation of the discoveries expected from the massive investment in molecular biology research being made worldwide will depend heavily on the development of engineering and technological solutions in the so-called downstream processes of separation, purification and preservation. In the next decades of food innovation in which the focus of consumer food manufacturers on development of value-added products with specific health claims is expected to increase along with the requirement of dairy ingredient suppliers to develop value-added ingredients with novel and substantiated health claims, it seems inevitable that many more opportunities for the application of membrane technology solutions will emerge. Foremost among those opportunities is likely to be the isolation and purification of bioactive peptides for the supplementation of food and beverage products targeted towards population groups at risk from oral and gastric infections, hypertension, suppressed immune function, obesity and a range of other syndromes. In this context, the yield may even be measured in activity units rather than dry matter recovered.

To meet these challenges, advances in membrane chemistry, membrane fabrication and process optimisation will be required. In particular, current shortcomings with regard to selectivity will have to be resolved. Research to develop membranes with a narrow pore size distribution is ongoing, based on asymmetric pore construction, track etching and the introduction of ceramic and metal membranes (Saxena *et al*., 2009). Furthermore, advances in the development of charged membranes and ultimately the development of industrial-scale membrane chromatography may enable the selective isolation of biomolecules by affinity chromatography from complex mixtures (Brandt *et al*., 1988). Chromatography membranes are usually hollow-fibre membranes, which have biospecific ligands immobilised on the pore surface that absorb target biomolecules during cross-flow of the permeate stream that can be subsequently eluted in high purity (Fig. 11.10). By using highly specific ligands, this affinity chromatography approach could lead to the purification of biopolymers based on biological functions rather than individual physical or chemical properties (Saxena *et al*., 2009). In addition, engineering innovations in the superimposition of electrical charge to improve the selectivity of membrane separation processes can be anticipated, potentially leading to industrial-scale electro-membrane separation systems, such as ED using microporous membranes in the NF, UF or even MF range (Bargeman *et al*., 2000; Poulin *et al*., 2006), or continuous flow electrophoresis (Galier & Balmann, 2004).

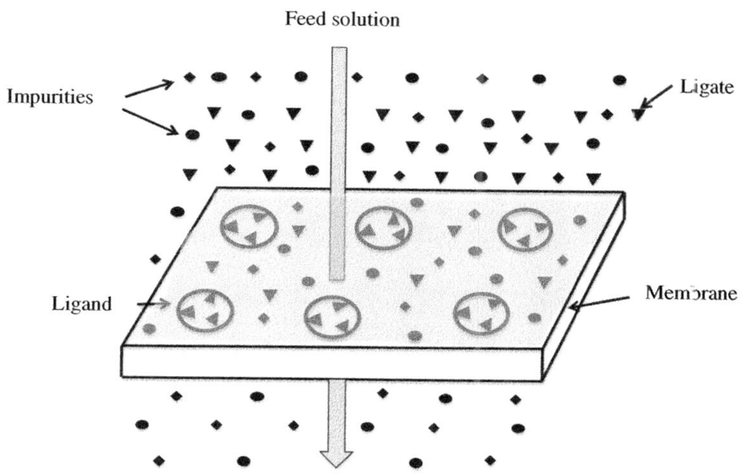

Fig. 11.10 Schematic representation of the principle of membrane chromatography (After Saxena *et al.*, 2009).

To whatever innovations in individual membrane technologies that emerge in the future must be added to the challenge of successful integration into large-scale dairy processing plant operations with all the attendant constraints of volume throughput and energy or water consumption.

References

Al-Akoum, O., Ding, L.H. & Jaffrin, M.Y. (2002) Microfiltration and ultrafiltration of UHT skim milk with a vibrating membrane module. *Separation and Purification Technology*, **28**, 219–234.

Allen, I.M., McAuliffe, A.G. & Donnelly, W.J. (1985) Simplified approaches to casein fractionation. *Irish Journal of Food Science and Technology*, **9**, 85–85.

Amundson, C.H., Watanawanichakorn, S. & Hill Jr., C.G. (1982) Production of enriched protein fractions of β-lactoglobulin and α-lactalbumen from cheese whey. *Journal of Food Processing and Preservation*, **6**, 55–71.

Anderson, M. & Cawston, T. E. (1975) The milk-fat globule membrane (review). *Journal of Dairy Research*, **42**, 459–483.

Anonymous (2004) *Sterile Drug Products Produced by Aseptic Processing – Current Good Manufacturing Practice*, Guidance for Industry published by United States Food and Drug Administration, Washington DC.

Anonymous (2011) *Guidelines for Lactase filtration in Tetra Aldose*, Available from Tetra Pak Dairy and Beverage Systems AB, High Wycombe, UK.

Astaire, J.C., Ward, R., German, J.B. & Jimenez-Flores, R. (2003) Concentration of polar MFGM lipids from buttermilk by microfiltration and supercritical fluid extraction. *Journal of Dairy Science*, **86**, 2297–2307.

Azuma, N. & Yamauchi, K. (1987) A glyco-phosphoprotein in human milk. *Journal of Dairy Research*, **54**, 199–205.

Bacher, T. & Konigsfeldt, P. (2000) WPI by microfiltration of skim milk. *European Dairy Magazine*, **11**, 14–16.

Ballard, F.J., Francis, G.L. & Regester, G.O. (1999) Milk protein mixture for promoting growth of animal cells or treating wounds and a method of making and methods employing the mixture, United States of America Patent Application No. 5866418.

Bargeman, G., Dohmen-Speelmans, M., Recio, I., Timmer, M. & van der Horst, C. (2000) Selective isolation of cationic amino acids and peptides by electro-membrane filtration. *Lait*, **80**, 175–186.

Baruah, G.L., Nayak, A. & Belfort, G. (2006) Scale-up from laboratory microfiltration to a ceramic pilot plant: design and performance. *Journal of Membrane Science*, **274**, 56–63.

Bayless, K.J., Davis, G.E. & Meininger, G.A. (1997) Isolation and biological properties of osteopontin from bovine milk. *Protein Expression and Purification*, **9**, 309–314.

Bazinet, L., Ippersiel, D., Gendron, C., Beaudry, J., Mahdavi, B., Amiot, J. & Lamarche, F. (2000) Cationic balance in skim milk membrane acidification. *Journal of Membrane Science*, **173**, 201–209.

Bazinet, L., Ippersiel, D. & Mahdavi, B. (2004) Fractionation of whey proteins by bipolar membrane electroacidification. *Innovative Food Science and Emerging Technologies*, **5**, 17–25.

Bazinet, L., Lin Teng Shee, F., Angers, P. & Ben Qunis, W. (2008) Method for extracting lipids from biological sources, United States of America Patent Application No. 2008/0171118 A1.

Beckman, S.L., Zulewska, J., Newbold, M. & Barbano, D.M. (2010) Production efficiency of micellar casein concentrate using polymeric spiral-wound microfiltration membranes. *Journal of Dairy Science*, **93**, 4506–4517.

Bird, J. (1996) The application of membrane systems in the dairy industry. *Journal of Society of Dairy Technology*, **49**, 16–23.

Brandsma, R.L. & Rizvi, S.S.H. (1999) Depletion of whey proteins and calcium by microfiltration of acidified skim milk prior to cheese making. *Journal of Dairy Science*, **82**, 2063–2069.

Brandt, S., Goffe, R.A., Kessler, S.B., O'Connor, J.L. & Zale, S.E. (1988) Membrane based affinity technology for commercial scale purifications. *Biotechnology*, **6**, 799–782.

Brans, G., Schroen, C.G.P.H., van der Sman, R.G.M. & Boom, R.M. (2004) Membrane fractionation of milk : state of the art and challenges. *Journal of Membrane Science*, **243**, 263–272.

Britten, M. & Pouliot, Y. (1996) Characterisation of whey protein isolate obtained from milk microfiltration permeate. *Lait*, **76**, 255–265.

Brulé, G., Roger, L., Fauquant, J. & Piot, M. (1980) Procédé de traitment d'une matiere a base caséine contenant des phosphocaséinates de cations monovalents ou leurs derives, produits obtenus et applications, French Patent Application No. 2 474 829.

Corredig, M. & Dalglesigh, D.G. (1997) Isolates from industrial buttermilk: Emulsifying properties of materials derived from the milk fat globule membrane. *Journal of Agricultural and Food Chemistry*, **45**, 4595–4600.

Corredig, M., Roesch, R.R. & Dalgleish, D. (2003) Production of a novel ingredient from buttermilk. *Journal of Dairy Science*, **86**, 2744–2750.

Costa, M.R., Elias-Argote, X.E., Jimenez-Flores, R. & Gigante, M.L. (2010) Use of ultrafiltration and supercritical fluid extraction to obtain a whey buttermilk powder enriched in milk fat globule membrane phospholipids. *International Dairy Journal*, **20**, 598–602.

Damodaran, S. (1995) Removing lipids from cheese whey using chitosan, United States of America Patent Application No. 5436014.

Danthine, S., Blecker, C., Paquot, M., Innocente, N. & Deroanne, C. (2000) Progress in milk fat globule membrane research: A review. *Lait*, **80**, 209–222.

Dinkov, K., Andreev, A. & Panaoitov, P. (1997) Separation of casein and whey proteins of cow milk using two types of ultrafiltration. *Milchwissenschaft*, **52**, 127–130.

Donnan, F. G. (1995) Theorey of membrane equilibria and membrane potentials in the presence of non dialysing electrolytes – A contribution to physico-chemical physiology. *Journal of Membrane Science*, **100**, 45–55.

Eckhardt, E.R.M., Wang, D.Q.H., Donovan, J.M. & Carey, M.C. (2002) Dietary sphingomyelin suppresses intestinal cholesterol absorption by decreasing thermodynamic activity of cholesterol monomers. *Gastroenterology*, **122**, 948–956.

Famelart, M.H. & Surel, O. (1994) Caseinate at low temperatures: calcium use in β-casein extraction by microfiltration. *Journal of Food Science*, **59**, 548–553, 587.

Famelart, M.H., Hardy, C. & Brule, G. (1989) Etude des facteurs d'extraction de la caséine β. *Lait*, **69**, 47–57.

FAO (2010) Standard for whey powders (CODEX Standard 289–1995). Available from Food and Agriculture Organization of the United Nations, Rome, Italy.

Fauquant, J., Maubois, J. L. & Pierre, A. (1988) Microfiltration du lait sur membrane minérale. *Technique Laitière*, **1028**, 21–23.

Fauquant, C., Briard, V., Leconte, N. & Michalski, M.C. (2005) Differently sized native milk fat globules separated by microfiltration: fatty acid composition of the milk fat globule membrane and triglyceride core. *European Journal of Lipid Science and Technology*, **107**, 80–86.

FitzGerald, R.J. & Meisel, H. (2003) Milk protein hydrolysates and bioactive peptides. *Advanced Dairy Chemistry* (eds. P.F. Fox and P.L.H. McSweeney), Vol. 1, 3rd edn, pp. 675–698, Kluwer Academic/Plenum Publishers, New York.

Fox, P.F. (2001) Milk proteins as food ingredients. *International Journal of Dairy Technology*, **54**, 41–55.

Fox, P.F. and McSweeney, P.L.H. (1998) *Dairy Chemistry and Biochemistry*, Chapman and Hall, London.

Francis, G.L., Regester, G.O., Webb, H.A. & Ballard, F.J. (1995) Extraction from cheese whey by cation-exchange chromatography of factors that stimulate the growth of mammalian cells. *Journal of Dairy Science*, **78**, 1209–1218.

Freudenstein, C., Keenan, T.W., Eigel, W.N., Sasaki, M., Stadler, J. & Franke, W.W. (1979) Preparation and characterisation of the inner coat material associated with fat globule membranes from bovine and human milk. *Experimental Cell Research*, **118**, 277–294.

Galier, S. & Balmann, H.R. (2004) Study of biomolecules separation in an electrophoretic membrane contactor. *Journal of Membrane Science*, **241**, 79–87.

Garcin, I. & Paviot, G. (1991) Cosmetical or pharmaceutical composition for cutaneous application and process for the manufacture of such a composition, European Patent Application No. EP0334776B1.

Garem, A., Léonil, J., Daufin, G. & Maubois, J.-L. (1996) Nanofiltration d'acides amines sur membranes organiques: influence des parameters physico-chemiques et de la pression transmembranaire sur la sélectivité. *Lait*, **76**, 267–281.

Giachelli, C.M. & Steitz, S. (2000) Osteopontin: A versatile regulator of inflammation and biomineralization. *Matrix Biology*, **19**, 615–622.

Goudédranche, H., Maubois, J.-L. & Fauquant, J. (1998) Produits, en particulier laitiers, comprenant des fractions selectionnees de globules gras, obtention et applications, French Patent Application No. FR/2/776/208/A1 and International Patent Application No. PCT/FR 99/00632.

Goudédranche, H., Fauquant, J. & Maubois, J.-L. (2000) Fractionation of globular milk fat by membrane microfiltration. *Lait*, **80**, 93–98.

Gourley, L., Britten, M., Gauthier, S.F. & Pouliot, Y. (1994) Characterisation of adsorptive fouling on ultrafiltration membranes by peptide mixtures using contact angle measurements. *Journal of Membrane Science*, **97**, 283–289.

Gourley, L., Gauthier, S.F. & Pouliot, Y. (1995) Separation of casein hydrolysates using polysulfone ultrafiltration membranes with pH and EDTA treatments applied. *Lait*, **75**, 259–269.

Gourley, L., Gauthier, S.F., Pouliot, Y. Molle, D., Leonil, J. & Maubois, J.-L. (1998) Identification of casein peptides interacting with polysulfone ultrafiltration membranes. *Lait*, **78**, 633–646.

Govindasamy-Lucey, S., Jaeggi, J.J., Johnson, M.E., Wang, T. & Lucey, J.A. (2007). Use of cold microfiltration retentates produced with polymeric membranes for standardisation of milks for manufacture of pizza cheese. *Journal of Dairy Science*, **90**, 4552–4568.

Grosvenor, C.E., Picciano, M.F. & Baumrucker, C.R. (1993) Hormones and growth factors in milk. *Endocrine Reviews*, **14**, 710–728.

Heino, A.T., Uusi-Rauva, J.O., Rantamaki, P.R. & Tossavainen, O. (2007) Functional properties of native and cheese whey protein concentrate powders. *International Journal of Dairy Technology*, **60**, 277–285.

Hewedi, M.M., Mulvihill, D.M. & Fox, P.F. (1985) Recovery of milk protein by ethanol precipitation. *Irish Journal of Food Science and Technology*, **9**, 11–23.

Holst, H.H., Gunther, W.S., Mogensen, M.T. & Jorgensen, A.S. (2007) Concentrate derived from a milk product enriched in naturally occurring sialylasctose and a process for preparation thereof, United States of America Patent Application No. 2007/0104843 A1.

Horton, B. (1997) Microfiltration – where is it headed? *Australian Journal of Dairy Technology*, **52**, 73–75.

Houldsworth, D.W. (1980) Demineralisation of whey by means of ion exchange and electrodialysis. *Journal of the Society of Dairy Technology*, **33**, 45–51.

Huppertz, T., Hennebel, J.-B., Considine, T., Ur-Rehman, S., Kelly, A.L. & Fox. P.F. (2006) A method for the large-scale isolation of β-casein. *Food Chemistry*, **99**, 45–50.

Hwang, D.-C. & Damodaran, S. (1995) Selective precipitation and removal of lipids from cheese whey using chitosan. *Journal of Agricultural and Food Research*, **43**, 33–37.

Iordache, M. & Jelen, P. (2003) High pressure microfulidization treatment of heat denatured whey protein for improved functionality. *Innovative Food Science and Emerging Technologies*, **4**, 367–376.

Jelen, P. (2009) Dried whey, whey Proteins, lactose and lactose derivative products. *Dairy Powders and Concentrated Products* (ed. A.Y. Tamine), pp. 255–267, Blackwell Publishing, Oxford.

Jono, S., Peinado, C. & Giachelli, C.M. (2000) Phosphorylation of osteopontin is required for inhibition of vascular smooth muscle cell calcification. *Journal of Biological Chemistry*, **275**, 20197–20203.

Karasu, K., Glennon, N., Lawrence, N.D., Stevens, G.W., O'Connor, A.J., Barber, A.R., Yoshikawa, S. & Kentish, S.E. (2010) A comparison between ceramic and polymeric membrane systems for casein concentrate manufacture. *International Journal of Dairy Technology*, **63**, 284–289.

Keenan, T.W. & Dylewski, D.P. (1994) Intracellular origin of milk lipid globules and the nature and structure of the milk lipid globule membrane. *Advanced Dairy Chemistry* (ed. P.F. Fox), Vol. 2, pp. 89–130, Chapman and Hall, London.

Kelly, J. & Kelly, P. (1995) Nanofiltration of whey: quality, environmental and economic aspect. *Journal of the Society of Dairy Technology*, **48**, 20–25.

Kelly, P.M., Kelly, J., Mehra, R., Oldfield, D.J., Raggett, E. & O'Kennedy, T. (2000) Implementation of integrated membrane processes for pilot scale development of fractionated milk components. *Lait*, **80**, 139–153.

Kernohan, E.A., Wadsworth, J.C. & Lascelles, A.K. (1971) Changes in the composition of bovine milk fat during milking. *Journal of Dairy Research* **38**, 65–68.

Kimura, S. & Tamano, A. (1986). Separation of amino acids by charged ultrafiltration membranes. *Membranes & Membrane Processes* (eds. E. Droli and M. Nahagaki), pp. 191–197, Plenum Press, New York, NY.

Kopf, H.B. & Kopf III, H. (2005) Method & apparatus for separation of milk, colostrum and whey, United States of America Patent Application No. 20050053707A1.

Kunz, C., Rudloff, S., Baier, W., Klien, N. & Strobel, S. (2000) Oligosaccharides in human milk; Structural, functional and metabolic aspects. *Annual Review of Nutrition*, **20**, 699–722.

Lawrence, N.D., Kentish, S.E., O'Connor, A.J., Barber, A.R. & Stevens, G.W. (2008) Microfiltration of skim milk using polymeric membranes for casein concentrate manufacture. *Separation and Purification Technology*, **60**, 237–244.

Le Berre, O. & Daufin, G. (1994) Fouling and selectivity of membranes during separation of β-casein. *Journal of Membrane Science*, **88**, 263–270.

Le Magnen, C. & Maugas, J.J. (1995) Method for obtaining β-casein, United States of America Patent Application No. 5397577.

Liaw, L., Skinner, M.P., Raines, E.W., Ross. R., Cheresh, D.A., Schwartz, S.W. & Giachelli, C.M. (1995) The adhesive and migratory effects of osteopontin are mediated via distinct cell surface integrins: Role of $\alpha_v\beta_3$ in smooth muscle cell migration to osteopontin in vitro. *Journal of Clinical Investigation*, **95**, 713–724.

Lin Teng Shee, F., Angers, P. & Bazinet, L. (2007) Delipidation of a whey protein Concentrate by electroacidification with bipolar membranes. *Journal of Agricultural and Food Chemistry*, **55**, 3985–3989.

Lucey, J.A. & Horne, D.S. (2009) Milk salts. *Advanced Dairy Chemistry* (eds. P.L.H. McSweeney and P.F. Fox), Vol. 3, pp. 351–390, Springer Science, New York, NY.

Lutin, F. & Bar, D. (2007) Keep it natural! Adjusting the pH of food products without chemical additives thanks to bipolar membrane electrodialysis. *Paper presented at the Proceedings of the 20th Annual Meeting North American Membrane Society*, Orlando. (Available from authors at Ameridia, Somerset, NJ 08873, USA).

Marcelo, P.A. & Rizvi, S.S.H. (2008) Physicochemical properties of liquid virgin whey protein isolate. *International Dairy Journal*, **18**, 236–246.

Martinez-Ferez, A., Rudloff, S., Guadix, A., Henkel, C.A., Pohlentz, G., Boza, J.J., Guadix, E.M. & Kunz, C. (2006) Goat's milk as a natural source of lactose-derived oligosaccharides: Isolation by membrane technology. *International Dairy Journal*, **16**, 173–181.

Mather, I. H. (2000). A review and proposed nomenclature for major proteins of the milk-fat globule membrane. *Journal of Dairy Science*, **83**, 203–247.

Maubois, J.-L. (1997) Current uses and future perspectives of MF technology in the dairy industry. *Implications of Microfiltration on Hygiene and Identity of Dairy Products/Genetic Manipulation of Dairy Cultures – Report of Subject F23*, Document No. 320, pp. 154–159, International Dairy Federation, Brussels.

Maubois, J.-L. (2002) Membrane microfiltration: a tool for a new approach in dairy technology. *Australian Journal of Dairy Technology*, **57**, 92–96.

Maubois, J.-L. & Leonil, J. (1989) Peptides du lait à activité biologique. *Lait*, **69**, 245–269.

Maubois, J.-L., Loic, R., Brulé, G. & Piot, M. (1984) Total enzymatic hydrolysate from whey proteins and process of obtaining the same, United States of America Patent Application No. 4427658.

Maubois, J.-L., Pierre, A., Fauquant, J. & Piot, M. (1987) Industrial fractionation of main whey proteins. *Trends in Whey Utilization*, Document No. 212, pp. 154–159, International Dairy Federation, Brussels.

Maubois, J.-L., Fauquant, J., Jouan, P. & Bourtourault, M. (2003). Method for obtaining a TGF-beta enriched protein fraction in activated form, protein fraction and therapeutic applications. Patent Application No. PCT/WO 03/006500.

McPherson, A.V. & Kitchen, B.J. (1983) The bovine milk-fat globule membrane-its formation, composition, structure and behaviour in milk and dairy products (review). *Journal of Dairy Research*, **50**, 107–133.

Mehra, R. & Kelly, P. (2006) Milk oligosaccharides: Structural and technological aspects. *International Dairy Journal*, **16**, 1334–1340.

Michalski, M.C., Cariou, R., Michel, F. & Garnier, C. (2002) Native vs damaged milk fat globules: membrane properties affect the viscoelasticity of milk gels. *Journal of Dairy Science*, **85**, 2451–2461.

Michalski, M.C., Gassi, J.F., Famelart, M.H., Leconte, N., Camier, B., Michel, F. & Briard, V. (2003) The size of native milk fat globules affects physico-chemical and sensory properties of Camembert cheese. *Lait*, **83**, 131–143.

Michalski, M.C., Ollivon, M., Briard, V., Leconte, N. & Lopez, C. (2004) Native fat globules of different sizes selected from raw milk: thermal and structural behaviour. *Chemistry and Physics of Lipids*, **132**, 247–261.

Michalski, M.C., Briard, V. & Juaneda, P. (2005) CLA profile in native fat globules of different sizes selected from raw milk. *International Dairy Journal*, **15**, 1089–1094.

Michalski, M.C., Leconte, N., Briard-Bion, V., Fauquant, J., Maubois, J.-L. & Goudedranche, H. (2006) Microfiltration of raw whole milk to select fractions with different fat globule size distributions: Process optimisation and analysis. *Journal of Dairy Science*, **89**, 3778–3790.

Michalski, M.C., Camier, B., Gassi, J.-Y., Briard-Bion, V., Leconte, N., Famelart, M.-H. & Lopez, C. (2007) Functionality of smaller vs control native milk fat globules in Emmental cheeses manufactured with adapted technologies. *Food Research International*, **40**, 191–202.

Molochnikov, V. & Zetteir, K. H. (1994) Process for separation of natural milk proteins. German Federal Republic Patent Application No. 4227666A1.

Morel, F. (1991) Production de caseine 'native' pour la première fois à l'echelle industrielle. *SAFIR ourvre la voie. Process*, **1058**, 53–56.

Morin, P., Jimenez-Flores, R. & Pouliot, Y. (2004) Effect of temperature and pore size on the fractionation of fresh and reconstituted buttermilk by microfiltration. *Journal of Dairy Science*, **87**, 267–273.

Morin, P., Pouliot, Y. & Jimenez-Flores, R. (2006) A comparative study of the fractionation of regular buttermilk and whey buttermilk by microfiltration. *Journal of Food Engineering*, **77**, 521–528.

Morin, P., Britten, M., Jimenez-Flores, R. & Pouliot, Y. (2007). Microfiltration of buttermilk and washed cream buttermilk for concentration of milk fat globule membrane components. *Journal of Dairy Science*, **90**, 2132–2140.

Morr, C.V. & Ha, E.Y.W. (1991) Off-flavours of whey protein concentrates: a literature review. *International Dairy Journal*, **1**, 1–11.

Mulvihill, D.M. (1991) Trends in the production and utilisation of dairy protein products: functional properties and utilisation. *Food Research Quarterly*, **51**, 65–73.

Mulvihill, D.M. & Ennis, M.P. (2003) Functional milk proteins: production and utilisation. *Advanced Dairy Chemistry* (eds. P.F. Fox and P.L.H. McSweeney), Vol. 1, 3rd edn, pp. 1175–1227, Kluwer Academic/Plenum Publishers, New York, NY.

Mulvihill, D.M. & Fox, P.F. (1994) Developments in the production of milk proteins. *New and Developing Sources of Food Proteins* (ed. B.J.F. Hudson), pp. 1–30, Chapman and Hall, London.

Murphy, J.M. & Fox, P.F. (1991) Fractionation of sodium caseinate by ultrafiltration. *Food Chemistry*, **39**, 27–38.

Nelson, B.K. & Barbano, D.M. (2005) A microfiltration process to maximise removal of serum proteins from skim milk before cheese making. *Journal of Dairy Science*, **88**, 1891–1900.

Noel, R. (1992) Process for separation of calcium phosphocaseinate and whey from skim milk, and more generally separation of a protein from a biological liquid, French Patent Application No. 2 671 945 A1.

Norddahl, B. (2001) Fermentative production and isolation of lactic acid, United States of America Patent Application No. 6319382B1.

Norddahl, B., Eriksen, S. & Pedersen, F.M. (2004) Method for producing lactic acid, United states of America Patent Application No. 20040033573.

O'Mahony, J.A., Lucey, J.A. and Smith, K.E. (2007) Purification of β-casein from milk, United States of America Patent Application No. 2007/0104847A1.

O'Mahony, J.A., McSweeney, P.L.H. & Lucey, J.A. (2008) Observations on the rheological and functional properties of model cheeses made using milk protein concentrate solutions with different ratios of α_{s1}-:β-casein. *Milchwissenschaft*, **63**, 145–148.

O'Mahony, J.A., McSweeney, P.L.H. & Lucey, J.A. (2009) Rheological properties of rennet-induced skim milk gels made from milk protein concentrate solutions with different ratios of α_s-:β-casein. *Milchwissenschaft*, **64**, 135–138.

Olofsson, M., Buhler, M. & Wood, R. (1981) Process for the preparation of a purified protein hydrolysate, United States of America Patent Application No. 4293571.

Olsson, E. (2008) Sterile filtration of heat sensitive products with the Tetra Aldose, Master of Chemical Engineering Thesis, University of Lund, Sweden.

Papadatos, A., Neocleous, M., Berger, A.M. & Barbano, D.M. (2003) Economic feasibility evaluation of microfiltration of milk prior to cheesemaking. *Journal of Dairy Science*, **86**, 1564–1577.

Pearce, G. (2007) Introduction to membranes: manufacturers' comparison: part 1. *Filtration-Separation*, **44**, 35–37.

Pearce, R.J., Marshall, S.C. & Dunkerley, J.A. (1992) Reduction of lipids in whey protein concentrates by microfiltration – effect on functional properties. *New Applications of Membrane Processes – Monograph by an IDF Group of Experts*, Special Issue 9201, pp. 118–129, International Dairy Federation, Brussels.

Petersen, R.J. (1993) Composite reverse osmosis and nanofiltration membranes. *Journal of Membrane Science*, **83**, 81–150.

Pfeuffer, M. & Schrezenmeir, J. (2002) Dietary sphingolipids: Metabolism and potential health implications. *Kieler Milchwirtschaftliche Forschungsberichte*, **53**, 31–42.

Pierre, A., Fauquant, J., Le Graet, Y., Piot, M. & Maubois, J.-L. (1992) Préparation de phosphocaséinate natif par microfiltration sur membrane. *Lait*, **72**, 461–474.

Piot, M., Faquant, J., Madec, M.N. & Maubois, J.-L. (2004) Preparation of serocolostrum by membrane microfiltration. *Lait*, **84**, 333–341.

Poulin, J.F., Amiot, J. & Bazinet, L. (2006) Simultaneous separation of acid and basic bioactive peptides by electrodialysis with ultrafiltration membrane. *Journal of Biotechnology*, **123**, 314–328

Pouliot, Y. (2008) Membrane processes in dairy technology – From a simple idea to worldwide panacea. *International Dairy Journal*, **18**, 735–740.

Pouliot, Y. & Gauthier, S.F. (1990) Effect of selected physico-chemical parameters on the flux decline of casein hydrolysate on polysulfone ultrafiltration membranes. *Proceedings of ICOM'90*, pp. 273–275, Chicago, IL.

Pouliot, Y. & Gauthier, S.F. (2006) Milk growth factors as health products: some technological aspects. *International Dairy Journal*, **16**, 1415–1420.

Pouliot, Y., Gauthier, S.F. & Bard, C. (1993) Fractionation of casein hydrolysates using polysulfone ultrafiltration hollow fiber membranes. *Journal of Membrane Science*, **80**, 257–264.

Pouliot, Y., Gauthier, S.F. & Bard, C. (1995) Skim milk solids as substrate for the preparation of casein enzymatic hydrolysates. *Journal of Food Science*, **60**, 111–116.

Pouliot, M., Pouliot, Y. & Britten M. (1996) On the conventional cross-flow microfiltration of skim milk for the production of native phosphocaseinate. *International Dairy Journal*, **6**, 105–111.

Pouliot, Y., Wijers, M.C., Gauthier, S.F. & Nadeau, L. (1999) Fractionation of whey protein hydrolysates using charged UF/NF membranes. *Journal of Membrane Science*, **158**, 105–114.

Pourcelly, G. & Bazinet, L. (2009) Developments of bipolar membrane technology in food and bio-industries. *Handbook of Membrane Separations* (eds. A.K. Pabby, S.S.H. Rizvi and A.M. Sastre), Section 2, pp. 581–634, CRC Press, Boca Raton, FL.

Rasanen, E., Nystrom, M., Sahlstein, J. & Tossavainen, O. (2002) Comparison of commercial membranes in nanofiltration of sweet whey. *Lait*, **82**, 343–356.

Rizvi, S.S.H. & Brandsma, R.L. (2002). Microfiltration of skim milk for cheesemaking and whey proteins, United States of America Patent Application No. 6485762B1.

Rogers, M,-L., Goddard, C., Regester, G.O., Ballard, F.J. & Belford, D.A. (1996). Transforming growth factor B in bovine milk: Concentration stability and molecular mass forms. *Journal of Endocrinology*, **151**, 77–86.

Rombaut, R. & Dewettinck, K. (2007) Thermocalcic aggregation of milk fat globule membrane fragments from acid buttermilk cheese whey. *Journal of Dairy Science*, **90**, 2665–2674.

Rombaut, R., Dejonckheere, V. & Dewettinck, K. (2006) Microfiltration of butter serum upon casein micelle destabilization. *Journal of Dairy Science*, **89**, 1915–1925.

Rombaut, R., Dejonckheere, V. & Dewettinck, K. (2007) Filtration of milk fat globule membrane fragments from acid buttermilk cheese whey. *Journal of Dairy Science*, **90**, 1662–1673.

Rosenberg, M. (1995) Current and future applications for membrane processes in the dairy industry. *Trends in Food Science and Technology*, **6**, 12–19.

Saboya, L.V. & Maubois, J.-L. (2000) Current developments of microfiltration technology in the dairy industry. *Lait*, **80**, 541–553.

Sachdeva, S. & Buchheim, W. (1997) Recovery of phospholipids from buttermilk using membrane processing. *Kieler Milchwirtschaftliche Forschungsberichte*, **49**, 47–68.

Sarney, D.B., Hale, C., Frankel, G. & Vulfson, E. (2000) A novel approach to the recovery of biologically active oligosaccharides from milk using a combination of enzymatic treatment and nanofiltration. *Biotechnology and Bioengineering*, **69**, 461–467.

Saxena, A., Tripathi, B.P., Kumar, M. & Shahi, V.K. (2009) Membrane-based techniques for the separation and purification of proteins: An overview. *Advances in Colloid and Interface Science*, **145**, 1–22.

Schack, L., Lange, A., Kelsen, J., Agnholt, J., Christensen, B., Petersen, T.E. & Sorensen, E.S. (2009) Considerable variation in the concentration of osteopontin in human milk, bovine milk and infant formulas. *Journal of Dairy Science*, **92**, 5378–5385.

Schmelz, E.M. (2004) Sphingolipids in the chemoprevention of colon cancer. *Frontiers in Bioscience*, **9**, 2632–2639.

Schuck, P., Piot, M., Méjean, S. Le Graet, Y., Fauquant, J., Brulé, G. & Maubois, J.-L. (1994) Déhydratation par atomisation de phosphocaséinate natif obtenu par microfiltration sur membrane. *Lait*, **74**, 375–388.

Senger, D.R., Perruzzi, C.A., Papadopoulos, A. & Tenen, D.G. (1989) Purification of a human milk protein closely similar to tumor-secreted phosphoproteins and osteopontin. *Biochimica et Biophysica Acta*, **996**, 43–48.

Singh, S., Viotto, L.A. & Rizvi, S.S.H. (1995) Fractionation of skim milk by ceramic membranes: II. Separation of caseins and evaluation of their functional properties. *Journal of Dairy Science*, **78**(Supplement 1), 147–147.

Slack, A.W., Amundson, C.H. & Hill, C.G. (1986) Production of enriched β-lactoglobulin and α-lactalbumin whey fractions. *Journal of Food Processing and Preservation*, **10**, 19–30.

Smith, K.E. (2004) Membranes 101. *Proceedings of Membrane Processing of Dairy Products Short Course*, Wisconsin Centre for Dairy Research and University of Wisconsin, Madison.

Sorensen, E.S. & Petersen, T.E. (1993) Purification and characterisation of three proteins isolated from the proteose peptone fraction of bovine milk. *Journal of Dairy Science*, **60**, 189–197.

Sorensen, E.S., Horup, P. & Petersen, T.E. (1995) Posttranslational modification of bovine osteopontin: Identification of twenty-eight phosphorylation and three O-glycosylation sites. *Protein Science*, **4**, 2040–2049.

Sorensen, E.S. Justesen, S.J. & Johnsen, A.H. (2003). Purification and characterisation of osteopontin from human milk. *Protein Expression and Purification*, **30**, 238–245.

Sorensen, E.S., Ostersen, S., Chatterton, D., Holst, H.H. & Albertsen, K. (2007) Process for isolation of osteopontin from milk, United States of America Patent Application No. 7259243B2.

Steventon, A.J., Donald, A.M. & Gladen, L.F. (1994) Thermal aggregation of whey proteins under fluid shear conditions. *Biochemistry of Milk Products* (eds. A.T. Andrews and J. Varley), pp. 133–142, Royal Society of Chemistry, Cambridge.

St-Gelais, D., Passey, C. A., Haché, S. & Roy, P. (1997) Production of low-fat Cheddar cheese from low and high mineral retentate powders and different fractions of milkfat globules. *International Dairy Journal*, **7**, 733–741.

Surel, O. & Famelart, M.H. (1995) Ability of ceramic membranes to reject lipids of dairy products. *Australian Journal of Dairy Technology*, **50**, 36–40.

Timmen, H. & Patton, S. (1988) Milk fat globules: Fatty acid composition, size and *in vivo* regulation of fat liquidity. *Lipids*, **23**, 685–689.

Tobin, J.T., Fitzsimons, S.M., Kelly, A.L., Kelly, P.M., Auty, M.A.E. & Fenelon, M.A. (2010). Microparticulation of mixtures of whey protein and inulin. *International Journal of Dairy Technology*, **63**, 32–40.

Tolkach, A. & Kulozik U. (2005) Fractionation of whey proteins and caseinomacropeptide by means of enzymatic crosslinking and membrane separation techniques. *Journal of Food Engineering*, **67**, 13–20.

Tomasula, P.M. Craig, J.C. & Boswell, R.T. (1997) A continuous process for casein production using high-pressure carbon dioxide. *Journal of Food Engineering*, **33**, 405–419.

Tomasula, P.M., Craig, J.C., Boswell, R.T., Cook, R.D., Kurantz, M.J. & Maxwell, M. (1995) Preparation of casein using carbon dioxide. *Journal of Dairy Science*, **78**, 506–514.

Tsuru, T., Shutou, T., Nakao, S.I. & Kimura, S. (1994) Peptide and amino acid separation with nanofiltration membranes. *Separation Science and Technology*, **29**, 971–984.

Turgeon, S.L. & Gauthier, S.F. (1990) Whey peptide fractions obtained from a two step ultrafiltration process: production & characterization. *Journal of Food Science*, **55**, 106–110, 157.

Walstra, P. (1994) Physical chemistry of milk fat globules. *Advanced Dairy Chemistry* (ed. P.F. Fox), Vol. 2, pp. 131–178, Chapman and Hall, London.

Walstra, P. & Jenness, R. (1984) *Dairy Chemistry and Physics*, John Wiley and Sons, New York.

Ward, L.S. & Bastian, E.D. (1996) A method for isolating β-casein. *Journal of Dairy Science*, **79**, 1332–1339.

Wijers, M.C., Pouliot, Y., Gauthier, S.F., Pouliot, M. & Nadeau, L. (1998) Use of nanofiltration membranes for the desalting of peptide fractions from whey protein enzymatic hydrolysates. *Lait*, **78**, 621–632.

Woychik, J.H. (1992) Preparation of simulated human milk protein by low temperature microfiltration, United States of America Patent Application No.5169666.

Woychik, J.H., Cooke, P. & Lu, D. (1992) Microporous ultrafiltration of skim milk. *Journal of Food Science*, **57**, 46–48, 58.

12 Fruit Juices

A. Cassano

12.1 Introduction

12.1.1 *General Background*

Epidemiological studies have shown that there is a significant positive association between the intake of fruits and vegetables and a reduced risk of chronic diseases, such as cancer, cardiovascular disease, diabetes, osteoporosis, Alzheimer's and immune disorders (Kaur & Kapoor, 2001). Consequently, the consumers' interest in these products and their derivatives, such as fruit juices and fruit-based drinks (nectars, cocktails and drinks), is expected to increase in the future. In particular, the global market for fruit juices was valued at approximately US$79 billion in 2009. This market is projected to grow at an annual growth rate of 3.4% during 2009–14 to reach a value of US$93 billion in 2014. Fruit juice blends of traditional juices (e.g. orange, grape fruit, apple and grape juices) and non-traditional fruits (e.g. tropical fruits) contribute significantly to the fruit juice market and have received wide consumer acceptance. In addition, the use of juices as carriers of nutrients will further contribute to the expansion of functional foods.

Although the raw material selection, cultivar and maturity of the fruits affect the quality of the produced juices, different treatments involved in juice processing (extraction, pasteurisation, concentration, packaging, etc.) have a significant impact on the organoleptic and nutritional properties of the processed juices.

12.1.2 *Background to manufacturing practice*

Conventional methods of producing clarified juice involve several batch operations that are labour- and time-consuming. A typical traditional scheme for producing clear apple juice is shown in Fig. 12.1. After preliminary washing, sorting and grinding, the juice is extracted through a juice press. For example, the apple juice collected from the press contains coarse particulate matter which can be removed by using a cider screen (e.g. cylindrical stainless steel screen of 100–150 mesh) or through a centrifugation step. The juice is then heated to 72°C for 16 s to inhibit yeast growth and prevent fermentation. The following steps aim to remove suspended solids and colloids (including proteins, pectins, starch, phenolic compounds and other carbohydrates) from the freshly pressed

Membrane Processing: Dairy and Beverage Applications, First Edition. A. Y. Tamime.

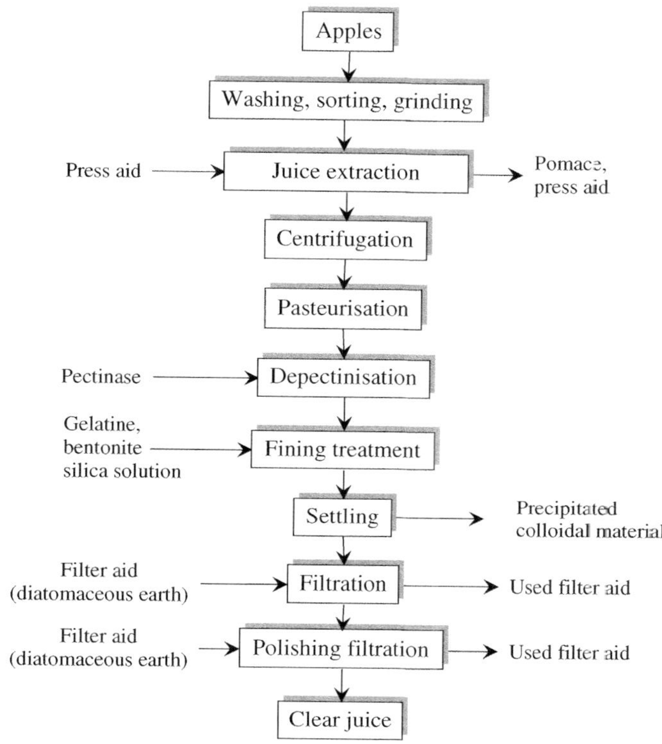

Fig. 12.1 Conventional process for the production of clear apple juice.

juice, responsible for haze formation in the final product. The juice is initially treated with pectinase in order to hydrolyse the pectin. This reduces the juice viscosity making the product easier to filter. Enzyme treatment takes place at temperatures between 40°C and 50°C and, for fruit of average size, the decomposition of pectin is concluded in less than 1 h. Fine materials contained in the juice precipitate during the subsequent settling as sediment. Before settling, fining agents, such as gelatine, bentonite or silica solution, are added to the juice to improve the effect of settling. The positively charged gelatine interacts with negatively charged proteins, tannins and other colloids causing aggregation and precipitation of these materials. The juice is typically held overnight before decanting the supernatant. After decanting, the juice is submitted to a precoat filtration using filter aids, i.e. diatomaceous earths and perlite to remove suspended solids, colloidal particles, proteins and polyphenols. In general, the filtered juice contains less than 0.1 g 100 g^{-1} suspended solids. It can be submitted to a second polishing filtration in order to produce a sparkling clear juice. The final product having a turbidity of 1.5–3 nephelometric turbidity units (NTU) can be pasteurised before packaging.

The use of traditional fining agents is characterised by several drawbacks: (a) significant costs (the cost of diatomaceous earth filters for small and medium-sized fruit juice manufacturers has been estimated between 60 000 and 100 000 Euro per year); (b) risks of dust inhalation with consequent health problems due to handling and disposal; and (c) environmental problems because of the deposition of sludge or filter cake after use.

Membrane operations, such as microfiltration (MF) and ultrafiltration (UF), are competitive and attractive alternatives to the conventional filtration of fruit juices. The advantages of using membrane technology in the beverage industry are related to economy, working conditions, environment and product quality (Koseoglu *et al*., 1990; Hägg, 1998). Typical advantages of membrane technology over conventional fining agents are:

- low energy requirements and costs
- possibility of avoiding the use of gelatines, adsorbents and other filtration aids (avoiding dust formation and sludge deposition
- possibility of lower temperature processing (hence reduction of thermal damage to food during processing)
- simpler process design
- reduction of waste products
- possibility of avoiding use of chemical agents
- increased juice yield (juice recovery of 96–98% can be obtained in contrast to conventional processes)
- reduction in enzyme utilisation (depending on the juice enzyme usage is reduced from 50% to 75%)
- easy cleaning and maintenance of the equipment
- better product quality
- possibility of operating in a single step reducing working times (traditional methods requiring fining agents may take 12–36 h; in membrane cross-flow filtration a clear juice can be obtained in 2–4 h)
- possibility of avoiding pasteurisation (at 60–65°C) and sterilisation (at about 110°C) (MF can be considered to be cold sterilisation, as it removes the bacteria and other live organisms present in the juice).

In addition, the cost savings associated with the clarification of fruit juices by MF or UF processes can be quite substantial. Mondor & Brodeur (2002) estimated that 79% reduction in annual production costs for a UF plant producing 250 m^3 of juice per day in comparison with conventional plants. This saving was attributed to the elimination of products required for the old method and to lower labour costs.

All these features are becoming very important factors in the production of new additive-free fruit juices with natural fresh taste (Cassano *et al*., 2004). Table 12.1 compares the characteristics of conventional processes and MF in the clarification of fruit juices.

In this chapter, a general overview on the use of MF and UF operations in fruit juice clarification is given. Basic principles, membrane module configurations and process

Table 12.1 Clarification of fruit juices – comparison of traditional processes with microfiltration.

Processing method	Operating time	Fibres	Sensory and nutritional quality	Operating cost
Decanting	Very long	Poor	Poor	Intermediate
Dead-end filtration	Long	High	Intermediate	Intermediate
Centrifugation	Very short	Poor	High	High
Microfiltration	Short	High	High	Low

designs involved in these operations are described. Membrane fouling and other parameters affecting the performance of MF and UF membranes in the clarification of fruit juices are also discussed. Finally, the impact of MF and UF processes on the quality of clarified juices and their integration on small- and large-scale operations is reviewed.

12.2 Fruit juice clarification by microfiltration and ultrafiltration

Membrane operations, such as MF and UF were introduced to the fruit juice industry in the late 1970s to replace holding, filtration and decantation steps.

12.2.1 *Microfiltration*

MF is based on the use of symmetric or asymmetric membranes having pore sizes in the range 0.05–10 μm. Basically, the separation principle is a sieving mechanism and transport through the membrane occurs as a result of a driving force acting on the individual components in the feed. In addition, MF is the membrane process which most closely resembles conventional coarse filtration, and it can be used to separate juices into a fibrous concentrated pulp and a clarified fraction free of spoilage micro-organisms. Also, MF is replacing a number of traditional clarification and sterilisation operations in a wide variety of industries. The increase in disposal and purchase costs will make it economically feasible to invest in alternative filtration methods for several industry sectors with a large use of filter aids, such as the brewing, pharmaceutical and beverage industries.

12.2.2 *Ultrafiltration*

In contrast, UF membranes have been extensively studied during the last 25 years for the clarification of fruit and vegetable juices. They are able to retain large components in the product, such as micro-organisms, lipids, proteins and colloids while small solutes, for example vitamins, salts and sugars, pass through the membrane together with water (i.e. permeate). Thus, the UF process substitutes the fining step in the traditional process. The possibility of microbial contamination in the permeate stream is minimised avoiding any thermal treatment and, as a consequence, a loss of volatile aroma substances will occur (Tallarico *et al*., 1998). In order to achieve high yield, high capacity and excellent quality, an enzyme treatment and proper pre-filtration must be carried out before the UF system is utilised.

Fruit juices are a colloidal mixture of soluble and suspended solids containing both microsolutes, such as sugars, organic acids, pigments, vitamins, mineral salts, and macrosolutes, such as proteins, enzymes, pectic substances, polyphenols, micro-organisms, etc. (Table 12.2). Cell wall debris, consisting principally of cellulose and pectin with small amounts of hemicelluloses and hydroxyproline-rich protein, is the main component in the suspended solids. Moreover, dietary fibres are generally retained in the retentate fraction of the MF and UF processes. In order to have a juice enriched in fibres, the retentate can be thermally sterilised and mixed in the appropriate proportion with the clarified juice. This product is characterised by improved qualities in terms of natural aroma, flavouring

Table 12.2 Typical fruit juice compounds and size.

Component	Molecule size	Product
Water (70–90 g 100 g^{-1})	–	–
Soluble matter	<1 kDa	Sugars, organic acids, vitamins, aroma compounds, pigments, simple phenols, salts
	>1 kDa (macromolecules)	Polysaccharides, proteins, tannins, polyphenols
Insoluble material	–	Cells, cell walls, fibres, crystals, starch granules, micro-organisms (1–200 μm)

agents, food values, etc., if compared with thermally treated juice. Fibres tend to settle with time in the transparent packing bottle, and the product loses its consumer appeal. Therefore, clarified juices depleted in fibres are preferred by consumers at the moment.

12.2.3 *Selection of microfiltration and ultrafiltration membranes*

The selection of MF or UF membranes for fruit juice clarification should take in consideration the viscosity of the raw juice, the sterilisation ability and the investment costs of the membranes, the retention of specific compounds and the membrane performance. For pulpy juices, with high solids content and viscosity, large-bore tubular modules or plate and frame modules with large spacers are preferred. On the other hand, particle clogging can occur in the narrow flow path of hollow fibre or spiral-wound membrane modules reducing permeate flux and interfering with subsequent chemical and mechanical cleaning. Consequently, the use of thin channel devices requires additional pre-treatments of the raw juice to reduce the overall content of solids.

Polysulphone membranes have been used extensively for ultrafiltered juice; compared with cellulose acetate membranes, they can withstand short exposures to hypochlorite during periodic cleaning cycles. Polyvinylidenefluoride (PVDF) and polyamide membranes are also largely used in some juice UF systems. Ceramic membranes have greater resistance to chemical degradation, much longer life and higher costs than organic membranes. Retention and performance should be tested at the pilot-plant level.

The most used configurations for the clarification of fruit juices at industrial level are tubular (inner diameter 5–10 mm), capillary (1–1.5 mm) and plate-and-frame membrane modules. The tubular configuration is associated with low packing density and high membrane replacement costs. A new concept, which has recently been developed, combines a high-speed separator with spiral-wound UF modules to overcome these limitations (Lipinzki, 2010).

12.3 Membrane fouling and membrane cleaning

12.3.1 *Membrane fouling*

As in other applications of membrane technology, the main problem in the use of MF and UF membranes for the clarification of fruit juices is membrane fouling. Membrane fouling manifests itself as a decline in flux with the time of operation, reducing the membrane permeability. The degree of membrane fouling determines the frequency of cleaning, the lifetime of the membrane, the membrane area needed and consequently costs, design and operation of membrane plants.

Membrane fouling is a complex phenomenon due to the accumulation of colloidal particles present in the juice on the membrane surface, in the form of a cake layer, or blocking of the membrane pores (Todisco *et al*., 1996). Fruit juice fouling materials are mainly composed of cell wall polysaccharides and macromolecules, such as pectins (polyuronic acids mostly derived from D-galacturonic acid, mainly plentiful in apple and citrus fruits), cellulose, lignin and hemicelluloses (Yu *et al*., 1986).

In addition to the membrane–solute interactions, process parameters, such as temperature, flow rate, feed concentration and pressure, as well as overall equipment design, have great influence on membrane fouling. Basically, the complete retention of pectin and an acceptable value of permeate flux are strictly considered for the selection of a membrane with an appropriate molecular weight cut-off (MWCO).

12.3.2 *Methods of reducing membrane fouling*

The pre-treatment of fruit juices before membrane filtration is an essential step to enhance filtration performance. In particular, mixtures of enzymes, known as pectinases, are used to hydrolyse pectin into poly-D-galacturonic acid fragments, reducing the viscosity of the juice with relatively low pulp leading to an increasing in permeate fluxes and yield recovery (Alvarez *et al*., 1998). UF of depectinised juices can achieve more than 95% recovery of the product. In the case of pulpy juices, a more aggressive enzymatic degradation is needed in order to liquefy insoluble cell wall polysaccharides retained by the membrane. Vaillant *et al*. (1999) reported that a synergistic effect between pectinase and cellulase activities enhanced permeate fluxes in the MF of passion fruit with ceramic membranes having a pore size of 0.2 μm.

Minimum viscosity, high clarity and minimum concentration of pectin are optimum depectinisation conditions required for juice clarification. Usually, at the industrial level, the depectinisation is carried out in batches and then the depectinised juice is used as feed for the UF system. However, pectinase can also be added directly to the feed tank and recycled with the retentate stream. This reduces enzyme addition by about 60–70%.

Other factors affecting the filtration efficiency of juice are proteins, fibres, suspended solids, etc. Therefore, in addition to depectinisation, various treatments can be used to improve the performance of MF and UF membranes. They include treatment with proteases for the removal of proteins, centrifugation (before or after depectinisation) and use of fining agents, such as gelatine and bentonite.

Rai *et al*. (2006, 2007) evaluated the effect of various pre-treatment methods on the performance of UF of mosambi juice in order to select an appropriate process for clarification with high permeate flux and low membrane fouling. The pre-treatment methods investigated included centrifugation, fining by gelatine, fining by bentonite and gelatine, enzymatic treatment using pectinase, enzymatic treatment followed by centrifugation and enzymatic treatment followed by fining with bentonite. Among the various pre-treatment methods, maximum permeate flux (23 L m^{-2} h^{-1} at 0.41 MPa) was obtained in the enzymatic treatment followed by adsorption using bentonite. Compared with the only enzymatic treatment, an increase in flux of about 77% was observed.

In addition to the pre-treatment of the feed solution, other flux enhancement methods have been proposed in order to reduce membrane fouling in juice filtration, and include ultrasonic vibration, periodic backwashing with air or N_2 (Su *et al*., 1993) and backwashing by pulsating entry flow (Padilla-Zakour & McLellan, 1993). The use of turbulence promoters (Millward *et al*., 1995) also reduces fouling, although fluidised bed systems and rotary modules seem not very flexible for large-scale application because of their high costs. In general, the limit to processing fruit juices by UF is due to the pulp content and viscosity of concentrate. For high pulp content, vibrating membrane systems are more appropriate because of the higher shear rate created at the membrane interface resulting in better control of fouling phenomena (Strathmann *et al*., 2006).

Membrane surface treatments, aimed at reducing the interaction between proteins and other foulants with the membrane, can be also useful to reduce fouling phenomena. Although fouling is more severe in porous membranes, such as MF and UF modules, a narrow pore size distribution can reduce the fouling. Moreover, hydrophilic membranes are less fouled than hydrophobic membranes. Negatively charged colloids in the juice can be rejected by negatively charged membranes reducing membrane fouling.

Sarkar *et al*. (2008) investigated the effect of an external electric field on the enhancement of permeate flux during clarification of mosambi (*Citrus sinensis* (L.) *Osbeck*) juice by UF. Experiments were performed in a cross-flow electro-UF cell equipped with a flat-sheet polyethersulphone membrane of 50 kDa under laminar flow regime for a wide range of operating conditions. The use of an electric field resulted in a 32% enhancement of permeate flux.

12.3.3 *Methods of fouling treatment*

Fouling of membrane systems can be minimised to some degree by using specific cleaning methodologies. The frequency with which membranes need to be cleaned can be estimated from the process optimisation. Basically, cleaning methods can be summarised in three different categories:

- chemical cleaning
- mechanical cleaning
- hydraulic cleaning.

The selection of the best cleaning method depends on the module configuration, the chemical resistance of the membrane and the type of foulant encountered. Chemical cleaning is largely used to reduce membrane fouling. Chemical detergents can be used in combination or separately. The frequency of cleaning and chemical concentration will depend on the chemical resistance of the membrane.

Mechanical cleanings can only be applied for tubular membrane systems by using oversized sponge balls. Hydraulic cleanings include back-flushing, which consists in a periodic reversing of the transmembrane flow. This system aims to remove the fouling layer from inside the membrane and from the membrane surface by using the permeate, another liquid or a gas as the back-flushing medium. The 'back-shock' technique was introduced by Wenten *et al*. (1995) in order to reduce the loss of permeate during back-flushing and optimise the operation time during the filtration process. It consists in reversing the permeate flow for a short period (less than a second) by using low linear velocities in order to reduce concentration polarisation and fouling phenomena.

12.4 Performance of microfiltration and ultrafiltration membranes

Permeate fluxes in the clarification of fruit juice by MF and UF are strongly affected by the operating conditions, such as transmembrane pressure, cross-flow velocity, temperature and volume reduction factor (VRF), nature of the membrane and nature of the feed solution (Cheryan, 1998).

In general, the permeate flux increases with membrane pore size whereas the retention of sugars decreases as membrane pore size is increased (Girard & Fukumoto, 2000). Laorko *et al*. (2010) studied the clarification of pineapple juice using both MF and UF membranes with pore size of 0.1 and 0.2 μm and MWCO of 30 and 100 kDa, respectively. The MF process with the 0.2 μm membrane produced the highest permeate fluxes and antioxidant compound recovery in the clarified juice and highest values of irreversible fouling due to the presence of large pores causing the penetration of particles in the internal structure of the membrane. The adsorption of colloidal material is strongly affected by the membrane material. Furthermore, the surface structure and flux resistance of the fouling layer is affected by membrane morphology. Dense fouling layers were detected on smooth surfaced membranes. In contrast, rough membranes produced a looser surface fouling layer with a lower flow resistance per unit thickness of foulant (Riedl *et al*., 1998).

Mathematical models employed to analyse flux decays involved in the UF of pineapple juice established that complete pore blockage and cake formation are the main mechanisms involved in the filtration process in ceramic tubular membranes and polysulphone hollow fibre membranes, respectively (de Barros *et al*., 2003). Analyses of permeate flux decline according to the resistance-in-series model (Jiraratananon & Chanachai, 1996) in the UF of kiwifruit juice with a tubular PVDF membrane revealed that the cake layer and irreversible fouling resistances gave a minimum contribution to the total resistance (2.23% and 2.75%, respectively), whereas the contribution of reversible fouling was more significant (29.4%). Good restoration of the hydraulic permeability of the membrane (about 96% of the initial one) was observed after cleaning using alkaline and acid detergents (Cassano *et al*., 2007a).

Many researchers investigated the effects of these operating conditions on the performance of MF and UF processes employed to clarify different fruit juices. Table 12.3 summarises the membrane type and the steady-state permeate flux values obtained under selected operating conditions for different fruit juices. Fig. 12.2 shows the effect of transmembrane pressure (TMP), feed flow rate and temperature on permeate flux in the clarification of blood orange juice by using a tubular PVDF membrane with a MWCO of 15 kDa. At low pressures, the permeate flux is proportional to the applied pressure. As the pressure is increased, flux shows a deviation from linear flux pressure behaviour, and it becomes independent of pressure (Fig. 12.2a). In these conditions, a limiting flux is reached and further pressure increases do not determine any significant increase in the permeate

Table 12.3 Steady-state permeate fluxes obtained in the clarification of fruit juices by microfiltration (MF) and ultrafiltration (UF) in selected operating conditions.

Fruit juice	Membrane	Operating conditions	Permeate flux ($L\ m^{-2}\ h^{-1}$)	References
Acerola	MF, tubular, PES, 0.3 μm	100 kPa, 30°C	150	Matta *et al.* (2004)
Apple	UF, hollow fibre, PS, 50 kDa	73 kPa, 20 L min^{-1}, 50°C	55	Constela & Lozano (1997)
	MF, flat-sheet ceramic, 0.2 μm	414 kPa, 8 m s^{-1}, 50°C	200	Fukumoto *et al.* (1998)
	UF, tubular, ceramic, 50 kDa	150 kPa , 7 m s^{-1}, 50–55°C	82.5	De Bruijn *et al.* (2003)
	UF, tubular, ceramic, 30 kDa	200 kPa, 800 mL min^{-1}, 55°C	30	Vladisavljević *et al.* (2003)
	UF, hollow fibre, PS, 30 kDa	150 kPa, 200 mL min^{-1}	12	Youn *et al.* (2004)
Black currant	MF, tubular, ceramic, 0.45 μm	390 kPa, 500 L h^{-1}, 25°C	30	Kozák *et al.* (2008)
Blood orange	UF, tubular, PVDF, 15 kDa	85 kPa, 800 L h^{-1}, 25°C	12	Cassano *et al.* (2007b)
Kiwi	UF, hollow fibre, 50 kDa	1375 kPa, 20°C	15	Wilson & Burns (1983)
	UF, tubular, PVDF, 15 kDa	85 kPa, 500 L h^{-1}, 25°C	7	Cassano *et al.* (2004)
	UF, hollow fibre, PEEK-WC, >69 kDa	75 kPa, 40 L h^{-1}, 25°C	25	Tasselli *et al.* (2007)
	UF, tubular, PVDF, 15 kDa	90 kPa, 700 L h^{-1}, 25°C	8	Cassano *et al.* (2007a)
Lemon	MF, flat-sheet, PVDF	60 kPa, 1 m s^{-1}, 20°C	20	Espamer *et al.* (2006)
Melon	MF, tubular, ceramic, 0.2 μm	150 kPa, 7 m/s, 35°C	70	Vaillant *et al.* (2005)
Mosambi	UF, flat-sheet, composite PA, 50 kDa	414 kPa, 30 ± 2°C	13	Rai *et al.* (2006)
Orange	MF, tubular, PVDF, 0.3 μm	50 kPa, 972 L h^{-1}, 19°C	11.8	Todisco *et al.* (1996)
	MF, tubular, ceramic, 0.2 μm	400 kPa, 7 m s^{-1}, 20°C	62	Cisse *et al.* (2005)
Passion fruit	MF, tubular, ceramic, 0.2 μm	150 kPa, 7 m s^{-1}, 36°C	50	Vaillant *et al.* (1999)
Pineapple	MF, tubular, PES, 0.3 μm	100 kPa, 6 m s^{-1}, 25°C	110	Carneiro *et al.* (2002)
	UF, tubular, ceramic, 0.01 μm	200 kPa, 4.17 m s^{-1}, 40°C	102	de Barros *et al.* (2003)
	MF, hollow fibre, PS, 0.2 μm	70 kPa, 3.4 m s^{-1}, 20±2°C	36	Laorko *et al.* (2010)
Pomegranate	MF, flat-sheet, PVDF, 0.45 μm	50 kPa	6	Mirsaeedghazi *et al.* (2010)

PS, polysulphone; PES, polyethersulphone; PVDF, polyvinylidenefluoride; PA, polyamide; PEEK-WC, sulphonated poly(oxa-p-phenylene-3,3-phtalido-p-phenylene-oxa-p-phenileneoxy-phenylene).

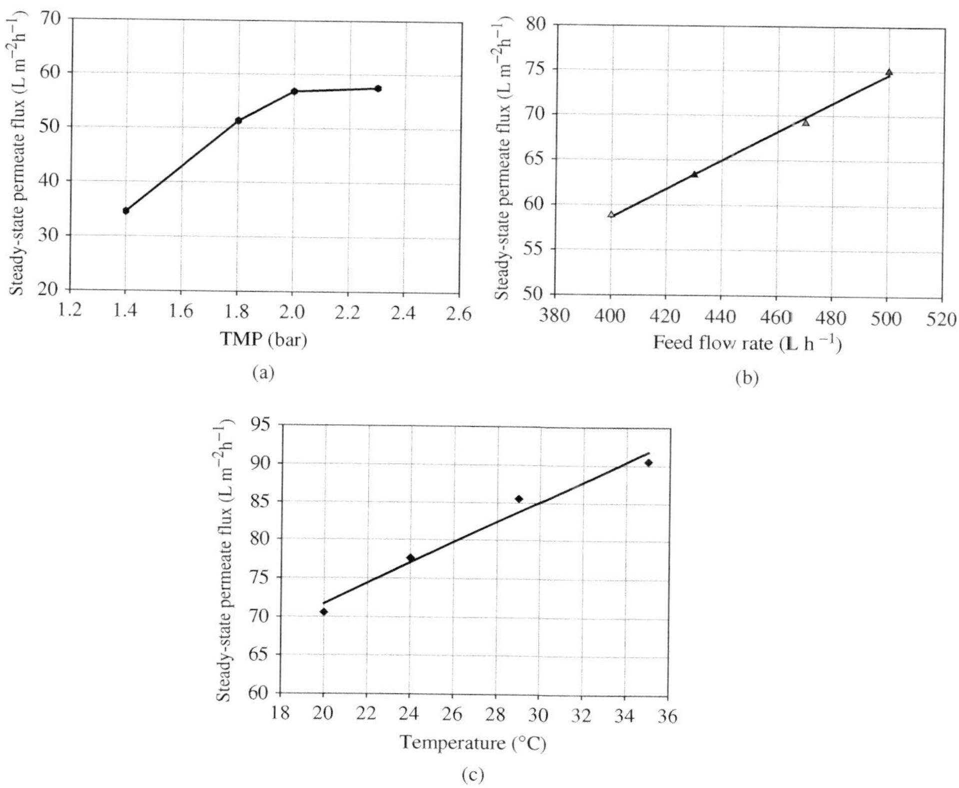

Fig. 12.2 Clarification of cactus pear juice with a flat-sheet microfiltration (MF) membrane. (a) Effect of transmembrane pressure (TMP) on steady-state permeate flux (temperature, 20°C; feed flow rate, 460 L h^{-1}); (b) effect of feed flow rate on steady-state permeate flux (temperature, 26°C; TMP, 0.2 MPa); and (c) effect of temperature on steady-state permeate flux (TMP, 0.2 MPa; feed flow rate, 500 L h^{-1}) (10 bar = 1 MPa).

flux. The existence of a limiting flux can be related to the concentration polarisation phenomenon that arises as the feed solution is convected towards the membrane where the separation of suspended and soluble solids from bulk solution takes place. A concentration profile from bulk solution to membrane surface is generated by the rejected material accumulated on the membrane. The formation of a viscous and gelatinous-type layer is responsible for additional resistance to the permeate flux in addition to that of the membrane. The TMP limiting value (TMP_{lim}) depends on physical properties of the suspension and feed flow rate. The gel-polarised layer is assumed to be dynamic, and higher flow rates tend to remove the deposited material, reducing hydraulic resistance through the membrane and increasing the permeate flux (Cheryan, 1998).

An increase in cross-flow velocity enhances wall shear stress on the membrane surface, reducing concentration polarisation and reversible fouling on the membrane surface; consequently, the mass transfer coefficient is enhanced, leading to an increase in permeation flux (Fig. 12.2b) (Vadi & Rivzi, 2001). When the operating temperature is raised, the feed viscosity is reduced and the diffusion coefficients of macromolecules increase. The effect of these two factors is to enhance mass transfer and to increase the permeation rate according to the film model (Fig. 12.2c) (Nilsson, 1990). However, results obtained by Jiraratananon & Chanachai (1996) during the clarification of passion fruit juice by UF showed that the flux rate increased with temperature from 30°C to 40°C and then decreased at 50°C. This phenomenon was attributed to the probable formation of a true gel on the membrane surface caused by cross-linking between deposited entangled pectin and starch molecules when the temperature was raised. This cross-linked gel increases resistance to flow. In general, it is recommended to operate at the highest possible temperature that is consistent with the limits of the feed solution and the membrane.

According to the film model theory, a decrease in permeate flux is observed when the juice concentration is increased. In addition, the method used for extracting the juice affects the suspended solids in the raw juice and, consequently, the productivity of the process. The VRF is another parameter influencing the productivity of MF and UF membranes in the clarification of fruit juices. VRF is defined as the ratio of initial feed to final retentate volume. High VRF (>10) can be easily reached in juices with low pulp content, such as grape and apple juices. For very pulpy juices, such as tropical fruits, optimum VRF values are less than 5. The yield (1 – (1/VRF)) of recovered juice is a crucial item in the economics of the process. Fig. 12.3 shows a typical variation of the permeate flux and VRF with the filtration time in the clarification of orange juice by UF. The permeate flux decreases rapidly at the initial stage and gradually thereafter. The rapid flux decay can be attributed to the deposition and growth of a polarised gel-type layer formed by high-molecular-weight compounds present in the juice (the leftover pectin, protein, cellulose, hemicelluloses, etc.). The growth of such a polarised layer is limited by the shear rate, and, finally, flux decline is arrested until a steady-state value is reached.

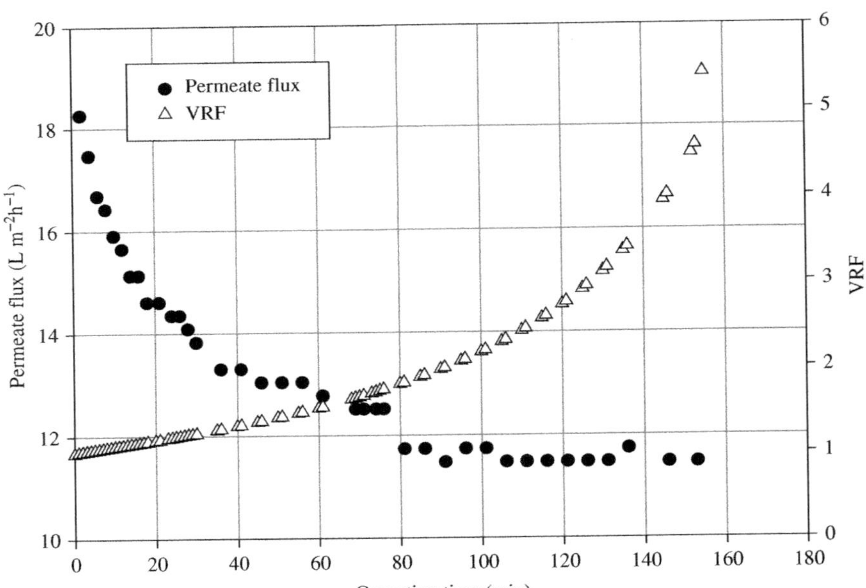

Fig. 12.3 Clarification of blood orange juice with a tubular ultrafiltration (UF) membrane. Evolution of permeate flux and volume reduction factor (VRF). Operating conditions: transmembrane pressure, 0.085 MPa; feed flow rate, 800 L h^{-1}; temperature, 21°C.

12.5 Process configurations

The most common MF/UF process configurations in the clarification of fruit juices are: (a) batch concentration; (b) feed and bleed; and (c) diafiltration (DF). In the batch concentration operation with full recycling of retentate (Fig. 12.4a), commonly used with laboratory- and pilot-scale units, the retentate is returned to the feed tank for recycling through the membrane module, and the permeate is collected separately. Since fruit juices contain low levels of retained solids, the optimum mode of operation is the modified batch operation with partial recycle of retentate (Fig. 12.4b) in which the permeate is collected separately and the feed is added continuously at the same rate as the flux. The bulk of the retentate is within the recycle loop and a small portion is recycled to the feed tank. Compared with the classic batch filtration with full recycling of retentate, this configuration reduces foaming formation in the feed tank and energy consumption of the recirculating pump.

DF (Fig. 12.4c) can be used in the manufacture of clear single-strength juices to recover sugars and low-molecular-weight compounds from the retentate and maximise the process yield. It involves the addition of water to the retentate and may be either sequential or continuous. The residue of the retentate contains potentially valuable pectins and enzymes.

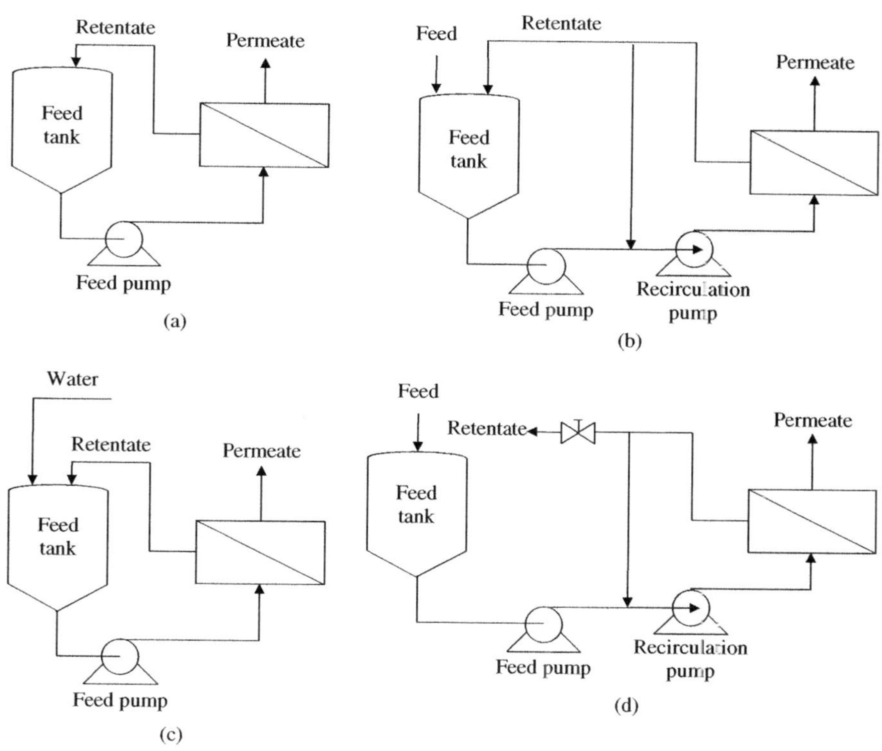

Fig. 12.4 Schematic illustration of different operation modes in microfiltration (MF) and ultrafiltration (UF) processes for fruit juice clarification. (a) Batch concentration with full recycle of retentate; (b) batch concentration with partial recycle of retentate; (c) diafiltration; and (d) feed-and-bleed operation.

The UF permeate, being sterile, could be submitted directly to storage or bottling so avoiding heat treatments.

The feed-and-bleed configuration (Fig. 12.4d) is generally used for continuous full-scale operations. The feed pump provides the required transmembrane pressure while the recirculation pump provides the cross-flow. The main advantage of this configuration is related to the possibility to immediately obtain the final concentration as the feed is pumped into the loop. Furthermore, the feed tank is not necessary for this operation. However, when the concentration factor is equivalent to the final concentration of the batch system, the permeate flux is lower than the average flux in the batch mode. Consequently, the membrane area required is higher.

12.6 Quality of the clarified juices

The most important effect in the production of clarified juice by using MF or UF is the removal of haze-forming components, such as suspended solids, colloidal particles, proteins and polyphenols, resulting in the production of superior quality juice. Turbidity values less than 0.1–0.3 NTU are typically obtained compared with the 2–5 NTU from conventional methods with fining agents (Cheryan, 1998). The removal of pectins, tannin–protein complexes and large carbohydrates produces a clear sparkling juice free from sediments and cloudiness. Membranes with MWCO greater than 25 kDa may produce a brownish permeate (or clouding of the filtered product during storage) because of the passage of tannins through the membrane pores.

Polysulphone hollow fibre MF and UF membranes, which have pore size of 0.1 and 0.2 μm and MWCO of 30 and 100 kDa, respectively, were able to remove yeasts, moulds and coliforms from pineapple juice (Laorko *et al*., 2010). The MF membrane with a pore size of 0.2 μm gave the highest recovery of phytochemical and antioxidant compounds. These results indicated that membrane filtration with a pore size of 0.2 μm permits clarification and sterilisation in one step.

Table 12.4 summarises the physico-chemical characteristics of samples obtained after clarification of depectinised kiwifruit juice using a cellulose acetate UF membrane with a MWCO of 30 kDa. Suspended solids were totally removed from the juice, and almost all total soluble solids (TSS) can be recovered in the clarified juice. The slight decrease in TSS can be attributed to the removal of suspended solids. As reported by different researchers (Vaillant *et al*., 1999; Matta *et al*., 2004) the presence of suspended solids and soluble pectin in fruit juices can interfere with refractive index measurements. Therefore, after the UF treatment, the nutritional value of the juice remains unchanged. The UF membrane showed a 13.5% rejection of total phenolic compounds while the rejection of organic acids and glutamic acid was in the range 0–4.3% (Cassano *et al*., 2008). Although the molecular weight of the protein fractions identified in the pressed kiwifruit juice ranged between 14 000 and 32 000, most proteins (about 61 g 100 g^{-1}) were retained by the UF membrane. This phenomenon was attributed to the formation of a cake layer on the membrane surface affecting the rejection characteristics of the membrane. In addition, other factors, such as the molecular conformation of the proteins and their interaction with the membrane and other compounds of the juice, cannot be excluded. The removal of most

Table 12.4 Composition of depectinised and clarified kiwifruit juice.

Characteristics	Depectinised juice	Clarified juice	Rejection(%)
pH	3.69	3.69	–
Suspended solids (g 100 g^{-1})	10.08	0.00	100.0
Pectins (mg L^{-1})	30.00	18.00	40.0
Total soluble solids (°Brix)	7.50	7.00	6.6
Folic acid (mg L^{-1})	460.00	440.00	4.3
Citric acid (mg L^{-1})	5.20	5.10	1.9
Ascorbic acid (mg L^{-1})	280.00	270.00	3.5
Glutamic acid (mg L^{-1})	3.98	3.98	0.0
Total phenols (mg L^{-1})	944.00	816.00	13.5
Proteins (mg L^{-1})	871.00	336.50	61.3

Source: Adapted from Cassano *et al*. (2008).

proteins from the juice prevents any subsequent protein precipitation caused by heat or any other factor. Also, the concentration of papain-like enzymes and actinidin as by-products could be of commercial significance (Wilson & Burns, 1998).

However, a 7.8% reduction in the total antioxidant activity was measured in the clarified kiwifruit juice obtained from UF of depectinised juice using a PVDF tubular membrane with a MWCO of 15 kDa; this reduction was attributed to a partial degradation of the initial content of ascorbic acid (Cassano *et al*., 2007a). The loss of ascorbic acid was probably due to oxidation of this component caused by the continual recycling of the juice around the UF pilot-plant loop. A reduction in total antioxidant activity of the same order (about 5%) was found in the clarified fraction of blood orange juice using the same UF membrane (Cassano *et al*., 2003).

Since elevated temperatures can be avoided in the production of a clear juice by UF, potential heat-induced flavour changes can be minimised. However, some studies have indicated some losses in flavour when clarifying apple juice by means of UF compared with MF (Wu *et al*., 1990). Rao *et al*. (1987) observed greater flavour losses in the UF of apple juice with polysulphone and polyamide membranes than with rack-and-frame filter presses, but lower than that using rotary vacuum precoat filtration.

Hydrophobic UF membranes employed in the clarification of kiwifruit juice showed a higher rejection (80–90%) of esters, such as methyl butanoate, ethyl butanoate, ethyl benzoate and methyl benzoate, and a lower rejection of more polar compounds, such as alcohols (3-hexen-1-ol, (E)-2-hexen-1-ol and 1-hexanol) (Cassano *et al*., 2006). UF membranes having a MWCO of 25 kDa retained over 80% of the flavour compounds in the passion fruit juice (Yu *et al*., 1986).

Cissè *et al*. (2005) found that some aroma compounds of orange juice, mainly terpenic hydrocarbons, were largely retained by MF ceramic membranes with a pore size of 0.2 μm because of their apolar properties and association with the insoluble solids of the juice. Oxygenated compounds, such as alcohols, esters, aldehydes and terpenols, were mainly recovered in the clarified juice. In particular, more than 60% of the ethanol, hexanal, decanal, benzaldehyde, linalool and terpineol from the fresh juice was recovered in the permeate, whereas more than 75% of limonene, terpinolene and valencene were found in the retentate. In addition, the content of vitamin C in the clarified juice and in the retentate was similar to that of the single-strength juice. Therefore, neither retention by the membrane nor oxidative damage affected this compound. Carotenoids, being associated with cell fragments, were retained by the membrane along with the pulp. As a consequence, the clarified juice was almost colourless, whereas the retentate was more intensely yellow than the initial juice.

12.7 Integrated processes

UF or MF can be combined with adsorbent resins to remove bitter compounds, such as limonin, naringin and hesperidin, present in the aqueous phase of citrus juices. The fresh juice is pre-treated by UF to separate suspended solids and pulp. The clarified juice, containing sugars and bitter compounds, is then treated by an adsorbent resin specifically designed to remove these compounds. The debittered juice is then recombined with the UF retentate to give a product containing less than 5 mg kg^{-1} limonin (Cheryan, 1998).

Clarified juice processed in MF and UF plants can be commercialised or submitted to a concentration process in order to reduce storage volumes and transport costs and improve the microbiological stability of the product. Membrane processes, such as reverse osmosis (RO) and osmotic distillation (OD), can be used for this purpose. Therefore, the role of UF for juice clarification is still technically important. A reconstituted juice with improved sensory properties can be obtained combining the pasteurised pulp (UF retentate) with the concentrated juice.

Different integrated membrane processes have been proposed to obtain concentrated fruit and vegetable juices (e.g. apple, carrot, kiwifruit, orange, cactus-pear, pineapple, melon, grapes, pomegranate and blackcurrant) by using UF or MF processes as clarification step (Alvarez *et al*., 2000; Cassano *et al*., 2003, 2004, 2006, 2007c, 2011; Cissè *et al*., 2005; Vaillant *et al*., 2005; Hongvaleerat *et al*., 2008; Kozák *et al*., 2008). The clarified juice was concentrated by using thermal evaporation or membrane processes, such as RO, membrane distillation or osmotic distillation (OD). Higher fluxes were obtained in membrane processes when processing the clarified juice indicating an effect of pulp content on performance (Calabrò *et al*., 1994; Hongvaleerat *et al*., 2008). In particular, UF of grape juice using membranes with pore diameters of 0.1 μm or less resulted in appreciable OD flux increases over that observed for unclarified juice. The increase in flux was attributed to a reduction in viscosity of the concentrated juice–membrane boundary layer as a result of protein removal. UF also resulted in a small increase in juice surface tension with a consequent reduction in the tendency for membrane wet-out to occur (Bailey *et al*., 2000).

An integrated membrane process for the production of concentrated orange juice is shown in Fig. 12.5. It involves UF as a clarification step, RO as a pre-concentration step of the clarified juice and OD as a concentration step of the pre-concentrated juice. Since the retentate composition is less sensitive to heat than the small aroma molecules, vitamins and sugars, it can be pasteurised to inactivate the enzymes and micro-organisms and then added in adequate proportions to the concentrated juice from the OD process.

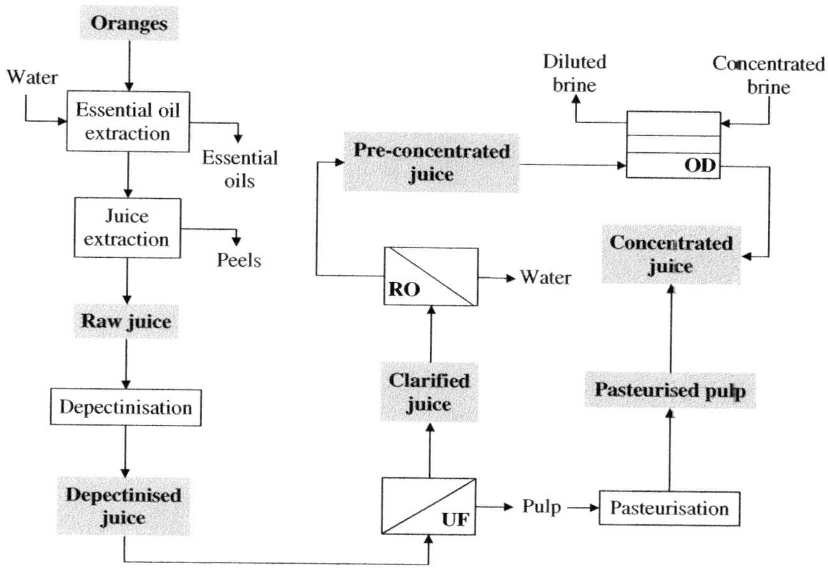

Fig. 12.5 Integrated membrane process for the production of concentrated orange juice UF, ultrafiltration; RO, reverse osmosis; OD, osmotic distillation.

12.8 Conclusions and future development

The continuous expansion of the fruit juice market offers significant opportunities for fruit juice processors and food scientists to develop processing technologies to retain as much as possible the peculiarity of fresh fruit, as well as colour, aroma, nutritional value and structural characteristics.

However, safety aspects, stability, convenience, waste prevention, formulation and sustainability are other important factors which have to be considered, and the potential of membrane technologies in the food and beverage industries is widely recognised today.

The key issue limiting the growth of membrane operations, such as MF and UF in the clarification of fruit juices, are basically related to the improvement of permeate fluxes, large amounts of energy used to circulate the feed solution, frequent cleaning of membrane modules and short membrane life time. Ceramic membranes offer many advantages over polymeric membranes, but they are limited to small-scale applications because of their high costs.

Membrane filtration technologies leading to the development of new material, structure designs and filtration operation technologies (module design, cleaning technology, etc.) are important areas of investigation for the sustainable growth of MF and UF operations in the clarification of fruit juices and beverages.

References

Alvarez, S., Riera, F.A., Alvarez, R., Coca, J., Cuperus, F.P., Bouwer, S., Boswinkel, G., van Gemert, R.W., Veldsink, J.W., Giorno, L., Donato, L., Todisco, S., Drioli, E., Olsson, J., Trägårdh, G., Gaeta, S.N. & Panyor, L. (2000) A new integrated membrane process for producing clarified apple juice and apple juice aroma concentrate. *Journal of Food Engineering*, **46**, 109–125.

Alvarez, S., Alvarez, R., Riera, F.A. & Coca, J. (1998) Influence of depectinization on apple juice ultrafiltration. *Colloids and Surfaces A: Physicochemical and Engineering Aspects*, **138**, 377–382.

Bailey, A.F.G., Barbe, A. M., Hogan, P.A., Johnson, R.A. & Sheng, J. (2000) The effect of ultrafiltration on the subsequent concentration of grape juice by osmotic distillation. *Journal of Membrane Science*, **164**, 195–204.

de Barros, S.T.D., Andrade, C.M.G., Mendes, E.S. & Peres, L. (2003) Study of fouling mechanism in pineapple juice clarification by ultrafiltration. *Journal of Membrane Science*, **215**, 213–224.

Calabrò, V., Jiao, B. & Drioli, E. (1994) Theoretical and experimental study on membrane distillation in the concentration of orange juice. *Industrial & Engineering Chemistry Research*, **33**, 1803–1808.

Carneiro, L., dos Santos Sa, I., dos Santos Gomes, F., Matta, V.M. & Corrêa Cabral, L.M. (2002) Cold sterilization and clarification of pineapple juice by tangential microfiltration. *Desalination*, **148**, 93–98.

Cassano, A., Drioli, E., Galaverna, G., Marchelli, R., Di Silvestro, G. & Cagnasso, P. (2003) Clarification and concentration of citrus and carrot juices by integrated membrane processes. *Journal of Food Engineering*, **57**, 153–163.

Cassano, A., Jiao, B. & Drioli, E. (2004) Production of concentrated kiwifruit juice by integrated membrane processes. *Food Research International*, **37**, 139–148.

Cassano, A., Figoli, A., Tagarelli, A., Sindona, G. & Drioli, E. (2006) Integrated membrane process for the production of highly nutritional kiwifruit juice. *Desalination*, **186**, 21–30.

Cassano, A., Donato, L. & Drioli, E. (2007a) Ultrafiltration of kiwifruit juice: operating parameters, juice quality and membrane fouling. *Journal of Food Engineering*, **79**, 613–621.

Cassano, A., Marchio, M. & Drioli, E. (2007b) Clarification of blood orange juice by ultrafiltration: analyses of operating parameters, membrane fouling and juice quality. *Desalination*, **212**, 15–27.

Cassano, A., Conidi, C., Timpone, R., D'Avella, M. & Drioli, E. (2007c) A membrane-based process for the clarification and the concentration of the cactus pear juice. *Journal of Food Engineering*, **80**, 914–921.

Cassano, A., Donato, L., Conidi, C. & Drioli, E. (2008) Recovery of bioactive compounds in kiwifruit juice by ultrafiltration. *Innovative Food Science and Emerging Technologies*, **9**, 556–562.

Cassano, A., Conidi, C. & Drioli, E. (2011) Clarification and concentration of pomegranate juice (*Punica granatum L.*) using membrane processes. *Journal of Food Engineering*, **107**, 366–373.

Cheryan, M. (1998) *Ultrafiltration and Microfiltration Handbook*, Technomic Publishing Company, Lancaster.

Cissè, M., Vaillant, F., Perez, A., Dornier, M. & Reynes, M. (2005) The quality of orange juice processed by coupling crossflow microfiltration and osmotic evaporation. *International Journal of Food Science and Technology*, **40**, 105–116.

Constela, D. T. & Lozano, J. E. (1997) Hollow fibre ultrafiltration of apple juice: macroscopic approach. *LWT-Food Science and Technology*, **30**, 373–378

De Bruijn, J.P.F., Venegas, A., Martinez, J.A. & Bórquez, R. (2003) Ultrafiltration performance of Carbosep membranes for the clarification of apple juice. *LWT-Food Science and Technology*, **36**, 397–406.

Espamer, L., Pagliero, C., Ochoa, A. & Marchese, J. (2006) Clarification of lemon juice using membrane process. *Desalination*, **200**, 565–567.

Fukumoto, L.R., Delaquis, P. & Girard, B. (1998) Microfiltration and ultrafiltration ceramic membranes for apple juice clarification. *Journal of Food Science*, **63**, 845–850.

Girard, B. & Fukumoto, L. (2000) Membrane processing of fruit juices and beverages: a review. *Critical Reviews in Biotechnology*, **20**, 109–175.

Hägg, M. (1998) Membranes in chemical processing: a review of applications and novel developments. *Separation and Purification Methods*, **27**, 51–168.

Hongvaleerat, C., Cabral, L.M.C., Dornier, M., Reynes, M. & Ningsanond, S. (2008) Concentration of pineapple juice by osmotic evaporation. *Journal of Food Engineering*, **88**, 548–552.

Jiraratananon, R. & Chanachai, A. (1996) A study of fouling in the ultrafiltration of passion fruit juice. *Journal of Membrane Science*, **111**, 39–48.

Kaur, C. & Kapoor, H. C. (2001). Antioxidants in fruits and vegetables – the millennium's health. *International Journal of Food Science and Technology*, **36**, 703–725.

Koseoglu, S.S., Lawhon, J.T. & Lusas, E.W. (1990) Use of membranes in citrus juice processing. *Food Technology*, **44**, 90–97.

Kozák, A., Bánvölgyi, S., Vincze, I., Kiss, I., Békássy-Molnár, E. & Vatai, G. (2008) Comparison of integrated large scale and laboratory scale membrane processes for the production of black currant juice concentrate. *Chemical Engineering and Processing*, **47**, 1171–1177.

Laorko, A., Zhenyu, L., Tongchitpakdee, S., Chantachum, S. & Youravong, W. (2010) Effect of membrane property and operating conditions on phytochemical properties and permeate flux during clarification of pineapple juice. *Journal of Food Engineering*, **100**, 514–521.

Millward, H.R., Bellhouse, B.J. & Walker, G. (1995) Screw thread flow promoters: an experimental study of ultrafiltration and microfiltration performance. *Journal of Membrane Science*, **106**, 269–279.

Lipinzki, F. (2010) Cross-flow membrane applications in the food industry. *Membrane Technology* (eds. K.V Peinemann, S.P. Nunes and L. Giorno), Vol. **3**, pp. 1–24, Wiley-VCH Verlag GmbH & Co. KGaA, Weinheim.

Matta, V.M., Moretti, R.H. & Cabral, L.M.C. (2004) Microfiltration and reverse osmosis for clarification and concentration of acerola juice. *Journal of Food Engineering*, **61**, 477–482.

Mirsaeedghazi, H., Emam-Djomeh, Z., Mousavi, S.M., Ahmadkhaniha, R. & Shafiee, A. (2010) Effect of membrane clarification on the physicochemical properties of pomegranate juice. *International Journal of Food Science and Technology*, **45**, 1457–1463.

Mondor, M. & Brodeur, C. (2002) *Le Monde Alimentaire. Membrane Filtration*, **March/April**, 1–4.

Nilsson, S.L. (1990) Protein fouling of UF membrane: causes and consequences. *Journal of Membrane Science*, **52**, 121–142.

Padilla-Zakour, O. & McLellan, M.R. (1993) Optimization and modelling of apple juice cross-flow microfiltration with a ceramic membrane. *Journal of Food Science*, **58**, 369–388.

Rai, P., Majumdar, G.C., Jayanti, V.K., Dasgupta, S. & De, S. (2006) Alternative pretreatment methods to enzymatic treatment for clarification of mosambi juice using ultrafiltration. *Journal of Food Processing Engineering*, **29**, 202–218.

Rai, P., Majumdar, G.C., Dasgupta, S. & De, S. (2007) Effect of various pretreatment methods on permeate flux and quality during ultrafiltration of mosambi juice. *Journal of Food Engineering*, **78**, 561–568.

Rao, M.A., Acree, T.E., Cooley, H.J. & Ennis, R.W. (1987) Clarification of apple juice by hollow fibre ultrafiltration: fluxes and retention of odor-active volatiles. *Journal of Food Science*, **52**, 375–377.

Riedl, K., Girard, B. & Lencki, R.W. (1998) Interactions responsible for fouling layer formation during apple juice microfiltration. *Journal of Agricultural and Food Chemistry*, **46**, 2458–2464.

Sarkar, B., DasGupta, S. & De, S. (2008) Cross-flow electro-ultrafiltration of mosambi (*Citrus sinensis* (L.) *Osbeck*) juice. *Journal of Food Engineering*, **89**, 241–245.

Strathmann, H., Giorno, L. & Drioli, E. (2006) Membranes and membrane processes applications. *An Introduction to Membrane Science and Technology*, pp. 332–336, Consiglio Nazionale delle Ricerche, Roma.

Su, S.K., Liu, J.C. & Wiley, R.C. (1993) Cross-flow microfiltration with gas backwash of apple juice. *Journal of Food Science*, **58**, 638–641.

Tallarico, P., Todisco, S. & Drioli, E. (1998) Use of ultrafiltration in the preventing of orange juice bitterness and its effect on the aroma compounds distribution. *AgroFood Industry Hi-Tech*, **33**, 739–756.

Tasselli, F., Cassano, A. & Drioli, E. (2007) Ultrafiltration of kiwifruit juice using modified poly(ether ether ketone) hollow fibre membranes. *Separation and Purification Technology*, **57**, 94–102.

Todisco, S., Peña, L., Drioli, E. & Tallarico, P. (1996) Analysis of the fouling mechanism in microfiltration of orange juice. *Journal of Food Processing and Preservation*, **20**, 453–466.

Vadi, P.K. & Rizvi, S.S.H. (2001) Experimental evaluation of a uniform transmembrane pressure cross-flow microfiltration unit for the concentration of micellar casein from skim milk. *Journal of Membrane Science*, **189**, 69–82.

Vaillant, F., Millan, P., O'Brien, G., Dornier, M., Decloux, M. & Reynes, M. (1999) Crossflow microfiltration of passion fruit juice after partial enzymatic liquefaction. *Journal of Food Engineering*, **42**, 215–224.

Vaillant, F., Cissè, M., Chaverri, M., Perez, A., Dornier, M., Viquez, F. & Dhuique-Mayer, C. (2005) Clarification and concentration of melon juice using membrane processes. *Innovative Food Science and Emerging Technology*, **6**, 213–220.

Vladisavljević, G. T., Vukosavljević, P. & Bukvić, B. (2003) Permeate flux and fouling resistance in ultrafiltration of depectinized apple juice using ceramic membranes. *Journal of Food Engineering*, **60**, 241–247.

Wenten, I.G., Koenhen, D.M., Roesnik, H.D.W., Rasmussen, A. & Jonsson, G. (1995) Method for removal of components causing turbidity from a fluid by means of microfiltration. United States of America Patent Application No. 5560828.

Wilson, E.L. & Burns, D.J.W. (1983) Kiwifruit juice processing using heat treatment techniques and ultrafiltration. *Journal of Food Science*, **48**, 1101–1105.

Wu, M.L., Zall, R.R. & Tzeng, W.C. (1990) Microfiltration and ultrafiltration comparison for apple juice clarification. *Journal of Food Science*, **55**, 1162–1163.

Youn, K.S., Hong, J.H., Bae, D.H., Kim, S.J. & Kim, S.D. (2004) Effective clarifying process of reconstituted apple juice using membrane filtration with filter-aid pretreatment. *Journal of Membrane Science*, **228**, 179–186

Yu, Z.R., Chiang, B.H. & Hwang, L.S. (1986) Retention of passion fruit juice compounds by ultrafiltration. *Journal of Membrane Science*, **51**, 841–844.

13 Beer and Cider

J. Bergin and J.J. Tuohy

13.1 Introduction

Since the beginning of the 21st century, rapid consolidation within the brewing industry has resulted in almost 50% of global beer sales being supplied by the top four brewing companies (Anheuser-Busch InBev, SAB Miller, Heineken and Carlsberg). Although growth is beer sales tailed off in 2009 because of economic conditions, growth had been steady in previous years and is predicted to grow a further 8% by 2014 (Annonymous, 2010). Although the overall global figures show growth, there are different pictures emerging from different markets. The main areas of growth are in emerging markets, such as China, Eastern Europe and Africa, whereas Western Europe continues to decline. China is now the largest beer market in the world, with Europe in second place. In 2014, the total beer sales volume is predicted to exceed 160 billion litres. The beer market is very competitive and brewing is big business with tight margins.

Cider generally refers to fermented beverages made from apples and a similar product made from pears is termed perry. Although the term cider refers to a diverse range of beverages native to a number of countries (principally European; UK, France (*Cidre*), Germany (*Apfelwein*), Spain (*Sidra*) and Ireland), with a range of alcohol contents, carbonation levels and having characteristic flavours and colours, this chapter will deal only with cider characteristic of that produced on a large scale in or for the United Kingdom (UK) market. The UK is the world's leading cider consumer at approximately 1 billon litres per annum (NACM, 2011). UK cider is a 'clear, light-flavoured, carbonated product, with varying degrees of sweetness and alcohol (2–8.5 mL 100 mL^{-1}) content, which is served chilled' (Lea & Drilleau, 2003; NCAM, 2008).

Cider is broadly defined by the National Cidermakers Association (NACM) in the UK as 'a beverage obtained by the partial or complete fermentation of the juice of apples (and pears if desired, provided that no more than 25% of the mixed juice is pear) ... or concentrated apple juice ... with or without the addition before or after fermentation of sugar or potable water ... without at any time adding alcoholic liquor ... without at any time adding any substance which gives colour or flavour (other than certain specifically permitted ingredients)' (NACM, 2008).

Ironically, while in the UK cider and lager beer compete in the same market sector (~25% of cider consumed in UK is draught), from a technological perspective cidermaking has more in common with winemaking than with beer brewing.

Membrane Processing: Dairy and Beverage Applications, First Edition. A. Y. Tamime.

In the past, the brewing and cider industries have been quite traditional and slow to move to new technologies. However, nowadays, brewers and cidermakers are continually looking to innovate in both product development, to drive growth, and process technology development, to reduce costs, and make the process more efficient. There is also greater environmental pressure on all industries to reduce their carbon footprint. Brewing and cidermaking have quite high energy costs and high water usage. Both industries are now more open than ever to technology innovations, which can help to reduce both these costs. Membrane filtration applications can address some of these issues, but must be cost effective in terms of both capital and running costs without having an adverse impact on quality. Membranes can be expensive; they have a limited life and can be difficult to clean. However, the use of membranes has the potential to provide not only novel solutions to processing problems, but also to replace existing unit operations with more efficient alternatives. In order to best appreciate the opportunities for membrane applications within brewing and cidermaking, both processes need to be briefly described.

13.2 Beer brewing process

Beer production (Fig. 13.1) brings together many well-known physical, chemical and biological food-processing operations, including size reduction, extraction, separation,

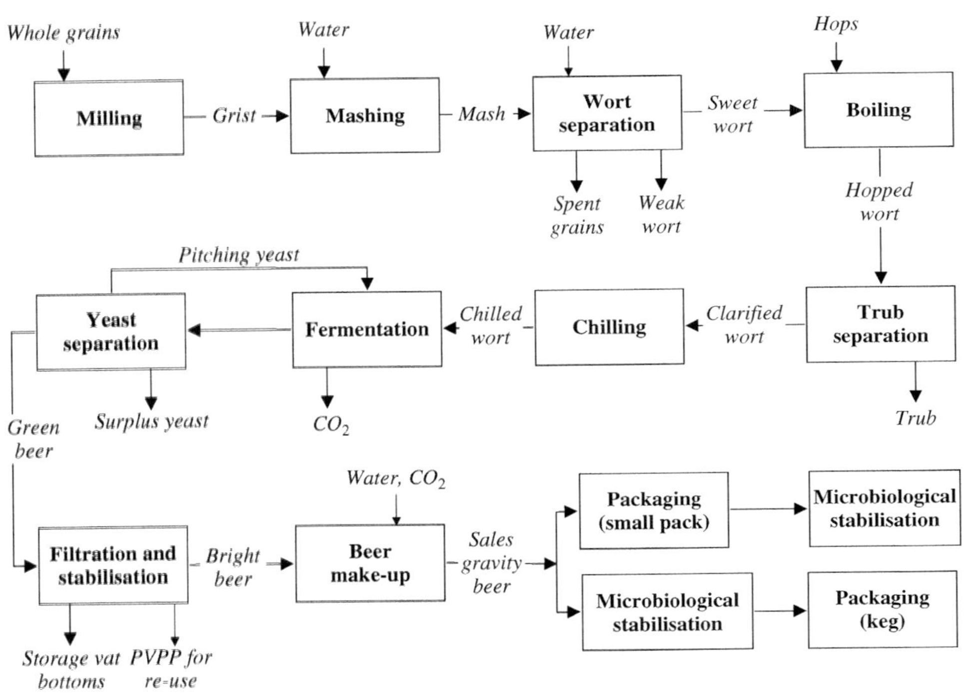

Fig. 13.1 Flow chart for generic brewing process. Summary of the key steps in the brewing process showing product and side streams.

heating, fermentation, clarification, microbiological stabilisation and packaging (Briggs *et al*., 2004).

The principle ingredients of beer are malt, hops, yeast and water. Apart from water, malt is the single biggest ingredient. Malt is made from barley through the process of malting, in which the proteins and starch within the barley grain are made available to the brewer for subsequent extraction. Malting is basically a controlled germination process consisting of three stages: (a) steeping; (b) germination; and (c) kilning. In the past, many brewers operated their own malting plants, but both now exist as separate industries with the malster supplying the brewer with malt to the brewers specifications. However, in recent years, due to cost and availability, brewers are using other cereals as starch sources. This is particularly the case in Africa where barley is not a native crop. Cereals, such as sorghum, maize and wheat, are commonly used to supplement or indeed fully replace barley in the brewing process. Nevertheless, beer remains a natural product made from natural materials through a natural fermentation process. The composition of beer is complex and varied (Briggs *et al*., 2004). Table 13.1 shows a simplified composition of a typical 5 mL alcohol 100 mL^{-1} beer.

13.2.1 *Milling*

The first step in brewing is milling where the malt is crushed using a variety of mill types. Milled grain is referred to as grist. The purpose of milling is to expose the starch in the malt endosperm for subsequent enzyme degradation during mashing. The mill type will depend on the separation method used subsequently to separate the liquid extract (wort). The most popular mill types are roller, hammer and disc mills. The roller and disc mills are available as either wet or dry, whereas the hammer mills are dry mills. Wet mills are generally cheaper and safer because the risk of dust explosion is reduced and, thus, requires no special explosion protection. This results in a smaller footprint and the possibility of locating the mill directly in the brewhouse. Wet milling may also result in improved quality due to reduced air contact with the grain as the water present during milling displaces air.

Roller mills provide a gentle crushing, and are designed to produce a grist of fine grits and flour from the starchy endosperm while at the same time leaving the malt husk virtually

Table 13.1 Typical composition of a 5 mL 100 mL^{-1} alcohol in beer.

Component	Concentration (g L−1)
Ethanol	39.0
Carbohydrate (mainly maltodextrins)	26.0
Protein	2.8
Inorganic matter	2.4
Volatile flavour compounds (e.g. esters, higher alcohols)	0.3
Calorific content (kcal L^{-1})	450.0

The composition of beer can vary significantly and is dependent on beer type, raw materials, yeast and production conditions.

intact. The husk is required to form a filter bed for mash separation when a Lauter Tun is used for separation. Hammer mills produce finely milled grist, and are employed when mash filters are used as the separation method. Hammer mills are also used if high levels of non-malt grains are used.

13.2.2 *Mashing*

After milling, the grist is mixed with water (liquor) to a set volume and at a specific temperature dependent on the beer type and malt quality. This mixture is referred to as the mash. Again, dependent on materials and brew type, the mash is taken through a number of temperature changes and rest times. The function of mashing is to convert the starch in the malt to fermentable sugars, which the yeast can subsequently convert to alcohol during fermentation. Mashing is essentially a biochemical process where the natural malt enzymes, which are formed during malting, convert the starch to sugars. The enzymes involved in this conversion are α- and β-amylases. Before the enzymes can be fully effective, the starch in the malt must be gelatinised, where water is adsorbed and the starch granules erupt opening the starch up to amylase activity. For barley malt, the gelatinisation temperature is $>60^{\circ}$C, and the optimum temperature for amylase activity is between 60°C and 65°C. The mash is continuously stirred during this process to maintain homogeneity. At the end of mashing, all starch will have been converted to simple sugars and maltodextrins. The fermentable sugars consist of mainly glucose, maltose and maltotriose, and make up approximately 70 g 100 g^{-1} of the total solids in the soluble liquid phase (wort). The wort will also be rich in other nutrients, such as amino acids, minerals and vitamins that the yeast will need to grow and ferment. Once the mash cycle is complete, the next stage is wort separation.

13.2.3 *Wort separation*

The purpose of wort separation is to remove the liquid wort from the spent grains. The objective being to achieve this quickly and efficiently to minimise any extract losses in the process. Nowadays, the two principle processes for mash separation are Lauter Tun and mash filter. Before the mash is transferred from the mash vessel to either of these, it is first heated to ~78°C. This arrests all enzyme activity and reduces the viscosity of the mash and wort to aid more efficient 'run off'.

Lauter Tun separation is the more traditional method. It is essentially another vessel with an integral perforated (false) floor just above the base, which acts as a strainer, and enough space underneath to allow the removal of wort. The mash is transferred into the Lauter Tun through bottom filling just above the strainer to minimise air incorporation. The vessel is sized based on brew size and raw material loading. The mash bed sits on the false floor, and the wort drains out of the bed through the slits in the false floor and is collected either directly into a boiling kettle or an intermediate holding vessel known as an underback. The bed depth is ~0.5 m. The collected wort is generally recycled back into the Lauter Tun until it has good clarity. The mash bed itself acts as the primary filter, which is why it is important during the milling to keep the husk material intact. The husk keeps the mash bed open and allows the wort to drain more effectively from the bed. Run-off of the wort is driven purely by gravity and is monitored by flow rate and differential pressure between the top of the bed and beneath the false bottom. The Lauter Tun is equipped with rakes positioned at the top of the mash bed which, while not totally disrupting the filter bed, can be operated in such a way as to keep the bed open and allow faster filtration. When the initial 'strong worts' are collected, hot water at 78°C is sprayed on the top of the bed in a process known as sparging. This keeps the bed wet and serves to flush out the entrained wort. As 'run-off' continues, the wort collected becomes more diluted. Sparging and wort collection continues until the bulk wort collected is at the required specific gravity and further collection of 'weak wort' would dilute the bulk gravity. However, the weak worts still contain valuable extract and would represent a significant loss to the brewer. Many brewers continue collecting these worts into a weak worts vessel. The weak worts are stored hot, and are used as part of the mashing liquor for subsequent brews. Weak wort collection is terminated when the specific gravity drops below a set value, which will vary from brewer to brewer (average between 1.002 and 1.005). The remaining liquid is drained from the mash bed in order to reduce the moisture level of the spent grains. The spent grains are discharged from the Lauter Tun, and are generally sold as animal feed.

A significant innovation within the brewing industry was the introduction of the mash filter for wort separation (Fig. 13.2). Mash filters have been used in breweries for over 30 years, and consist of a number of filtration chambers. Each chamber has a filter sheet (polypropylene) and an inflatable membrane. Mash filter design combines a large surface area, a very thin mash bed (3–4 cm) and the ability to apply an additional pressure to the filter bed by inflating the membranes with compressed air thereby squeezing the grain to expel the wort. Because of the filter sheets, the thin mash bed and the ability to squeeze the mash bed, the grain can be milled very finely and a hammer or disc mill is used for this

Fig. 13.2 Typical mash filter in-situ (Reproduced with permission from Meura S.A.).

purpose. This has significant advantages as fine milling and the squeezing effect recovers more extract for the brewer. This needs to be balanced with the potential for extracting unwanted lipids and polyphenols, which can have a damaging effect on the quality and long-term stability of the subsequent beer. Mash filters are also ideally suited when a high level of non-malt grains is used and fine milling is necessary for efficient extraction (Briggs *et al.*, 2004).

The transfer of the mash to the filter is identical to the process for the Lauter Tun. The mash is heated to 78°C prior to transfer. However, because of the filter sheets, wort collection can begin before the chambers are completely full with no wort circulation, as the clarity of the collected worts will be excellent from the start. In general, the filling takes ~5 min, and is followed by the main filtration stage where the bulk of the strong wort is transferred to the boiling kettle. Afterwards, pre-compression is applied, where the membranes are gently inflated to squeeze the remaining strong wort from the grain, and is followed by sparging. As the membranes are slowly deflated, sparge liquor is pumped into the chambers and, when full, the membranes are again inflated for final compression where the sparge liquor is pushed through the grain bed removing the weaker extract and leaving a relatively dry filter cake on the filter sheets (Fig. 13.3). The filter can then be opened to discharge the dry spent grain, which again is used as animal feed.

Fig. 13.3 Discharge of filter cake (spent grain) from a single chamber of mash filter. Note uniform thickness and dryness of filter cake (Reproduced with permission from Meura S.A.).

13.2.4 *Boiling*

During wort collection, the wort is either stored directly in the boiling kettle or in an underback (an intermediate storage vessel). Before boiling, the wort is referred to as 'sweet wort'. When all the wort has been collected to the specific gravity required, the wort is now ready for boiling. At this point hops are added to achieve the required hop character and bitterness. Depending on the beer type, the hops can be added at the beginning, middle or end of the boiling step. The wort in the kettle is heated to just over 100°C, and is boiled for a set time to deliver a required evaporation rate, which varies depending on the brewer's recipe and the beer type. The boil time can range from 30 min to 2 h and the evaporation rate from 4% to 10%. The boiling step can represent over one-third of the total energy required for the complete brewing process. Although energy recovery systems are very efficient, it is an area where brewers strive to reduce energy costs through reduction of the evaporation rate. Improvements in kettle design and emerging technologies are geared towards this. It is also believed that reducing the heat load during brewing has a beneficial effect on quality and long-term flavour stability.

The primary functions of wort boiling are:

- wort sterilisation to ensure that no spoilage of the wort and subsequent beer occurs
- precipitation of protein/polyphenol complexes – high levels of both protein and polyphenols in beer can lead to very poor colloidal stability, as they tend to precipitate over time to form a haze; the precipitate is known as trub.
- removal of undesirable volatile wort compounds through evaporation – the main volatile removed is dimethyl sulphide (DMS), which leads to a distinctive sweetcorn type flavour in lager beers
- extraction of the desirable hop components which give beer its characteristic aroma and flavour
- destruction of any residual enzyme activity.

Other effects of boiling are: (a) the wort is concentrated due to the evaporation; (b) the pH level is reduced; and (c) increase in colour due mainly to Maillard reactions and carmelisation. At the end of the boil, the wort is known as 'hopped wort'.

13.2.5 *Trub separation*

On completion of the boil before wort cooling, the precipitated trub needs to be removed to give wort of good clarity. The most common system for wort clarification is the whirlpool system. This is a circular vessel with a side inlet through which the hot wort is pumped from the kettle. This tangential entry of the wort causes a whirling action in the vessel, and the consequent centrifugal action forces the trub particles towards the centre of the vessel. The wort is left circulating for ~20 min. The trub eventually settles to form a cone at the centre of the vessel, which is designed with a series of run-off points or standpipes that allows removal of the clarified wort while leaving the trub cone undisturbed in the vessel. Although quite compact, the trub contains over 80% wort, at the same strength as the clarified wort, and would represent a significant loss to the brewer if not recovered.

A common recovery method is the addition of the trub to the mash vessel or to the mash separation system (Lauter Tun or mash filter). In this way, the liquid is recovered through the normal brewing process, and the solid trub is removed with the spent grains. Whatever method is used needs to be carefully controlled and monitored to achieve a balance between extract recovery and potential quality impact (Briggs *et al*., 2004).

13.2.6 *Fermentation*

En-route to the fermenter, the wort is first cooled using a plate heat exchanger (PHE). The temperature set point will vary depending on the yeast and beer type, and will generally be just below the fermentation temperature. After cooling, the wort is aerated using either air or oxygen, and yeast is added (pitched). The aeration and pitching are normally carried out inline over the duration of the wort transfer to ensure proper mixing. Depending on yeast type, aeration adds between 8–20 mg L^{-1} dissolved oxygen to the wort. Oxygen is used to achieve the higher levels due to limitations around oxygen saturation when air only is used. Yeast, which has been recovered from a previous fermentation, is pitched to achieve yeast levels of between 6 and 20 $\times$ 10^6 colony forming units (cfu) mL^{-1} in the wort. The fermentation process consists of two phases. Initially, due to the presence of dissolved oxygen, the yeast grows through three to five doublings, which, when the oxygen is depleted and growth has finished, results in approximately a 10-fold increase in yeast cell numbers. At this point fermentation proper begins, and the growth phase ensures that at least 90% of the yeast going into fermentation is new young cells, which is a prerequisite for a good fermentation.

During fermentation, the primary function is the conversion of fermentable sugars to ethanol and carbon dioxide (CO_2). Modern fermenters are cylindro-conical in shape, and this combined with the CO_2 evolved during fermentation ensures effective mixing of the fermenter contents during fermentation. However, fermentation is a biological step with many complex biochemical processes combining to contribute to the overall flavour and character of the final beer. Fermentation reactions of particular importance to beer flavour are listed below.

- The formation of esters and higher alcohols – yeast, wort composition and fermentation conditions are critical to achieve the correct balance for a particular beer type.
- The formation and subsequent breakdown of vicinal diketones – particularly diacetyl which has a distinctive buttery flavour that is not desirable in beer; the yeast produces the diacetyl precursor during growth, which over time is converted chemically to diacetyl and, once converted, the yeast can then break down diacetyl.
- The formation of compounds related to sulphur metabolism – in a properly controlled fermentation, these should not be a problem, but compounds like hydrogen sulphide and some mercaptans can lead to negative flavour issues.

Towards the end of the fermentation stage, the yeast cells start to settle out (flocculate) and collect in the cone of the fermenter. At the end of primary fermentation, there will need to be some form of flavour maturation to ensure diacetyl levels are below the flavour threshold for the beer. This is also referred to as lagering or conditioning, which can be carried out in the primary fermentation vessel or in a secondary conditioning vessel where the beer is transferred with the addition of a small amount of fresh wort and yeast. Whatever method is employed, the bulk of the fermentation yeast needs to be removed so as to prevent it autolysing and impacting the flavour. The beer can be chilled in the fermenter, which speeds up the flocculation process, and the yeast is removed (cropped) as a plug from the fermenter, or a centrifuge can be employed to achieve more rapid removal. Part of the recovered yeast is held and used to pitch subsequent brews, but the bulk of the yeast is surplus and is a waste stream. However, depending on the cropping method, this surplus stream can be a significant volume with low yeast solids. The low yeast solids make it unsuitable for other uses and also the yeast contains a high content of beer which would be a significant loss to the brewer. Add to this, the effluent costs for disposal and surplus yeast can be an expensive waste stream for the brewer to handle.

If the surplus yeast could be concentrated to a high yeast solids content

- the brewer could recover up to 75% of the volume as beer, which can be incorporated back into the process
- the concentrated yeast is now a saleable commodity with a range of uses from food to pharmaceutical applications
- the total effluent stream will be removed along with the disposal costs

Over the years, many processes have been used for this purpose, ranging from settling tanks, drum filters and yeast presses, and, while they all achieved the level of yeast concentration, the quality of the recovered beer suffered because of autolysis of the yeast during the long holding time associated with these processes. However, in recent years, the introduction of membrane technology has allowed yeast concentration with the recovery of excellent quality beer.

Once primary fermentation and flavour maturation are complete and the bulk of the yeast has been removed, the beer is now ready to be clarified as there will be still low levels of yeast and sediment present.

13.2.7 *Clarification*

Beer clarification normally refers to dead-end filtration using a filter aid. The most common being diatomaceous earth (more commonly called kieselguhr) depth filters. Kieselguhr is a powder made from the skeletal remains of diatoms, tiny sea creatures that inhabited the North Sea millions of years ago. It acts as a depth filter, with the incoming unfiltered beer being continuously dosed with the powder as a body feed. Thus, the filter bed is continuously built up, preventing blinding of the filter face with suspended material (Briggs *et al.*, 2004). Membrane technology has become more popular as a kieselguhr alternative in recent years, and this will be discussed at length later.

Filtering beer so that it remains bright and sparkling throughout its shelf-life requires two distinct processes: primary filtration to remove all suspended solids and haze stabilisation. A hazy beer is one that has an obvious cloudiness and lack of sparkle. A number of factors can lead to haze formation in beer, including the formation of calcium oxalate, residual carbohydrate, such as starch and β-glucan, and biological haze due to microbial contamination and growth. However, the most common type of haze in beer arises from the cross-linking of proteins and polyphenols. Weak cross-linking through hydrogen bonds between polyphenols and proteins causes a temporary or chill haze, which forms when beer is cooled to 0°C due to decreased solubility. This is a reversible association, and chill haze disappears when the beer is warmed. Over time, the polyphenols in beer tend to polymerise during storage, and these larger polyphenols can bind covalently to proteins to form an irreversible haze. The haze appears even at 20°C, and is known as permanent haze. Both chill and permanent hazes are a problem, but chill haze will arise first before developing into a permanent haze, and, so it is this which needs to be addressed. Most brewers approach the problem by reducing the level of haze-forming protein or haze-forming polyphenol or both in beer. While primary filtration removes all the visible suspended solids, it can also be used to remove potential haze-forming material if the beer is held below 0°C. It is common practice to hold beer for filtration in storage vessels below 0°C for a number of days prior to filtration. This holding period can result in further settling of suspended solids, which need to be removed (bottomed) before filtration to prevent the filter from blinding. These storage vat bottoms are another high-volume, high-value waste stream that the brewer has to deal with, and are analogous to the surplus yeast generated in primary filtration. Alongside standard cold filtration, other methods are commonly used to further improve the haze stability of beer. These methods involve the use of adsorbents, which can either adsorb polyphenols or proteins or a combination of both. The most common adsorbents are polyvinylpolypyrrolidone (PVPP), which adsorbs polyphenols, and silica gel, which adsorbs proteins. These can be used individually or together, and they are both powders that need to be slurried in water for addition. They can both be dosed into the beer en route to the filter and removed by the kieselguhr filter, but they both require a contact time of at least 5 min to be effective. They also need to be dispersed evenly throughout the beer to ensure good contact and maximum adsorption. PVPP is the more expensive of the two, and is available in both single-use and regenerable forms. When the regenerable form is used, it is added after primary filtration, held for the required contact time and removed through a dedicated PVPP filter, thus allowing the PVPP to be collected for regeneration. The regeneration process uses a caustic solution to desorb the polyphenols from the PVPP, and the caustic used in the regeneration process can also be recovered for reuse through the use of membrane technology.

After clarification and haze stabilisation, the beer is known as bright beer.

13.2.8 *Beer make-up*

Prior to packaging, final adjustments can be made to the beer. These can include dilution (if beer has been brewed at a higher strength), addition of post-fermentation bitterness and adjustment of gas levels (CO_2 and possibly nitrogen – N_2). If the beer is to be diluted, de-aerated water must be used with dissolved oxygen levels less than 0.1 mg L^{-1}. Beer is very sensitive to oxygen and the presence of oxygen can lead to off flavours, premature ageing and a shorter shelf-life. There are a number of ways to produce de-aerated water, including heating and cooling in an air-free environment, injection of CO_2 or N_2, and application of vacuum. There is a membrane technology that can also be used for this purpose. Indeed the same technology can also be used to add gas to beer. This will be discussed later.

13.2.9 *Packaging and microbiological stabilisation*

Packaging can be either in keg or in small pack (cans and bottles). At some point during the packaging process, the beer is microbiologically stabilised by pasteurisation. The degree of pasteurisation is measured in pasteurisation units (PU): 1 PU is defined as 1 min at 60°C. The PU scale is logarithmic and increases 10-fold every 7°. Therefore, 1 min at 67°C would be 10 PU, and 1 min at 74°C would be 100 PU, etc.

Depending on the pack type and packaging plant, there are two forms of pasteurisation.

- *High temperature short time (flash) pasteurisation* – This is used primarily for beer which is to be subsequently packaged in kegs, but can also be used for small pack provided that the subsequent filling operation is carried out aseptically or with a high degree of asepsis. Plate heat exchangers are used for this purpose, and a typical profile would be 74°C for 15 s delivering 25 PU followed by cooling to $<10°$C. Afterwards, the beer is packaged.
- *Low temperature long time (tunnel) pasteurisation* – This is only used for small pack applications, and consists of an enclosed tunnel equipped with water sprays. Tunnel pasteurisation is applied after packaging, and a typical holding temperature is between 60 and 63°C with a holding time between 10 and 20 min. The main advantage of tunnel pasteurisation is that the liquid and the package both get pasteurised. There are three distinct areas in the tunnel: heating, holding and cooling. Thus, the footprint of tunnels is quite large and the energy/water consumption is very high. Another disadvantage of tunnel pasteurisers is the fact that the beer is held at temperatures over 60°C for a considerable amount of time. If the beer prior to packaging has not been handled correctly and has picked up some oxygen, the pasteurisation process will cause rapid oxidation of the beer.

To minimise gas breakout during filling, the beer is packaged cold (close to 0°C). Packaging into keg is a closed fill operation as the keg can be fully cleaned, sterilised and pressurised before filling. There is no break between filling and sealing the keg as this happens automatically at the end of the filling process. Kegging is essentially aseptic filling with no exposure to air during filling.

During small pack filling, the bottles and cans are generally rinsed prior to filling but are not sterilised. Returnable bottles are more aggressively cleaned. The package is pressurised before the beer is filled to prevent gas breakout and subsequent foaming. The package must then disengage from the filling head for sealing (crowning for bottles or seaming for cans). The distance between filling and sealing is kept to a minimum, but this step does pose the risk for air pick-up, which can be detrimental to the beer. After filling and pasteurisation, the bottles are labelled and placed in secondary packaging for sale.

13.3 Cidermaking process

13.3.1 *Juice extraction and formulation*

Premium cider is made from true cider apple cultivars, but most modern UK ciders have a high proportion of dessert and culinary outgrades. Apples are stored for a few weeks after harvesting to allow for maximum conversion of starch to fermentable sugars. When apples retain the impression of a thumbprint after squeezing in the hand they are judged to be ready for cidermaking.

The first stage in cidermaking is the preparation of the juice by milling and pressing (Jarvis, 2001). Prior to this, apples are sorted to remove rotten fruit, wood and other orchard debris, and washed to remove any other foreign matter. Most modern cider producers use high-speed grinding mills which produce a pulp that is then fed to a semi-continuous high-pressure piston press (Fig. 13.4) or continuous belt press. This extracts the juice, leaving a pressed pulp, referred to as apple pomace, from which weaker juice can be extracted by addition of water and repressing. Juice yields of more than 75% with low levels of suspended solids are achievable.

Freshly pressed juice is screened and stored chilled in stainless steel blending vessels prior to fermentation. Fresh apple juice is commonly supplemented with apple juice concentrate (70°brix), and sugar syrups are also commonly added before fermentation, so that many UK ciders contain only the minimum 35% fresh juice content in the pre-fermentation mix required by legislation (SI, 2010).

With the addition of apple juice concentrate (AJC) and glucose syrup to fresh juice, cidermakers are able to increase the specific gravity up to 1.080–1.100. With this level of fermentable sugars it is possible to achieve a final alcohol level of 10–12 mL 100 mL^{-1} in the fermentation (Lea & Drilleau, 2003). Apple juices have low levels of amino nitrogen compared with beer worts or wine musts, so up to 25 mg kg^{-1} of ammonium sulphate or phosphate is commonly added to bring the amino nitrogen level up to 100 mg L^{-1} in order to promote yeast activity and optimise alcohol production. Other ingredients, such as water-soluble vitamins, are also added at this stage.

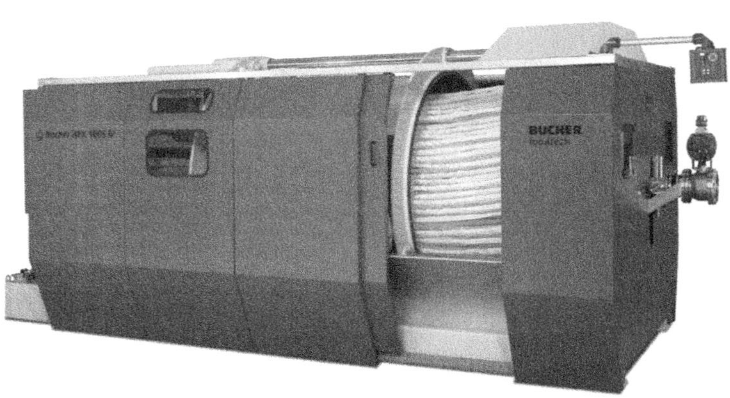

Fig. 13.4 High pressure piston press for apple juice extraction (Reproduced with permission of Bucher Unipektin AG).

13.3.2 *Fermentation*

The internal apple microflora, together with inocula from milling and pressing equipment, results in apple juice with 10^4-10^6 yeast cells mL^{-1} and, in traditional cidermaking, spontaneous fermentation is started within a few hours. However, in industrial-scale cidermaking, dried wine yeast (mixed strain) is hydrated and added to the blended juice to achieve more consistent, shorter fermentations with high alcohol yields (Beech, 1993). Fermentation temperatures vary in the range 15–25°C, and the fermentation is usually completed in 2–3 weeks.

13.3.3 *Racking and maturation*

When the fermentation is complete, flocculated yeast cells (lees) are separated from the cider in a clarification process termed 'racking' in the industry (Beech, 1972). In some factories, racking and clarification take place immediately post fermentation for immediate blending and packaging without any maturation, whereas in others the lees are not racked for several weeks post fermentation and are matured for months before blending and packaging.

Settling of well-flocculated yeast cells in the tank bottoms occurs naturally and, in modern plants, is removed by centrifugation, which has replaced the traditional 'fining' process that required the addition of flocculating agents.

13.3.4 *Blending, filtration and packaging*

In some cases, each fermentation batch is individually stored and matured for up to 12 months (Lea & Drilleau, 2003). The individual batches are then blended in a mixing tank, which can be up to 250 000 L in capacity, to achieve the uniform consistency of flavour and colour associated with different cider products or brands.

In modern high-capacity plants, blended cider base (10–12 mL 100 mL^{-1}) is generally microfiltered at this stage to remove residual yeast, adventitious bacteria and any suspended material, such as precipitated pectins. Hollow fibre membrane configurations are commonly used with nominal pore sizes in the range 0.1–0.5 μm.

After this primary filtration stage, potable water, together with sugar, other sweeteners, malic or other organic acids and SO_2 or other preservatives, are added to reduce the alcohol content to the correct final product strength (1.2–8.5 mL 100 mL^{-1}) and comply with compositional standards. An equilibrium level of 30 mg kg^{-1} SO_2 is standard for most ciders with the exception of cider which is to be canned, where the addition of SO_2 is avoided (or added at less than half this level) due to the risk of base metal corrosion (in the event of localised lacquer failure). The final strength cider is then microfiltered to give microbiological stability, cooled and stored for pasteurisation and/or carbonation and packaging.

Immediately prior to bottle filling, chilled cider is sterile filtered using in-line cartridge filter units (0.2 μm) and carbonated. Alternatively, high temperature, short time (HTST) heat treatment followed by hot-filling or in-bottle tunnel pasteurisation is used.

In addition to the traditional glass bottle, cider is also packed in PET bottles and cans. In the case of polyethylene terephthalate (PET) bottles either cold aseptic filling after in-line sterile filtration or ultra-clean filling following HTST heat treatment and cooling is used. Canned cider is tunnel pasteurised.

13.4 Membrane applications in the brewing process

Many potential applications that have been alluded to in the introduction to brewing will be described in more detail. Table 13.2 summarises process steps that have a membrane application, and a chart illustrating the range of membrane types required to separate out

Table 13.2 Summary of brewing processing steps, liquid streams and existing/potential membrane applications.

Processing steps	Liquid stream	Membrane application
Wort separation	Primary wort for boiling	Use of mash filters instead of more traditional Lauter Tun
	Weak worts	None commercially; potential for solids concentration to generate higher value liquid using nanofiltration (NF); hot water recovery
Trub Separation	Trub	None commercially; combine with weak worts stream
Yeast separation	Surplus yeast from fermentation	Recovery of beer from surplus yeast and concentration of yeast solids using cross-flow microfiltration (MF); future potential to generate higher flux rates and longer life membranes
	Storage vat bottoms	Recovery of beer from surplus yeast and concentration of yeast solids using cross-flow MF; combine with surplus yeast from fermentation for greater efficiency.
Filtration and stabilisation	Green beer	Primary beer filtration; using cross-flow MF; opportunities to reduce unit cost and combine filtration with haze stabilisation
	Caustic used in polyvinylpolypyrrolidone (PVPP) regeneration	NF of caustic stream for reuse
Beer make-up	Bright beer	Gas exchange membrane for addition or removal of gas to beer
	Deaerated water	Gas exchange membrane for removal of air from water
Microbiological stabilisation	Beer pre-packaging	MF of beer prior to packaging as an alternative to heat stabilisation
General process	Waste-water streams	Use of ultrafiltration/reverse osmosis (UF/RO) for waste water recovery
	Cleaning-in-place (CIP) liquids	Use of UF/RO for caustic recovery
Miscellaneous brewing applications	Bright beer	Application of NF for alcohol removal or reduction

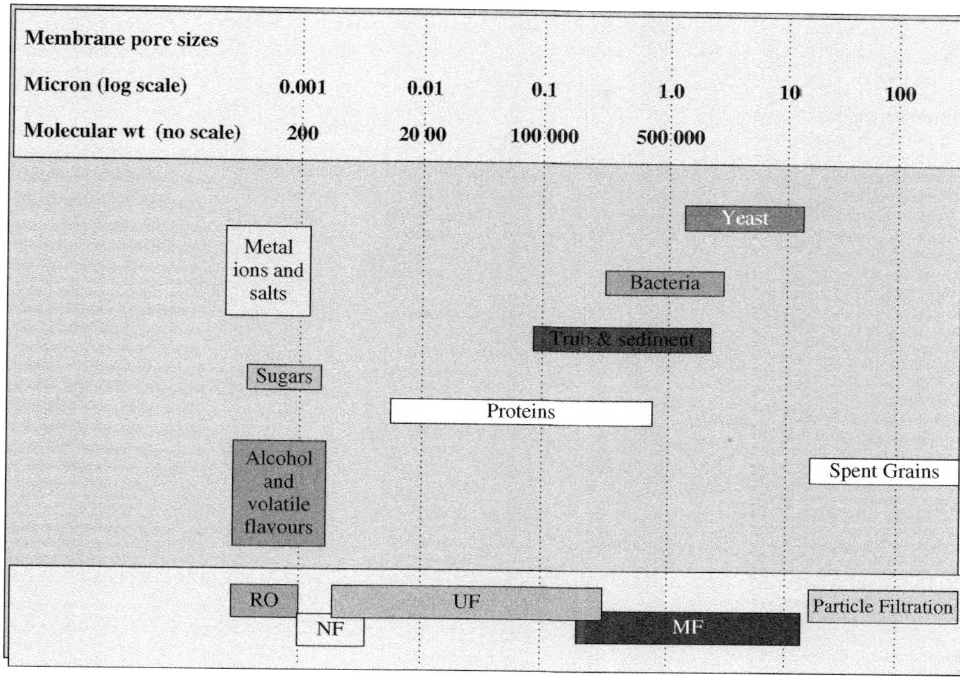

Fig. 13.5 Cut-off chart showing the relative particle size and molecular weight of the major components of beer and the relevant filtration technologies for their separation. MF, microfiltration; NF, nanofiltration; RO, reverse osmosis; UF, ultrafiltration.

the major components of interest in the brewing process, based on membrane pore size distribution and component molecular weight or particle size, is shown in Fig. 13.5.

13.4.1 *Wort separation*

As already explained, wort separation is an integral part of the brewing process. Mash filtration is a dead-end filtration process, and its introduction was probably one of the biggest innovations in brewing technology in the past 50 years. The mash filter provided an alternative to the traditional Lauter Tun, which relied on gravity to drive the filtration with the grain bed itself as the filter medium. The Belgian company Meura is credited with developing the mash filter for brewing applications, and still remains as the market leader in this area with the Meura 2001 Mash Filter (Fig. 13.6). Other brewery plant manufacturing companies, such as Ziemann (Germany) and Landaluce (Spain), also manufacture versions of the mash filter. The ability of mash filters to handle grist, which has high levels of non-malt cereals, has popularised their use in countries where barley is not locally grown (Braekeleirs *et al*., 2007). The common element in all is the use of filter sheets, which are generally woven from polypropylene and a thin mash bed for the wort filtration step.

Although mash filters provide many advantages to the brewer, some brewers still favour the traditional methods and would argue that mash filters carry with them some disadvantages. Table 13.3 gives a summary of the advantages and disadvantages of both systems.

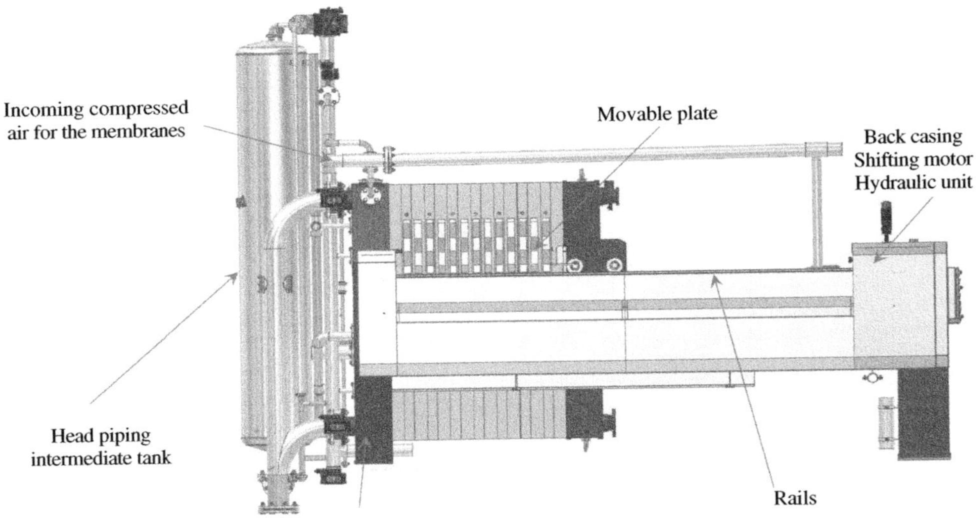

Fig. 13.6 Schematic of mash filter illustrating key components (Reproduced with permission from Meura S.A.).

Obviously, requirements will vary from brewer to brewer and selecting a system will need careful scrutiny, but the flexibility, size and rapid turnaround time of mash filters will certainly make their use attractive to many brewers. The selection of the best system for mash separation continues to be one of the most debated topics in brewing literature (Baumgärtner & Kantelberg, 2011a, b).

Future developments in wort separation are expected to be generally concerned with minor modification of existing systems or application of existing systems in novel ways. The use of mash filters is at the core of current developments to design a 'brewery of the future', which proposes continuous or semi-continuous brewing that could result in significant reduction in both capital and running costs (Harmegnies *et al*., 2003; Vandenbussche, 2009). Other research is looking at higher temperature separation (up to 95°C), which will speed up the separation time and reduce the total hot time (Harmegnies *et al*., 2005). Mash filters are continually being pushed to their limits to reduce turnaround time and increase efficiency. The typical number of brews per day from a mash filter is 12 (turnaround time of 2 h), but many brewers achieve more than this through manipulation of temperature, flow rate and sparge volume (Clifford, 2006). Another possible development would be to control or force the precipitation of protein/polyphenol trub during mashing extraction and use a mash filter to remove trub thereby removing the need for a trub separation step.

Limited research has been carried out on the use of cross-flow filtration for mash separation, which demonstrated that final wort and beer quality was comparable to Lauter Tun (Schneider *et al*., 2005). Although this research focused on quality, it is anticipated that both the capital and running costs associated with such systems on a commercial scale would make membranes unattractive for this purpose.

Table 13.3 Mash filter performance compared to traditional wort separation technologies.

Capital cost	Mash filters are generally slightly more expensive than Lauter Tun; however, as mash filters have a smaller footprint, this will be offset by reduced civil costs; this will only apply in a new build situation
Cycle time	Mash filters have a faster brew turnaround time; typically mash filters can produce up to 16 brews per day while Lauter Tun would process approximately 12
Flexibility	Owing to assisted filtration, mash filters are particularly suited to grists with a high degree of non-malt raw materials, which need to be finely ground; because of this, mash filters are very popular in hot countries where barley growing is difficult and local raw materials are used for brewing; most African breweries, who use high levels of sorghum and maize, use mash filters; mash filters are also more suitable for high gravity brewing and can produce more concentrated worts; it is also believed that mash filters are more robust in handling variability in malt/barley performance from year to year due to the fine milling of the grain
Running costs	Overall, the running costs are higher for mash filters; they require higher maintenance, and incur higher consumable costs such as filter cloth replacement; they also need more on operator attention than Lauter Tun
Extract recovery	Mash filters on average recover up to 2% more raw material extract than Lauter Tun; this generally compensates for the increased running cost resulting in overall cost neutral between mash filters and Lauter Tun
Effluent and water usage	Mash filters use less water and generate less effluent; the spent grains are dryer and are more marketable as an animal food source or suitable for absorbing other waste streams or for biomass/grains combustion in countries where sending spent grains for animal feed is not an option
Quality	There is no firm evidence to suggest that there is any significant difference in the quality of beers produced from either; Lauter Tun users will argue that the fine milling and mash filtration can extract undesirables from the grain which can adversely impact final product quality while mash filters reduce the overall hot time in separation which could improve flavour stability; however, as long as best practice is followed in both systems, there should be no quality differences

As mentioned earlier, the wort separation process can generate high levels of weak worts, some of which are reused whereas others contribute to the overall effluent stream from the brewery. These worts can contain up to 2 g 100 g^{-1} fermentable sugar, but cannot be used in the fermentation process because of their dilute state. These weak worts are also generated at temperatures close to 80°C. The availability of a thermostable cross-flow nanofiltration (NF) membrane, would make it possible to concentrate the dissolved solids in the worts and recover up to 90 mL 100 mL^{-1} as hot water (permeate stream), which could be reused in the brewing process. The concentrated wort (retentate) could be used for other purposes: (a) as a growth medium for industrial fermentation; (b) for the production of bio- or food ethanol through fermentation and distillation; and (c) fed directly into a waste-water treatment plant, but with a significantly reduced hydraulic load.

The trub formed during boiling and trub separation could also be included with the weak worts for hot water recovery and wort concentration. However, the wort contained within the trub is at full strength, and economically it would be better to recover this wort into the brewing process provided there was no adverse impact on quality.

A careful financial analysis for such a system would need to be carried out to ensure an adequate return on investment.

13.4.2 *Beer filtration and stabilisation*

In the early 1990s, brewers began to examine cross-flow microfiltration as an alternative to depth filtration with kieselguhr. Although the technology was shown to be capable of producing high-quality filtered beer, it was not until the early part of this century that it became technically and economically feasible. This was due in part to innovations in membrane materials and their regeneration, but also due to rising costs and concerns around kieselguhr usage (Buttrick, 2007; McCrorie, 2011).

As described previously, kieselguhr filtration has been used worldwide in the brewing industry for over 50 years. The base powder is composed of calcined diatoms, giving a highly porous filter bed and therefore an excellent medium for depth and chemical filtration of lagers and ales. Unfortunately, there are several drawbacks to this form of filtration.

- *Disposal* – After filtration, the spent kieselguhr has to be disposed of. The most common disposal method is to spread the used kieselguhr over agricultural land. However, this is not a satisfactory solution to the problem and in the long term will not be environmentally sustainable. The disposal costs increase every year, and there are also fears that in the future kieselguhr use may be banned for environmental and health reasons.
- *Handling* – Handling of kieselguhr creates a dust, which is of concern from a health and safety perspective due to risks of inhalation. The wearing of dust masks in the make-up area currently mitigates this.
- *Batch process* – Kieselguhr filtration is a batch process that prohibits the rapid changeover to a different brand of beer and reduces plant flexibility.

Cross-flow microfiltration addresses all these concerns, and also provides a number of advantages.

- *Increased fermentation/storage capacity* – Because some systems are designed to handle high levels of yeast, there is no requirement for long settling times thereby decreasing the vessel turnaround.
- *Fully automated* – There is minimum operator intervention required.
- *Guaranteed throughput* – Membrane systems are generally more consistent in performance.
- *Reduced beer losses* – Again due to the ability to handle high yeast solids. Also depending on configuration, these systems may reduce or indeed remove the requirement for secondary systems for beer recovery from yeast.
- *Future expansion* – All systems are modular and can be added to quite easily.

These systems may provide many more advantages, which are best realised when designed as part of a new brewery rather than a direct replacement for an existing kieselguhr filter.

There are three major manufacturers globally supplying cross-flow membrane filtration; Pentair Process Technology (formerly Norit), Pall and Alfa-Laval (currently known as Tetra Pak). The major differences between these systems are listed below.

- Pall and Westfalia have developed the Profi™ system (Fig. 13.7), which has a centrifuge prior to the membrane system. The centrifuge removes and concentrates the bulk of the suspended solids, leaving the membrane to do the final polishing. The membranes are polyethersulphone in a hollow fibre configuration. The system is configured in modules with spare capacity so that one module is either being cleaned or in standby, such that continuous filtration without stopping and with minimal changeover between brands is possible (Pickerell & Heeb, 2008; Snyder & Gaub, 2005).
- The Pentair Process Technology (formerly Norit) system (Fig. 13.8) has a similar membrane and configuration as the Pall system. The system does not have a centrifuge prior to the membranes resulting in significantly higher suspended solids loading on the membranes. However, the system utilises pulsing and backflushing, which interrupt filtration, within a run to maintain good flux rate. This is not as continuous process as the Pall system (Schuurman *et al*., 2005).
- Similar to the Profi™ system, Alfa-Laval and Sartorius have teamed up to offer the Alfabright™ system. This also has a centrifuge to pre-clarify the beer prior to the membrane. The membranes are polyethersulphone and configured as a cassette with thin flow channels between the membranes, which the manufacturers claim, reduces the need for big circulating pumps (Modrok *et al*., 2005).

The pore size of all the above membrane systems range from 0.5 to 0.65 μm and, although this pore size is close to sterilising grade, the filters are not guaranteed as sterile filters.

All the above systems while offering many opportunities to optimise the filtration process are still merely replacements for the traditional kieselguhr filter. They do not impact on

Fig. 13.7 Profi beer microfiltration system with clarifier in the foreground on the left (Reproduced with permission from GEA Westfalia Separator Group).

Fig. 13.8 Norit beer microfiltration system (Reproduced with permission from Pentair Process Technology B.V.).

the haze stability of the beer. Separate haze stabilisation as described previously will need to be carried out. Dosing of haze stabilisers, such as PVPP or silica gel, prior to membrane filtration and using the membranes to remove them may be risky due to the possible damage to the membranes because of abrasion from these particulate materials. Membrane systems fitted with a pre-clarifying centrifuge will be able to remove dosed powders through the centrifuge. Otherwise, traditional haze stabilisation methods will need to be used.

Novel haze stabilisation systems, such as the combined stabilisation system (CSS), have been developed which complement the continuous nature of membrane systems. These are based on the adsorbent material (insoluble agarose) being permanently held in a matrix within the processing modules. Proteins and polyphenols are adsorbed and removed during regeneration removing the need to dose in any stabilising powders (Jany, 2008).

13.5 Membrane applications in cidermaking

13.5.1 *Background*

Although the applications for membranes in cidermaking and the type of membrane technology used are in general quite similar to those already covered in the previous section on brewing, there are nonetheless some distinct requirements and performance issues associated with cider filtration

In cidermaking, apple juice is mechanically extracted from the pulp leaving the insoluble apple solids, apple flesh, seeds and stem, which is termed pomace. Some suspended solids is desirable in apple juice as it provides a physical support structure to keep yeast cells in suspension, by adhesion, throughout the fermentation vessel thereby promoting yeast activity. Thus, the pomace separation stage can be regarded as a coarse filtration operation, and does not represent an opportunity for membrane filtration.

For the production of naturally sweet ciders containing residual unfermented sugars, which are popular in France, juice is maintained at 5°C prior to fermentation to promote enzymatic demethylation, gelation and removal of pectin by decantation prior to fermentation in a process termed 'keeving'. The keeving procedure yields a pectin-reduced juice, which facilitates the production of a high-clarity cider. Keeving is not practised in industrial cidermaking in the UK (Lea & Drilleau, 2003), but merits consideration as a pre-filtration step to potentially improve flux rates later on in the process during clarification.

13.5.2 *Cider clarification*

Cross-flow MF is better suited to the clarification of cider than centrifugation for the following reasons.

- A high-clarity filtrate is obtained – cross-flow MF can separate fine particles like cell debris whereas with centrifugation the efficiency of recovery declines as the particle size gets smaller and particles less than 0.5 μm are generally not removed by centrifugation.
- High recovery of cider can be obtained as high volumetric concentration factors can be achieved by cross-flow MF.
- The difference in density between cells/particles and the fermentate liquor is not important for cross-flow MF compared to centrifugation.
- Cross-flow MF process can be more easily contained and no aerosols are produced.

The performance of cross-flow MF for cider clarification is primarily influenced by the following factors:

- porosity of the membranes used
- plant operating parameters and system hydrodynamics
- cider composition; concentration of yeast cells in the feed and presence of cell wall material and residual fruit cells
- physico-chemical properties of cider, such as pH and the relationship between cell concentration and viscosity
- filtration temperature.

In bench-top scale experiments with crude fermented cider broths, Scott (1988) found that, for a range of pore sizes and flow rates, the flux increased shortly after the start of cross-flow MF, declined again and then levelled off at or below the initial steady-state value until a critical cell concentration of 55–60 g L^{-1} was reached. Beyond this concentration range, the flux started to decline. The increase in cell concentration was reflected in an increase in feed inlet and outlet pressures (0.05 MPa) due to the increase in viscosity. Not surprisingly, filtration temperature, through its effect on viscosity, was found to a huge effect on flux – a 5°C increase in temperature (in the range 13–30°C) increased the steady state flux rate by approximately 50%.

The data for 0.2 and 0.45 μm MF membranes and a 100 kDa nominal molecular weight cut-off UF membrane evaluated by Scott (1998) show that, under steady-state conditions, there is a direct relationship between flux and pore size. Interestingly, a more rapid rate of initial flux decline was observed for the 0.45 μm membrane, which was attributed to progressive partial blockage by small cells and cell debris. They observed that fouling of a new MF membrane occurred in two stages: primary fouling with small particulate matter (that could not be subsequently removed by cleaning-in-place – CIP), which occurred before steady state flux was reached, and secondary fouling, which was a function of the cider characteristics and system hydrodynamics, which could be removed by CIP. Cross-flow velocity was critical to the arrest of secondary fouling through backflow diffusion and physical erosion of the deposits, and the data indicate that a critical cross-flow rate exists for a system below which a continual fall in flux occurs as the cell concentration increases.

The native pH of fermented cider, ~pH 3.1, appears to be the optimum for cross-flow MF. Scott (1991), again working with crude cider fermentation broth at bench-top scale, found that adjusting the pH in the range pH 2–3 had little impact on steady-state flux through PVDF MF membranes but above pH 3.5 a large decrease in flux occurred. Interestingly, the decrease in flux was greater in the case of a more open 0.45 μm MF membrane than with a 0.2 μm membrane. However, the flux through a 100 kDa molecular weight cut-off UF membrane was not affected at all in the pH range 2–8. Increasing the pH would have been expected to promote the spontaneous flocculation of spent yeast cells, and the large particles formed as a result of flocculation should be more easily removed from the surface of membranes at the velocities achieved in cross-flow MF and contributed to an improvement in flux.

In addition, Scott (1991) demonstrated that while the clean water flux of a new 0.45 μm membrane was not affected by pH, the cross-flow MF of one that had been used to clarify cider was lower than that of the new membrane and decreased with increasing pH. He attributed the observed reduction in flux with the 'used' MF membranes to the irreversible adsorption of organic material of subcellular dimensions on the membrane resulting in a reduction in the effective porosity of the membrane and the pH effect to a rising membrane hydrophobicity associated with the adsorbed organic material.

A recent article refers the use of a cross-flow MF system installed by Filtrox AG at the Bulmers Heineken cider plant in Hereford, and equipped with Pentair X Flow and Puresep membranes (McCrorie 2011). Pentair X Flow membranes are described by the company as capillary membranes with a pore size of less than 0.2 μm, made from a blend of polyethersulphone and polyvinylpyrrolidone having a double asymmetric pore structure which mitigates against in-depth pore clogging (Anonymous 2011a). Another recent article on the Bevisol (formerly Universal Beverages) factory in Ledbury refers to the use of a Puresep Pureflow hollow fibre cross-flow MF plant with 0.2 μm pore size for cider clarification (Anonymous, 2011b). One membrane equipment manufacturer, Filtrox Engineering AG, recommends their dual-flow ceramic MF plant (Fig. 13.9) equipped with 0.3 μm pore size ceramic elements with 8 mm diameter channels for cider clarification (Anonymous, 2011c).

13.6 Membrane applications common to brewing and cidermaking

13.6.1 *Yeast separation and product recovery*

Surplus yeast and tank bottoms represent a considerable waste stream for the brewer and cidermaker, leading to increased effluent and significant beer losses (up to 2–3% of total

Fig. 13.9 Cerinox MF 100 ceramic membrane plant for cider clarification (Reproduced with permission from Filtrox Engineering AG).

site output containing at least 50% saleable product (McCrorie 2011). Because of this, product recovery from yeast has always been an area of great interest (Buttrick, 2006). Over the last 10 years, the use of membranes for this purpose has become widely accepted in the drinks industry. Another advantage of membrane systems over older methods is the quality and suitability of the concentrated yeast for food grade applications as it is produced without any requirement for pre-processing, such as sedimentation, thickening, pressing, etc. The key to successful product recovery from yeast is to process the yeast as fresh as possible so that the recovered product is of similar quality to the bulk of the freshly fermented liquid. If possible, the yeast slurry should be processed within 24 h and should be stored at temperatures below 4°C prior to processing. Thus, the recovered liquid can be added back to the main product stream later in the process without any impact on quality.

Surplus yeast slurry is generally up to 8 g 100 g^{-1} of dry solids, and cross-flow systems are capable of concentrating this to 20 g 100 g^{-1} dry solids, recovering 60 g 100 g^{-1} of the beer contained within the yeast. Beer or cider recovery has a number of financial advantages: (a) removal of effluent or disposal costs (tank bottoms have a biological oxygen demand (BOD), of 140 000 mg kg^{-1}); (b) the value of the recovered liquid including all raw material and processing costs; and (c) the value of the concentrated yeast. At the original dry solids concentration, the brewer would have to pay for disposal, but after membrane processing the concentrated yeast is a revenue stream, which has many uses within the food, pharmaceutical and animal feed sectors. The combined savings could result in an average payback on investment of less than 2 years.

The most commonly available membranes suitable for this process are tubular ceramics with a pore size between 0.4 and 0.6 μm. These are long lasting, and can handle feed streams with high suspended solids. Owing to the increased levels of solids, high shear forces are required to prevent layer formation at the membrane surface and maintain good flux rates. To achieve this, flow velocities of up to 6 m s^{-1} are needed, which require quite large pumps and will have a significant impact on energy and maintenance costs.

Conventional cross-flow MF is less well suited to the removal of cider yeast lees (tank bottoms) because in addition to cell concentration potential being limited (due to the high viscosity of concentrated yeast cells), membrane fouling by yeast cells, cell wall material (autolysed cells), fruit cell debris and other colloidal or non-colloidal material present in the fermentate results in low flux rates and short run times. One membrane equipment manufacturer, Filtrox Engineering AG, recommends that cider must be racked at least once before cross-flow MF with their dual-flow ceramic MF plant (Anonymous, 2011c), although they claim that cider tank bottoms can be successfully filtered with a cross-flow MF plant fitted with specially designed large pore size ceramic membranes. However, for optimum filtration performance, some form of in-line prefiltration, such as the self-cleaning Russell Eco filter or centrifugal separation to remove abrasive particulate debris from yeast lees, would appear to be advisable before going to cross-flow MF with cider lees.

A clever innovation to reduce the pumping requirements while still maintaining high shear forces at the membrane was the introduction of the vibrating membrane cross-flow filtration launched by the Pall Corporation (Chalmers & Haughney, 1998; Sellick & Pratt, 1998). This is essentially a plate and frame type system using polysulphone membranes. Ceramic membranes are brittle and would not be suitable in a vibrating membrane configuration. The entire membrane head or pack is vibrated generating far higher shear forces than traditional pumping. This technology overcomes the problem of membrane fouling associated with conventional cross-flow plants and is ideally suited to the removal and concentration of yeast cells from beer and cider. The energy consumption is greatly reduced, but membrane replacement is more frequent because the polysulphone membranes have a shorter life span (typically 2 years, but will vary depending on throughput and cleaning cycles). Furthermore, it is known that the constant vibration of the membrane stack can lead to stress damage to the membrane housing resulting in high maintenance frequencies. It is believed that because of this, the vibrating membrane filter systems are not currently being supplied for this application.

13.6.2 *Microbiological stabilisation*

The microbiological stabilisation of beer and cider is required to protect product quality and guarantee the shelf-life. In the case of cider, SO_2 addition during the process contributes to stabilisation, but is generally followed by a final stabilisation treatment before or as part of the packaging process.

With the pressure to reduce capital and running costs and the environmental concerns over energy and water usage, there is no doubt that brewers will eventually abandon the use of tunnel pasteurisation. This means that both small pack and keg beers will be flash pasteurised using a PHE before packaging. This will put greater pressure on the small pack filling operation to ensure that no contamination is introduced into the beer between pasteurisation and filling. Both pasteurised and sterile filtered beer are required to be stored in a sterile buffer tank prior to packaging.

Cold sterile filtration through a filter with a pore size not greater than 0.4 μm is widely used by cidermakers and can be used as an alternative to flash pasteurisation of beer. Pasteurisation destroys all pathogenic and spoilage micro-organisms, which are in the vegetative state, but may not kill all dormant bacterial spores (although these can never grow or cause spoilage in beer). Sterile filtration on the other hand removes all viable microbial cells but, in addition, removes dead microbial cells and spores. The traditional filtration process is operated mainly as an in-line dead-end filtration rather than in cross-flow mode. The cider or beer will already have been processed through the normal beer filtration cycle so the sterile filter will only need to remove any micro-organisms present. However, a pre-filter is generally placed before the sterile filter to protect it and enable longer run times, less frequent cleaning cycles and ultimately extend the life time of the filter. The cider or beer is processed cold with no additional heating or cooling requirements. The primary reasons for considering sterile filtration over pasteurisation are the following.

- *Energy consumption* – Pasteurisers require heating and cooling, which is not required in sterile filtration. However, sterile filters do have additional running costs, such as membrane replacement, and the membranes may require more extensive cleaning regimens and specialised chemicals for cleaning. Also, all modern pasteurisers have heat recovery systems with up to 95% heat regeneration. Total energy costs need to be considered when comparing both systems.
- *Quality* – Because sterile filtration processes the liquid cold, there is no potential for thermal damage to the product, which could ultimately impact the flavour of the fresh cider or beer or the overall flavour stability resulting in reduced shelf-life. However, if pasteurisation is carried out using best practice with no air pick-up during the process, flavour should be unaffected.

Flash pasteurisers are relatively easy to maintain. They have a holding section, designed to ensure the correct holding time, at a constant set flow rate which is closely controlled. The pasteurisation temperature is tightly controlled and, if the temperature drops, the system automatically diverts to recirculation mode to ensure no unpasteurised product goes forward to filling or sterile feed tank in the case of beer. Sterile filter cartridges do need to be replaced, and as the system runs cold they need to be sterilised in place before each run. Integrity testing is carried out daily, before and after filtration, to test the system for cartridge damage to ensure maximum microbiological safety.

Small-scale sterile filtration systems have proven popular for microbreweries where standard pasteurisation at that scale would be expensive and difficult to control. However, in recent years, larger breweries have begun to embrace sterile filtration, and suppliers, such as Pall and Alfa-Lava Sartorius, are offering complete systems as pasteurisation alternatives (Krottenthaler *et al*., 2003). It is also possible to streamline the brewing process further through combining primary filtration and sterile filtration into one operation. The use of such systems could provide the brewer with a positive public relations/marketing message around low-temperature processing/cold stabilisation, etc.

13.6.3 *Gas standardisation using membranes*

There are a number of stages in the brewing and cidermaking process that may require addition or removal of gas, primarily:

- *Adjusting the CO_2 level* – All alcoholic fermentation is accompanied by CO_2 evolution. Thus, the fermented product will contain a natural level of CO_2 in solution. If the final packaged product is to be highly carbonated, then the presence of CO_2 will be advantageous. For a final product, which has low carbonation (e.g. nitrogenated beers, stouts), some dissolved CO_2 may need to be removed. Whatever the case, the level of dissolved CO_2 will more than likely need to be adjusted (either up or down) in the final product. Addition of CO_2 is relatively simple and easy to control. It involves injecting the gas into the liquid at the required rate and ensuring adequate mixing to aid dissolution. The rate of addition can be monitored using gas flowmeters and feedback control is achieved using on-line CO_2 meters. The fact that the gas is physically bubbled into the liquid can lead to foam formation. In beer, foam formation can lead to loss of head-forming potential, valuable hop flavour compounds and bittering compounds. Removal of CO_2 is less controllable, and various methods used to accomplish this goal include physical agitation and displacement using other gases (e.g. bubbling with nitrogen). Both of these methods are rather inelegant, and can generate large amounts of foam. Addition or removal of CO_2 is relatively simple and easy to control with membranes. Liquid flows on one side of the membrane and a vacuum and/or a gas is passed on the other side of the membrane (Fig. 13.10). By controlling the pressures of gasses in contact with the liquid a highly efficient method for gas control is possible (Annonymous, 2007).

- *Addition of nitrogen* – This mainly applies to beer. Some beers, such as stout and some ales, require addition of nitrogen to achieve the 'creamy' head associated with such products. Nitrogen gas has low solubility and can be difficult to get into solution. The same technology that is used for the addition of CO_2 can be used for nitrogen. Again membrane processes diffuse small N_2 molecules into the liquid, which promotes a consistent and stable gas level in the fluid.
- *Water deaeration* – Any water used for product dilution (in particular for beer) must be deaerated to ensure that no oxygen is added to the beer. Oxygen will negatively impact on the shelf-life and taste of the final product. Additionally, any water that is

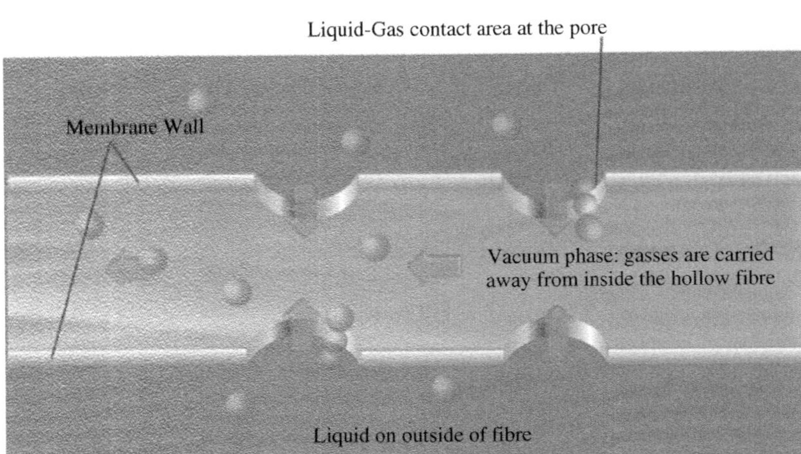

Fig. 13.10 Cutaway view of a hollow fibre gas exchange membrane. Liquid/beer flows on the outside of the hollow fibber while a strip gas and/or vacuum is used on the inside of the hollow fibre for gas removal. For carbonation or nitrogenation, CO_2 or N_2 is used (under pressure) on the inside of the membrane. Simultaneous oxygen removal teamed with carbonation or nitrogenation is also possible in one device (Illustration is courtesy of Membrana-Charlotte, A Division of Celgard, LLC).

used to flush product out of pipes and tanks also needs to be deaerated. If this water is not deaerated, the beer will absorb gasses present in the flushing water as it is being extracted from the system.

Gas exchange membrane contactors offer an alternative to traditional methods of adding or removing gases without foaming (Brown *et al*., 1999; Gill, 1997). The membranes used are hollow fibre microporous membranes that are hydrophobic. The membrane has a highly porous wall, which provides a substantial surface/contact area for gas transfer to occur with the liquid. These hollow fibre membranes are integrated into a gas transfer contactor that allows maximum surface area for the smallest footprint system where the liquid stream is on one side of the membrane and the gas stream is on the other (Fig. 13.11). Gas exchange occurs at the membrane pore but, unlike filtration, the liquid does not pass through the membrane. Gas is either transferred to or removed from the liquid by diffusion and is driven by partial pressure differences between both phases. Because the gas transfer is by diffusion, there is no bubbling and, thus, no foam formation. This technology can be applied to any gas exchange application, and it is also possible to simultaneously deaerate and carbonate or nitrogenate with one module.

For water deaeration, there are many advantages of using membrane gas exchange modules over sparging and steam deareators. For example, one method can simultaneously carbonate and deaerate, and systems have a small footprint and can easily be

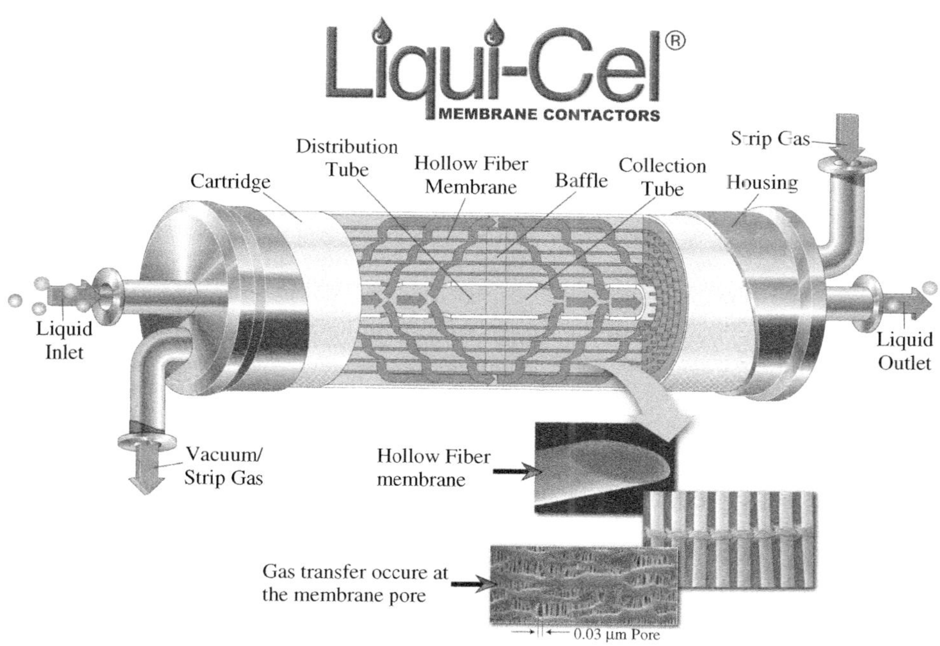

Fig. 13.11 Polypropylene microporous hollow fibre gas exchange membrane is knitted into a membrane array and assembled into a membrane contactor to maximise liquid/gas contact. The hydrophobic nature of the membrane allows gas transfer to occur at the membrane pore. The module construction forces the fluid to come in contact with all of the membrane for maximum efficiency (Illustration is courtesy of Membrana-Charlotte, A Division of Celgard, LLC).

added anywhere in the processing plant. There is also very low energy consumption with membrane systems and there is no need to heat-up and cool water back down as with a steam deareator. As with all membrane applications, one should review the broad advantages of the membranes and the needs of individual producers and bottlers. Membrane contactors have been operational for years without issues.

13.6.4 *Water recovery/cleaning-in-place systems*

The principal component of beer and cider is water ($\sim$92 g 100 g^{-1}) , but the production process itself is quite intensive in its water usage. Environmental concerns, water shortages and current/future costs associated with water usage are all drivers towards reducing the overall water volumes used in brewing through improved water management. This is particularly important in countries that are already water stressed. Membrane systems are fast becoming an efficient and affordable technology for water treatment and recovery (Brister & Morrow, 2010; Cahill, 2011). Current best practice is between three and four volumes of water per volume of final packaged beer. Recent developments by the Martens brewery in Belgium in streamlining the brewing process to define a 'brewery of the future' will drive water usage down to almost two volumes per volume of beer (Bergenhenegouwen, 2010; Nelson, 2009).

This is achieved through a combination of process changes: semi-continuous brewhouse, change to PET bottles for small pack; and significant cross-flow filtration technologies for primary filtration and waste-water recovery.

Membrane systems for waste-water recovery can be either stand alone or integrated into a waste-water treatment plant or as part of a membrane bioreactor for water treatment and purification. The application will depend on the water source and quality. The level of solids, both dissolved and suspended, will determine the configuration. For waste water with low levels of dissolved and suspended solids, the most simple configuration is UF followed by NF or reverse osmosis (RO). To protect the membranes, a rough screen filter would generally be placed before the membranes. The subsequent use of the recovered water will also determine membrane configuration. Recovered water for cleaning or rinsing may only require UF/NF, while RO would be needed for higher levels of purity. It is worth noting that while the recovered water is of excellent quality and more than suitable for brewing and cidermaking, many producers may be reluctant to use the water recovered directly as production or dilution water, particularly if the waste water from all sources is combined prior to recovery. The main concern is purely around public perception and the risk of adverse publicity through the use of 'recovered' water. This would be less of an issue if the waste water from a particular process step was processed at source and recycled back to the same process. As already described, water recovery from weak worts would be an example of this. However, placing small dedicated systems at each source of waste water would not be as economically attractive as a centralised system for collecting and processing all waste water.

Owing to the nature of brewing/cidermaking and the importance of maintaining cleanliness and sterility, at all steps in the process, the design and management of CIP systems are critical. Tank and pipe cleaning need to be carried out after every use. The main chemical used for cleaning purposes is a dilute caustic (NaOH) solution. Most modern production plants operate a CIP feed and return system where there is a central CIP plant, and the caustic is returned to the plant after use for potential reuse. The caustic is dumped periodically when the level of suspended and dissolved solids reaches a specified level. The waste caustic still has high alkalinity and any disposal requires neutralisation through pH balancing. Again driven by cost and environmental concerns, the recovery of waste caustic for reuse is now an attractive proposition and developments in membrane technology have enabled this to be commercially viable (Annonymous, 2009).

The membrane configuration will depend on the quality of the waste caustic and required quality of the recovered caustic. A combination of MF to remove suspended solids and/or NF to remove dissolved solids may be required. Traditionally, ceramic membranes are used for the MF step because of their high chemical resistance. Ceramics, however, are not generally produced or suited to low molecular separations like NF. In the late 1990s, polymeric membranes with high pH tolerance were developed that were suitable for NF applications. Depending on the quality of the waste caustic, recovery of between 80% and 90% is achievable. An excellent example of a typical application alluded to earlier is the recovery of caustic from PVPP recovery systems. In this instance, the caustic is used merely to desorb polyphenol compounds from used PVPP. The polyphenols are dissolved in the caustic, and are easily separated with a single NF process with 90% recovery of the caustic (Annonymous, 2008).

13.6.5 *Alcohol removal for non- or low-alcohol products and malt beverage production*

Although representing a small segment of the overall drinks market, low- and reduced-alcohol products are an important product category. Owing to social responsibility around misuse of alcohol and strict legislation on permitted alcohol levels for driving, most companies produce at least one low- or reduced-alcohol product. In a competitive market, the challenge is to produce a product with as much flavour as the full-alcohol counterpart. The definition of non-alcoholic and low-alcoholic drinks varies from market to market. Generally, non-alcoholic beer (NAB) or alcohol-free beer (AFB) refers to products with less than 0.5 ml 100 mL^{-1} alcohol by volume (ABV). Some definitions are as low as <0.05 ml 100 mL^{-1}. Low-alcohol beers tend to fall between AFB and full-alcohol products in terms of alcohol content.

There are numerous options available for the production of non-alcoholic or reduced-alcohol products:

- Use of unfermented wort/juice in formulating final product; for beer, this is more a malt beverage and has strong malty/worty flavours or excessive sweetness.
- Reduction in the level of fermentable material available to the yeast by using dilute worts or worts produced by high-temperature mashing to produce a wort with lower fermentability. This is mainly applicable to beer.
- Control of fermentation by either stopping fermentation at the required alcohol level or using yeast with limited fermentation capability.
- Cold contact of wort/juice and yeast, where this produces little or no fermentation. In brewing, contact with the yeast is believed to reduce worty flavours.

However, for authenticity, many no- and low-alcohol products are made from already existing product streams by removing or reducing the alcohol content. Although this appears wasteful and rather expensive, starting with a fully fermented product can result in a superior reduced-alcohol final product. The principal methods employed are:

- *Distillation/evaporation* – This is carried out under high vacuum conditions, which allows alcohol to be removed at low temperatures (35–50°C), thus minimising any heat damage to the beer. Falling film and spinning cone evaporators are the most commonly used. Evaporation removes both alcohol and water, resulting in concentration of dissolved solids. The degree of concentration depends on the desired NAB/low-alcohol content. The lower the required alcohol level, the greater the degree of concentration required on evaporation. This approach is very effective in removing alcohol, but it also removes the volatile flavour fractions, and can potentially cause heat damage to the remaining non-volatile fraction. Systems are available which minimise the degree of concentration while recovering some of the volatile flavour fraction. This recovered fraction can then be incorporated back into the final product.
- *Cross-flow membrane technology* – This utilises RO, NF or tight UF, and these systems can be operated cold, which minimises the risk of heat damage to the product. However, it is difficult to selectively remove alcohol without removing a high proportion of flavour components, which have similar molecular weights to ethanol. These compounds are mainly esters and higher alcohols, which are flavour active at low concentrations, and contribute significantly to overall flavour. Also, due to the nature of cross-flow filtration, it is difficult to remove all the alcohol from the feed liquid without some level of diafiltration. This would be only required for final products requiring alcohol levels less than 0.05 mL 100 mL^{-1}.

Both evaporation and membrane systems generate an alcoholic side stream. Because the alcohol recovered during evaporation is through a heat process and is more concentrated than the starting beer, it is deemed to be a distilled spirit. However, the alcohol removed using membranes is the same concentration as the starting beer, and can still technically be described as a beer. This is of great interest, particularly in the USA, where distribution and taxation of spirit-based products is different from malt-based products. This was first demonstrated in the early 1990s with the introduction of a colourless beer product called Zima™ by the Miller™ brewing company (Tripp *et al*., 1995). The process describes the production of two products from the same starting beer: a non-alcoholic beer and a clear beer in which the alcohol-rich permeate from the membrane step is used as a base to which various additions are made. The additions can include dextrins for body, hop constituents to give characteristic beer bitterness and natural foam enhancers. Fruit flavours may also be incorporated to give a fruit-flavoured malt beverage. Some equipment suppliers are marketing this technology primarily for the production of flavoured malt beverages (Blanchard, 2008).

13.7 Future opportunities

Membrane technology has become widely accepted within brewing in the last 20 years. It is difficult to see any further applications, but certainly the focus for developers and suppliers should be in the area of

- reduced capital costs
- reduced running costs
- extended membrane life
- reduced chemical cleaning
- low-pressure systems
- reduced pumping requirements

Fig. 13.12 Macro-view of the microsieve filter wafer manufactured using technology developed in the microelectronics industry (Reproduced with permission from Fluxxion B.V.).

- greater membrane functionality – selective rejection of specific components, e.g. for dealcoholisation, removal of alcohol only without removing the other flavour components would be of huge benefit
- smaller, more efficient systems – a novel membrane called a microsieve, which has been patented but not yet fully commercialised, promises to revolutionise beer microfiltration (van Rijn *et al*., 1998). The membrane is manufactured using 'wafer' technology from the microelectronics industry (Fig. 13.12). It has been demonstrated that this technology can produce uniform and precise pore sizes creating membrane wafers with very high pore density, which are capable of running at extremely high flux rates. Normal cross-flow MF systems would have average flux rates of ~100 L m^{-2} h^{-1}. Beer trials on a microsieve system have demonstrated fluxes of 4000 L m^{-2} h^{-1} and greater (Lommi *et al*., 2003; van Lersel *et al*., 2005). Filtration plants using this technology would occupy a significantly smaller footprint than traditional filtration plants (Fig. 13.13). A major barrier to full commercialisation of this technology is the initial set up costs for the reliable production of these membranes. The membranes are also quite delicate and sensitive to pressure changes requiring a high level of pressure control within the system. Like most new technologies, manufacturers and customers are reluctant to 'be the first' and would rather work with technologies proven at industrial scale. This technology could be a real 'disruptive' technology within filtration applications.

Fig. 13.13 Microsieve filtration rig. The membrane stack is located to the right of the drawing and occupies a small area due to the high porosity of the membranes (Reproduced with permission from Fluxxion B.V.).

References

Annonymous (2007) TechBrief 63: Liqui-Cel® Membrane Contactors are widely used in the soft drink and brewing industries to control dissolved gasses. Available from http://www.liqui-cel.com/uploads/documents/TB63 7–07 Beverage Market Summary.pdf.

Annonymous (2008) Spent Caustic Recovery Dundalk Brewery. Liquid Effluent Reduction using Membrane Technology. Available from http://www.scribd.com/doc/50425781/Spent-Caustic-Recovery-Dundalk-Brewery-Liquid-Effluent-Reduction-using-Membrane-Technology-Diageo-Global-Supply-2008.

Annonymous (2009) Caustic recovery using membrane filtration. 55th Annual Citrus Engineering Conference. Available from http://www.geafiltration.com/filtration_library/caustic_recovery_membrane_filtration.pdf Accessed 26/08/2011.

Annonymous (2010) Beer: Global Industry Guide. Datamonitor report summary. Available from http://marketpublishers.com/report/consumers_goods/food_beverage/beer_global_industry_guide.html.

Anonymous (2011a) Pentair Xflow Applications -Wine Filtration. Available from http://www.x-flow.com/applications/wine%20filtration/.

Anonymous (2011b) Bevisol – The new beverage industry solutions provider. *Food & Drink Business Europe*, **July**, 17–18.

Anonymous (2011c) Cider Filtration. [Internet Document] URL http://www.filtrox.com/depth-fitlration-products/beverage/cider/. Accessed 18/10/2011c.

Baumgärtner, Y. & Kantelberg, B. (2011a) Lauter Tun versus mash filter – a comparison (Part 1). *Brauwelt International*, **29**, 32–35.

Baumgärtner, Y. & Kantelberg, B. (2011b) Lauter Tun versus mash filter – a comparison (Part 2). *Brauwelt International*, **29**, 85–87.

Beech, F.W. (1972) English cidermaking – Technology, microbiology and biochemistry. *Industrial Microbiology* (ed. D.J.D Hockenhull), pp. 133–213, Churchill Livingstone, London.

Beech, F.W. (1993) Yeasts in cider making. *The Yeast* (eds. A.H. Rose & J.S. Harrisson), 2nd edn, Vol. **5**, pp. 169–213, Academic Press, London.

Bergenhenegouwen, S. (2010) The brewery of the future. *Brauwelt International*, **28**, 288–289.

Blanchard, B. (2008) *Cross-flow membrane filtration for producing neutral malt base*. *World Brewing Congress Proceedings*, CD available from World Brewing Congress, St. Paul.

Braekeleirs, R., Vandenbussche, J. & Harmegnies, F. (2007) Practical experiences with mash filtration on thin-bed filters from brews made with several kinds of raw materials. *Technical Quarterly Master Brewers Association of the Americas*, **44**, 121–126.

Briggs, D. E., Boulton, C. A., Brookes, P. A., Stevens, R. A. (2004) *Brewing: Science and Practice*, Woodhead Publishing, Cambridge.

Brister, R. & Morrow, N. (2010) The use of membrane systems for recovery of waste water to reduce brewery water footprint. *Proceedings of the 31st Asia Pacific Convention of the Institute of Brewing and Distilling (Queensland)*. Available from http://www.ibdasiapac.com.au/asia-pacific-activities/convention-proceedings/2010/Papers%20&%20Presentations/Brister%20R%20-%20%20Membrane%20Systems%20for%20Recovery%20of%20Waste%20Water%20.pdf.

Brown, J.W., Bott, N.J., Bowerman, R. & Smith, D. (1999) Control of dissolved gases in the brewery using a hydrophobic membrane. *Proceedings of the 27th European Brewery Convention Congress (Cannes)*, pp. 225–231.

Buttrick, P. (2006) Recovery of beer from tank bottoms – a review. *Brewer & Distiller*, **2**, 19–22.

Buttrick, P. (2007) Filtration – the facts. *Brewer & Distiller*, **3**, 12–19.

Cahill, G. (2011) Purification of brewery process water using membrane technology. *Proceedings of the 13th Africa Convention of the Institute of Brewing and Distilling (* Uganda*)*. (Available from the author through The Institute of Brewing & Distilling, 33 Clarges Street, Mayfair, London, W1J 7EE.)

Chalmers, S. & Haughney, H. (1998) Pallsep VMF – for the recovery of beer from barm/tank bottoms. *Proceedings of the 23rd Asia Pacific Convention of the Institute of Brewing (Perth)*, **25**, 165–169.

Clifford, D. (2006) 16 Brews/day mash filter: myth or reality? *Brewing & Distilling International*, **37**, 14.

Gill, C. (1997) Membrane technology in gas control systems. *Brewers Guardian*, **126**, 27–28.

Harmegnies, F., Bonacchelli, B. & Tigel, R. (2003) Continuous brewing. *Proceedings of the 29th European Brewery Convention Congress (Dublin)*, CD available from European Brewery Convention, Brussels.

Harmegnies, F., Marle, L. & Tigel, R. (2005) Mash filtration: influence of sparging parameters on wort quality and sparging efficiency. *Proceedings of the 30th European Brewery Convention Congress (Prague)*, CD available from European Brewery Convention, Brussels.

Jany, A. (2008) *Beer stabilization in combination with cross-flow filtration*. *World Brewing Congress Proceedings*, Hawaii, CD available from World Brewing Congress, St. Paul.

Jarvis, B. (2001) Cider, perry, fruit wines and other alcoholic fruit beverages. *Fruit Processing*, (eds. D. Arthey and P.R. Ashurst) , 2nd edn, pp. 111–148, Aspen Publishers, Gaithersburg.

Krottenthaler, M., Zanker, G., Gaub, R. & Back, W. (2003) Sterile filtration of beer by membranes – economical and physiological aspects. *Proceedings of the 29th European Brewery Convention Congress (Dublin)*, CD available from European Brewery Convention, Brussels.

Lea, A.G.H. & Drilleau, J-F. (2003) Cidermaking. *Fermented Beverage Production* (eds. A.G.H Lea and J.R. Piggott), 2nd edn, pp. 59–88, Kluwer Academic/Plenum Publishers, New York.

van Lersel, M., van den Bogaard, W., Loser, W., Dusamos, M. & van de Kerk, D. (2005) High throughput microfiltration of beer. *Proceedings of the 30th European Brewery Convention Congress, Prague*, CD available from European Brewery Convention, Brussels.

Lommi, H., Raspe, O.J.A., van Rijn, C.J.M. & Vos, J. (2003) New filter and method for beer clarification. *Proceedings of the 29th European Brewery Convention Congress (Dublin)*, CD available from European Brewery Convention, Brussels.

McCrorie, C. (2011) Recent developments in filtration technology. *Brewer & Distiller International*, **7**, 11–16.

Modrok, A., Weber, D., Diel, B. & Rodenberg, M. (2005) Cross-flow filtration of beer – a true alternative to diatomaceous earth filtration. *Proceedings of the 30th European Brewery Convention Congress (Prague)*, CD available from European Brewery Convention, Brussels.

NACM (2008) *Code of Practice for the Production of Cider and Perry*, National Association of Cidermakers, London.

NACM (2011) *Cider Market Statistics*, National Association of Cidermakers, London.

Nelson, L. (2009) Is this the brewery of the future. *Brewers Guardian*, **138**, 32–34.

Pickerell, A. & Heeb, W. (2008) Cross-flow membrane filtration at the Coors Shenandoah Brewery. *World Brewing Congress Proceedings (Hawaii)*, CD available from World Brewing Congress, St. Paul.

van Rijn, C., Johannes, M., Raspe, O., Johannes, A., van Hoof, S. & Cornelis, J. (1998) Device for beer filtration. European Patent Application EP0879635 (A1).

Schneider, J., Krottenthaler, M., Back, W. & Weisser, H. (2005) Study on the membrane filtration of mash with particular respect to the quality of wort and beer. *Journal of the Institute of Brewing*, **111**, 380–387.

Schuurman, R., Meijer, D., Broens, L. & Mepschen, A. (2005) Full scale results of (kieselguhr free) beer membrane filtration and latest developments on combined membrane filtration and stabilization. *Proceedings of the 30th European Brewery Convention Congress (Prague)*, CD available from European Brewery Convention, Brussels.

Scott, J.A. (1988) Application of cross-flow filtration to cider fermentations. *Process and Biochemistry*, **23**, 146–14.

Scott, J.A. (1991) Influence of sub-cellular material and pH on cross-flow membrane clarification of fermentation broths. *Journal of Institute of Brewing*, **97**, 291–296.

Sellick, I. & Pratt, J. (1998) Advances in the recovery of beer from tank bottoms. *Brewing & Distilling International*, **29**(2), 26–27.

SI (2010) Alcoholic Liquor Duties (Definition of Cider) Order 2010, No.1914, Statutory Instrument, HM Treasury.

Snyder, J. & Gaub, R. (2005) Profi process for diatomaceous earth replacement. *Technical Quarterly Master Brewers Association of the Americas*, **42**, 65–69.

Tripp, M.L., Rader, S., Rao, S.C. & Ryder, D.S. (1995) Flavoured Malt Beverages. International Patent Application WO9612788 (A1).

Vandenbussche, J. (2009) Continuous Brewhouse – First Industrial results. *Brewing and Beverage Industry – International*, **3**, 40–43.

14 Wine

K. Grainger

14.1 Background

The changes in the pattern of wine consumption of wine during the last 40 years have been dramatic. There has been spectacular growth in sales in many countries that have limited domestic production; for example, in the United Kingdom, the per capita consumption rose from just 2 L in the early 1960s to 22.2 L in 2009 (Anderson & Nelgen, 2011). However, during the same period many traditional European wine production countries have seen a decline in per capita consumption by some 50%, albeit mostly of very low-quality wines. Prior to the 1970s, only 10% of wines were exported from producing countries, including intra-Europe 'exports', and by 2009 over a third of wine consumed around the world was produced in another country (Anderson & Nelgen, 2011). This huge rise in exports has been made possible by considerable improvements in the stability of the products sold. 'It's a great wine to dink locally, but it doesn't travel' was an often heard expression when visiting wine regions in the 1970s and 1980s. Microfiltration (MF) and ultrafiltration (UF) using membrane technology have been important tools in ensuring clarity and stability of the product, particularly in the case of the output of large producers. Other membrane processes, including reverse osmosis (RO) have enabled producers to achieve product of reasonably consistent quality and in styles that the marketers and gatekeepers believe to be popular with a wide audience, including those who were traditionally not wine drinkers.

Several 'new' markets have emerged recently, particularly Asian countries, and in the last few years sales in China have surpassed the wine industry's expectations, especially for so called 'fine' wines – China purchased much of the 2010 production of the top châteaux of Bordeaux. The interest among wine lovers in the qualities and differences in styles in wines produced around the world is also high, as evidenced by the proliferation in wine societies and attendance at tutored tasting and other wine events, and the growth in wine 'tourism'. Global wine sales amount to the equivalent of 30 billion 75 cL bottles. However, the diversity retailers of wine has declined, particularly in the last 10 years, and the typical consumer of everyday wines is now most likely to make their purchase in a supermarket rather than in a specialist outlet, although recent years have seen a growth in web-based independent retailers, often offering high-quality wines from small, individualistic producers.

It is generally accepted that the quality and style of wines from any region or any producer, however, small or large will not be *totally* consistent from vintage to

Membrane Processing: Dairy and Beverage Applications, First Edition. A. Y. Tamime.

vintage – indeed it is these variations that makes individual wines so exciting to wine lovers. In other words, the properties of the finished product are not precisely predictable, unlike the product of other drink production technologies (Urkiage *et al*., 2002). There is constant debate as to the respective roles of nature and nurture, and as to how much wine production is, or should be, art or science. However, there is no doubt that the use of modern winemaking techniques has led to wines moving closer in style (Bird, 2010), and this is decried by passionate small producers and serious and educated wine lovers.

The origins of wine production date back some 8000 years, but the advances in the science and technology of production since the Second World War have outstripped those of the previous 79 centuries. These have included inter alia:

- grape sorting and selection systems to exclude unripe, rotten or damaged fruit – spectacular advances have been made in the technology of sorting equipment in the last five years;
- the introduction of temperature controlled fermentation vessels (generally constructed of AISI 316 or 304 grade stainless steel, although there is move today is 'back' to wood or concrete);
- development of numerous strains of cultured yeasts aiding control of fermentations and development of required flavours;
- a detailed understanding of malolactic fermentation and the development of suitable inoculums for use when deemed necessary;
- a wider range of extraction techniques including pre-fermentation cold soaks and rack and return (délestage), flash détente and thermo détente;
- the use of gases namely carbon dioxide, nitrogen and argon, primarily to avoid unwanted oxidation;
- micro-oxygenation in tank, useful to help polymerise long chain tannins in red wines, particularly if barrel ageing is not to take place;
- pre-bottling cold stabilisation, sometimes including the utilisation of the 'contact process', to precipitate crystals of potassium bi-tartrate and calcium tartrate.

Membrane processing applications relevant to the wine industry include

- front end MF
- cross-flow (tangential) MF
- UF
- RO
- electrodialysis (ED)
- pervaporation.

The amount of research into the oenological use of membranes has, in the last few decades, been immense. In addition to the institutes undertaking broadly based research into food and drink science and technology, there are several institutes specialising in wine research. There is a perception that these are mostly based in so-called 'New World' wine producing countries, and indeed the Australian Wine Research Institute, established in 1955, and the Department of Oenology of the University of California at Davis (and several other New World institutions) have produced a considerable amount of ground breaking studies. However, 'Old World' institutions have been equally active. The work of

Émile Peynaud, regarded as the father of modern oenology and the first to experiment with RO as a means of must concentration (Peynaud & Allard 1970), Jean and Pascal Ribéreau-Gayon, Denis Dubourdieu and, recently, Pascal Chatonnet's work on taints and Martine Mietton-Peuchot's work on membrane technology at the Université Bordeaux (2) Segalen have helped maintain that institution's position as a world leader. However, widespread use of membrane technologies in winemaking, particularly in small and medium-sized wineries, is not as universal and has not developed as rapidly as many authors suggest. Even as recently as the start of the current century the use of membrane filtration in the wine industry was mainly as a safeguard after pad filtration to ensure the complete removal of micro-organisms before bottling (Rankine, 2004).

Consumers require and expect wines that are clear and stable and, with the exception of bottle matured fine wines that constitute a tiny percentage of the total wine market, free from sediments. However, it is the so called gatekeepers, not the consumer or producer who have the power in the industry and who prescribe product specifications. In the United Kingdom, supermarkets account for over 75% of all 'off sales' of wine. Many of these specify a complete absence of all viable micro-organisms, which means that so-called sterile filtration regimes have to be used (Bird, 2010). Conversely, Robert Parker, undisputedly the most important wine critic in the world, whose influence is so great that even Cru Classé Chateaux in Bordeaux adjust their winemaking to suit his perceived palate, is well documented as having an aversion to filtration. He, and some other influential critics, decries the use of interventionist techniques and what they perceive as excessive manipulation. The websites of many producers of super premium wines boast that the wines are bottled without filtration or only after a light (pad) filtration. 'Good wine is made in the vineyard' is an oft heard expression, meaning that if top-quality, healthy grapes are produced from vines that are in balance, it is not necessary for the winemaker to carry out other than basic adjustments and manipulations. Careful fruit selection on a sorting table to exclude rotten, damaged or less than perfectly ripe fruit is now commonplace at quality conscious producers – 30 years ago all would go into the fermentation vats. Rackings of classic red wines during their time in barrel maturation have been reduced – 30 years ago four rackings in the first year in the chai was commonplace in the Bordeaux region; today many producers undertake a single racking.

Although membrane processing *per se* should perhaps have a good press as methods of improving and stabilising wines that do not involve chemical additions, in practice perceptions can be negative and the processes regarded as highly manipulative, interfering with nature and stripping the heart and soul from wines. However, producers of inexpensive and 'branded' wines need to bring product to market quickly, and the historic slow passage through the winery has been greatly reduced by using intervention, such as a single pass through a cross-flow membrane filter shortly after fermentation.

14.2 Clarification and filtration methods

14.2.1 *Traditional methods in common use*

There are several methods of processing in common use in wineries around the world for the clarification and filtration of must, wine and lees. Centrifugation, generally using a plate separator centrifuge, is effective for basic clarification. The must, prior to fermentation of white wines, needs to be reasonable free of solid matter to avoid development of off-flavours and, although natural setting (débourbage) may take place over 24 hours or so, centrifugation eliminates this waiting period. After fermentation, centrifugation may be used at various stages, including post-fining. It is often used pre-filtration; however, centrifugation is a violent method and many wineries avoid the process. Models of centrifuge are now available that can be flushed with inert gas to avoid the problem of oxygen uptake.

Immediately after the conclusion of the alcoholic fermentation, wines will contain considerable solid matter including grape solids and yeast hulls. Given time most of these solids will settle and the wine may be racked off, but alternatively the wine may undergo a depth filtration using Kieselguhr (diatomaceous earth) powder. Diatomaceous earth has been used as a filtration adjuvant since the late 19th century (Ribéreau-Gayon *et al.*, 2006a). The filtration may be undertaken using a mobile earth filter or rotary vacuum filter, which will cope with the filtration of lees. Diatomaceous earth is a carcinogen material, and presents issues regarding deposits of residue in landfill sites. Perlite, made from processed volcanic rock, may also be used for filtration of grape must and wine containing a high percentage of solids. However, it is very abrasive and can cause wear to injection pumps (Ribéreau-Gayon *et al.*, 2006a).

Although they will not cope with wines containing gross solids, sheet filters, also known as pad filters and plate and frame filters, are very versatile and extensively used in wineries of all sizes; in small wineries it may be the only type of filter available. The sheets are constructed of cellulose fibres, sometimes with the addition of granular components, such as Kieselguhr, perlite or polyethylene fibres, and sometimes also cation resins (with an electrostatic charge) to attract particles with the opposite electrostatic charge (Bird, 2010). Filter sheets are manufactured in several grades with uses ranging from a basic polishing to complete sterilisation. The large internal surfaces of filter pads enable the retention of considerable volumes of turbid liquid, up to 3 L m^{-2}. It is argued that such performance levels are unequalled by membranes (Ribéreau-Gayon *et al.*, 2006a). However, the filters gradually block, necessitating a labour-intensive taking part and reassembly, which also results in wine loss. The filters can also be messy – it is a very rare exception to see a drip-free example! The enclosed lenticular depth filter is a more hygienic and less labour-intensive piece of equipment.

All the above filtration equipment requires design and manufacture to high standards and careful operation by trained personnel to avoid negative consequences. In particular, ingress of oxygen is damaging to wine and is to be avoided – ideally filter equipment should be purged to evacuate air, but in practice this only happens in the most quality conscious wineries. Both earth filters and pad filters present an environmental challenge regarding the disposal of depleted filter material. However capital costs relatively low compared with most membrane filters.

14.2.2 *Membrane filtration*

Membrane filtration of wine can be carried out as front-end MF, which necessitates a previous clarification and filtration using the methods detailed above, or cross-flow filtration which will achieve the desired clarity in a single pass.

If a wine contains any residual sugar, and has a alcoholic degree of less than ~15.5% alcohol by volume (abv) (12.234 g 100 mL^{-1}), there is a considerable risk of a most undesirable refermentation taking place in the bottle (or other packaging) if any yeast colony-forming units are present. Thermotic bottling, heating the pre-bottled wine to 54°C is one method commonly used to guard against this risk, but it does present issues with fill heights as the wine cools. Other methods are flash pasteurisation, immediately prior to bottling, at a temperature of a approximately 75°C for 30 s and post-bottling tunnel pasteurisation at 82°C for ~15 min. However, MF immediately prior to bottling is the method preferred by many quality conscious producers. It avoids the flavour modification occasionally associated with pasteurisation (Jackson 2008). Membrane filtration is also a sensible precaution prior to bottling of wines that have not undergone malolactic fermentation, as the consequences of this taking place in-bottle would also result in wine spoilage. *Leuconostoc oenos*, *Pediococcus damnosus*, *Oenococcus oeni* and *Lactobacillus brevis* are the main bacteria involved and are all removed by a 0.45 μm membrane (Starbard 2008). *Lb. brevis* (lactic acid bacteria) can break down compounds in wine and result in, *inter alia*, a threefold increase in volatile acidity (Ribéreau-Gayon *et al*., 1976). Yeasts of the species *Saccharomyces cerevisiae* (the most dominant yeast in most fermentations) and the closely related *Saccharomyces bayanus* are removed using 0.8 μm membrane pore size (*Dekkera intermedia* – a sporulating form of *Brettanomyces*), which causes a defect recognisable on the nose as a pronounced 'burnt sugar' smell or a 'mousey' off flavour (Grainger, 2009) are removed at 1 μm. Unenclosed plate filters, of course, permit immediate reintroduction of yeasts and bacteria. Membranes with pore diameters of 0.2–0.45 μm are usually used for filtering white wine and 0.45–0.65 μm for reds (Lipnizki, 2010). In addition, it is claimed by the same author that MF eliminates by the use of fining substances, but many winemakers still undertake traditional fining to remove colloids, even if the wine is destined for MF.

14.2.3 *Cross-flow microfiltration*

Cross-flow (tangential) filtration was first used for the clarification of wine in the late 1980s. The system will normally comprise a mono-block containing a pre-filter unit (to prevent rapid fouling), and a membrane filtration unit arranged in series. The main advantage over other filtration methods is that it is capable of clarifying wine in a single operation. Even if the juice has an initial turbidity of 400 nephelometric turbidity units (NTU) the filtered wine can be of brilliant clarity, with a turbidity of 1 NTU. Cross-flow filters work best when the wine has a wide particle size distribution, which means that they are ideal for filtering wine shortly after fermentation when there is a high concentration of suspended solids. Other accepted advantages with cross-flow filtration include a reduction in labour costs compared with multistage filtration, less oxygen pick-up with a single operation and minimal wine waste – 2% or less. The economics of cross-flow membrane filtration for lees recovery compared with Kieselguhr can be particularly attractive to the larger producer. The colour, aroma and flavour of wine recovered from lees are generally regarded as little affected by the membrane filtration process. If cross-flow filtration is undertaken shortly after the completion of the alcoholic fermentation, the filtrate may be used as inoculums for other alcoholic or malolactic fermentations (Jackson 2008). There are valuable reasons for performing MF prior to wine going into vat or barrel for maturation, and cross-flow filtration really is the only cost and time efficient procedure at this stage. Spoilage of wine by *D. intermedia* has been mentioned above, but *Saccharomyces ludwigii* and other rogue yeasts that are known to spoil wine in tanks and several film forming yeasts including *Candida* spp., *Issatchenkia orientalis*, *Pichia membranaefaciens* can spoil wine in barrels. It must be said that with good cellar hygiene and meticulous topping up of barrels such problems are rare.

The automation aspect of cross-flow filtration is attractive to many winemakers – less chance of human error, but this is not accepted by all. Stefano Migotto, founder of wine specialist filtration company Winetech states (Cutler 2007): '... we removed all the automation from our systems. I couldn't trust the accuracy of the computer. Now, if I make a mistake, I make the mistake. I don't want some computer to open or close a valve because some high frequency device is passing by. Now, all of our cross-flow filtration machines are operated manually. That gives me better control of the process'.

When cross-flow filters were first introduced, there was a prevalent view amongst wine lovers, critics and many oenologues that cross-flow filtration was detrimental to the quality of wine (Rankine 2004). The technologies and membranes used had migrated from other industries, in particular the bio-pharmaceutical industry and stripped flavours and tannins from wine. Taints sometimes occurred because of the plastic nature of the membranes with a correspondent recommendation that membranes were conditioned with acidified hot water prior to use (Rankine 2004). There have been huge improvements membrane technologies in the last decade, including the introduction of ceramic (inorganic) and wine specific hollow-fibre (organic) membranes. Spiral-wound membranes are now generally falling out of use. Nevertheless, the negative perceptions, although not as universal, lingers today. On several occasions, recently, when the present author has visited wineries with professional colleagues, upon seeing the cross-flow filter, the colleagues have made remarks, such as 'clearly the focus here is not on quality'.

The apparently simple questions as to whether modern cross-flow filtration negatively impacts upon aromas, flavour and quality are perhaps still unanswered. There is a risk that high molecular weight carbohydrate and protein colloids may be removed during tangential filtration – these are an integral part of the wines composition, aroma and taste (Ribéreau-Gayon *et al.*, 2006a). Rektor *et al.* (2004) conducted sensory analyses on musts of Hungarian varieties (Kekfrankos, Furmint and a blend) that had undergone MF treatment using hollow-fibre and multitube membranes and control samples. Trained assessors considered that the appearance of the microfiltered musts was more attractive (there was a noted colour intensity decrease); although a certain extent of loss in flavour and odour intensity was observed in the treated samples, the tasters preferred the treated samples as they found the 'off-flavour' less intense. A decrease was also noted in the colour density of a Cabernet Sauvignon wine studied by Arriagada-Carrazana *et al.* (2005), who examined the effects of membrane filtration (0.65 μm final filter) on its aromatic and phenolic quality. The researchers noted that polyphenolics were also decreased, and there was a difference in some aroma compounds which they attributed to membrane adsorption. De Pinho (2010) reported that the application of MF (and UF) to wine clarification has been restricted by the lack of knowledge on the possible removal of polysaccharides and other macromolecules that may be an important contribution to the quality of wines.

14.3 Membrane fouling

In the last 20 years, there has been considerable advancement in the understanding of fouling of membranes during wine filtration. Boulton *et al*. (1999) attributed poor membrane life with certain wines to be mainly caused by colloidal complexes of proteins, phenols, other entities and polysaccharides. By 1999, it was established that polysaccharides with polyphenols are the main compounds responsible for the fouling of membranes in wine clarification (Vernhet *et al*., 1999). Koehler *et al*. (2000) reported that it is well established that membrane fouling by proteins leads to large flux decline; the strongest fouling occurring on account of hydrophobic interactions between protein and membrane material in combination with conditions where protein-protein interactions are also facilitated. Urkiage *et al*. (2002) found the best membrane material for filtering red wine to be cellulose acetate with pore sizes of 0.2 and 0.45 μm being suitable for the physico-chemical and flux reduction point of view, although today this would be regarded as too fine for a quality red wine. Vernhet *et al*. (2002) made significant contributions to the understanding of fouling build-up, and again Vernhet & Moutounet (2003) made huge advances in the understanding of the impact on fouling of wine polysaccharides and polyphenols. In a study by Ulbicht *et al* (2009), it was again noted that aggregates of polyphenols and polysaccharides present in red wine have a major contribution to adsorptive fouling. This study indicated that fouling is strong for polyethersulphone, but very weak for polypropylene membranes. The low adsorption tendency of wine ingredients to polypropylene membranes results in higher fluxes and longer service life of the respective filtration modules in wine clarification. However, membrane fouling remains a major limiting factor to wider use of the technology in the wine industry.

Wines made from grapes affected by *Botrytis cinerea*, which produces the polysaccharide β-glucan that has a molecular weight of 1 000 000 kDa, can cause serious membrane fouling. Filtration inhibition starts from 0.6 mg L^{-1} of β-glucan with a 0.45 μm membrane filter (a figure much less than previously believed), and concentrations of 1 mg L^{-1} and above lead to complete blocking of the membrane after a filtrate quantity of only 1.5 L (Schneider, 2011).

14.4 Must correction, wine correction and alcohol reduction using membrane technologies

14.4.1 *Reverse osmosis*

RO may be used for: (a) must correction (pre-fermentation); (b) wine correction (post-fermentation) including the removal of excessive acetic acid; and (c) reducing alcohol content. The process is highly suitable for must and wine treatments, taking place at low temperature (15°C), with a maximum turbidity of 400 NTU.

Must correction

If a must has a deficiency of sugar, which will result in an unbalanced wine having an inadequate degree of alcohol, several methods may be available to adjust this. These are chaptalisation (must enrichment), the addition of concentrated grape must or rectified concentrated must, the bleeding (saignée) of juice from vats containing crushed red grapes, cryo-concentration, vacuum evaporation and RO. Their usage will depend, *inter alia*, on the legal restrictions within the region in which the wine is produced and the territory in which it is sold. Traditionally in the more northerly producing regions in Europe, the potential alcoholic degree of wines would be increased by the process of chaptalisation (i.e. the addition of sucrose). This method has lost favour, partially on account of its impact on wine balance, and there has been discussion in the European Union (EU) on banning the practice, but this has not been implemented. The addition of concentrated grape must is occasionally practised in some countries, e.g. Italy. Cryo-concentration involves the cooling of the must until it begins to freeze. The unwanted water element, as ice, is removed by filtration. It is claimed by Bird (2010) that by this method there is virtually no loss of flavour and, in the case of a white must of the Macabeo variety, that the method comes closer to the ideal objective of removing the water without affecting the other components (Hernández *et al*., 2009). Vacuum evaporators have improved somewhat in the last decade, and now operate at acceptable temperatures (25–30°C) – aromas are retained in an aroma trap, condensed and added back to the wine.

The first experiments in using RO for must concentration were made back in 1970 (Peynaud & Allard, 1970), but it was to be nearly 20 years (following the advent of composite membranes) before further detailed experiments took place (Guimberteau *et al*., 1989). The water content of the must may be reduced by between 5% and 20%. The method is particularly valuable to reduce water content when there has been rain at harvest time. Jackson (2008) states that RO may be used for immature berries, if the ripeness of the fruit has been restricted due to cold weather its utilisation for must concentration is less successful, for although sugar will be concentrated, so will acidity and green tannins, resulting in an aggressive and unbalanced wine. Mietton-Peuchot *et al*. (2002) reported that the technique should not be used in palliation of lack of fruit maturity, but that good results are obtained with mature water inflated grapes. Fermentations can slow down when high sugar concentrations have been achieved by various methods, including RO (Ribéreau-Gayon *et al*., 2006b). The juice should be pre-filtered using membrane (20 μm), or finer prior to the operation. RO for the purpose of concentration of musts to be fermented into wine was only permitted in the EU from 1999 – Council Regulation (EC) No. 1493/1099 (EU, 1999), and since superseded by Council Regulations No. 479/2008 (EU, 2008) and Council Regulation No. 479/2009 (EU, 2009a), Commission Regulation No. 606/2009 (EU, 2009b, 2009c). The maxima set by the EU Regulations and International Organisation of Vine and Wine in the *International Code of Oenological Practices* (2012) for must concentration by RO are a maximum 20% decrease in must volume and a maximum 2% increase in potential alcohol. It is interesting that the EU authorities do not consider that this negates the concept of *terroir*, but when two Bordeaux properties covered their vineyards with plastic sheeting to deflect rainwater during the 2000 growing season the resulting wines were not accepted as Appellation Controlée, resulting in the properties marketing the wines as the world's most expensive Vins de Table (Grainger & Tattersall, 2005).

RO has also been used to produce concentrated juice destined for the making of icewine-like wines (Jackson, 2008), but experienced tasters will immediately recognise that the wine lacks the distinctive flavours that result from vineyard over-ripening and the freezing of the grapes on the vine, or even by cryoextraction.

As will be discussed further below, harvesting of wine grapes later than in the past in order to ensure phenolic ripeness, together with the impact of climate change, has resulted in an often unwelcome increase in must density and consequent high alcohol content after fermentation. UF plus nanofiltration (NF) is one effective method for sugar reduction, and alternative process being UF plus evaporation. It is noted by Massot *et al*. (2008) that the desired reduction in must sugars does involve a slight concentration of other components.

14.4.2 *Ultrafiltration*

UF operates at a lower pressures and larger porosities than RO. UF plus NF can be used for sugar removal from must, NF giving greater flow rates than RO (Mietton-Peuchot, 2010). UF can be a useful tool to adjust the style of red wine after fermentation. Tannins can, if required, be selectively removed based on molecular weight. The tannins and colour are held in the retentate, i.e. producing a red wine with deeper colour and firmer tannic structure than the base wine. There is then the possibility of adding these tannins and colour from the retentate to other red wines should they have a deficiency of these. Alternatively, UF can produce two separate wines that can stand alone: (a) the permeate a lighter, smoother red, or even a rosé, blush or blanc de noirs style; and (b) the retentate of red with full body and firm tannic structure. Bitterness resulting from the inclusions of hard press fractions can also be reduced. However, it should be noted that careful handling of sound fruit should have avoided these problems arising. UF can also be used to remove proteins and, therefore, can offer security against protein haze formation. The use of UF to help correct wine flaws can be particularly valuable – by way of examples in the case of white wines the removal of tint and pinking, and in the case of reds by removing oxidised phenolics that would otherwise cause premature browning. These can be selectively removed based upon molecular weight, and the precursors of browning can also be removed. In addition, UF may have beneficial effects in rejuvenating old wines (Jackson, 2008).

14.4.3 *Wine correction: reducing alcohol content*

The mean alcohol content of wines has increased considerably in the last 30 years. This is due to delaying harvesting until phenolic ripeness is reached (especially as the market and some critics now demand softer styles of red wines than in the past), the use of alcohol tolerant yeast and the impact of climate change (Grainger, 2009). In 1989 in the first edition of *Making Good Wine*, referring to Australian Cabernet Sauvignon Rankine (1989) wrote: 'A ripeness of 10°–12° Baumé (18–21.6° Brix) is usual, which results in wine containing between about 10 and 12% abv'; these figures respectively equate to alcohol of 7.893 and 9.471 g 100 mL^{-1}. Today the alcohol levels in excess of 14% abv (11.05 g 100 mL^{-1}) are not uncommon, and in some regions producers struggle to restrain alcoholic content and keep the wine balanced. A report from the Australian Wine Research Institute details that between 1984 and 2004 the average alcohol levels rose in wines made from all grape varieties, e.g. in all Australian regions red wines rising from an average of 12.3% to 13.9% abv (9.708 to 10.971 g 100 mL^{-1}) (Godden & Gishen, 2005). Wines from many countries, including California (United States of America – USA) and Chile have faced adverse criticism from wine experts and others on account of their perceived excessive alcohol. Apart from the obvious health and social issues raised by increase in average alcohol levels, excessive alcohol in any individual wine numbs the fruit, inhibits freshness and can even give a burning sensation on the palate. In simple terms, the wine is unbalanced, less pleasant and less drinkable in other than relatively small quantities.

There are several techniques available for alcohol reduction; all have organoleptic impacts and other positive and negative consequences. The addition of unfermented grape juice (süssreserve) is a process commonly used in Germany during the production of inexpensive wines, with an unfermented grape juice addition representing 10–15% of the total volume, resulting in a finished wine that is medium-sweet with a high level of primary fruit aromas, which is the real reason for the operation. Processes based on the use of high temperature (evaporation) cause strong alteration or loss of the wine aroma (Liguori *et al*., 2010), and are not generally used in the wine industry. One technique that is illegal in most production countries (including member states of the EU) is simply adding water to the must, fermenting juice or wine. Nevertheless, winemakers who use this technique have euphemisms for the technique – employing the 'white horse', or 'black snake' or 'flushing out the picking bins and tanks very well' are terms that might be heard in some New World producing countries.

Technical processes include spinning cones, RO, RO plus perstractive membrane and RO plus distillation. However, the use of RO for this purpose has remained controversial (Grainger, 2009). A recent study indicates that NF and pervaporation are effective for de-alcoholising wine and retaining original characteristics (Catarino & Mendes, 2011). A study by Agence Nationale de la Récherché (2009) looked at various methods of de-alcoholising wines, including treatment of must by the combination of UF–NF already noted, and de-alcoholisation of finished wine by blending after partial or total distillation, stripping, RO combined with distillation or membrane contactors, NF combined with membrane contractors and direct treatment by membrane contractors. The study found that in most cases, a treatment that eliminated 2 degrees of alcohol, obtained by blending of the original wine with the wine totally de-alcoholised by distillation or the combination of RO combined with membrane contactors results in products that cannot be distinguished from the a control wine. This study claims that in the case of the samples of Merlot, Syrah, Chardonnay and Sauvignon Blanc and with original pre-treatment alcohol degrees of 13–14% abv (10.26–11.05 g 100 mL^{-1}), reductions in alcohol of up to 3% abv (2.368 g 100 mL^{-1}) were not perceptible to the average consumer (even with sensory analysis training), although wine professionals rated the reduced alcohol wines lower, citing a loss of complexity.

Interesting tests in de-alcoholisation of red wine by direct RO with a pilot plant equipped with a polypropylene hollow-fibre membrane contractor have been recently conducted (Liguori *et al*., 2010). After five cycles, the alcohol content was reduced from 12.8% abv (10.103 g 100 mL^{-1}) to less than 0.5% abv (0.395 g 100 mL^{-1}), most of the reduction taking place in the first two cycles. The resultant de-alacoholised 'wine' was lacking in aroma, but the same authors stated that the process is suitable for use in the wine industry to reduce the alcoholic degree.

14.4.4 *Wine correction: removing acetic acid*

As with ethanol, the molecules of acetic acid are very small and can pass through RO membranes. An excess of volatile acidity is generally regarded as a fault in wines. If the level is too high, the wine may smell vinegary or of nail varnish remover (Grainger, 2009). Although other acids including carbonic, sulphurous, butyric and formic acids are volatile, acetic acid is the main culprit. Effective removal of volatile acidity can be achieved by RO plus ion exchange or RO plus adsorption. However, the level of volatile acidity that is acceptable in any wine is subjective, by way of examples high volatile acidity is regarded by connoisseurs as part of the intrinsic character of many reds from Italy, and some of the highest regarded Bordeaux wines from ripe years have demonstrated volatile acidity levels that might technically be regarded as a fault.

14.4.5 *Wine correction: removal of taints*

Brettanomyces

With the increase in gatekeepers demanding red wines with a few grams of residual sugar, the increase in alcohol content in most wines during recent years, and the drive by all involved in the industry including winemakers, regulatory authorities, the press and consumers to reducing the levels of sulphur dioxide in wine; it is not surprising that the instances of red wines tainted with *Brettanomyces* spp. (known as Brett taint) have increased during the last decade. Brett is recognisable on the nose by a pronounced smell of cheese, wet horse, farmyards, baked orange or the plasters used for small wounds (Grainger, 2009). The perception threshold is between 425 and 600 μg L^{-1}. However, the presence of Brett in wine remains a controversial topic, for although almost all purist oenologues regard the organism as a fault, many producers consider a limited presence, below perception thresholds, to add complexity and to be a part of their 'terroir'. Many wines from the southern Rhône show high levels of Brett, and it is perhaps a trademark of Château Musar, a very popular wine from Lebanon.

Calderón *et al*. (2004) studied wine filtration for the removal of *Brettanomyces* spp. and reported that using membranes with a pore size smaller than 0.45 μm resulted in effective removal of microbial cells. Ugate *et al*. (2005) obtained at 77% reduction in the total ethylphenols (4-ethylguaiacol and 4-ethylphenol) to what was considered to be acceptable levels. The 3 h process comprised a two-stage integrated process, a membrane with cross-flow filtration and a hydrophobic absorbent resin (Amberlite XAD-16 HP). However, a reduction in some (desired) aromatic compounds, ethyl and methyl vanillate and other esters was also reported. The process is not authorised under EU wine regulations, and the only company with a derogation to conduct trials (in association with *Inter-Rhône* is *Bucher Vaslin*.

Smoke taint

Australia, in particular, has had many instances of wines affected by smoke taint in the last 10 years, but other countries, including South Africa, have also had wines affected. Grapes are tainted, as a consequence of bush fires, or controlled burning taking place near vineyards. Affected wines may smell of burnt ash, smoked salmon or medicinal. The financial cost of such taints is huge – in 2006–2007, the loss in Australia was estimated at Australian \$75–90 million (Whiting & Krstic, 2007). As with Brett, high concentrations of volatile phenols are present in tainted wines – in this case the compounds are guaiacol and 4-methyl-guaiacol. It should be noted that wines may contain low concentrations of these as a result of oak maturation or treatments (Grainger, 2009). The sensory threshold is ~6 μg L^{-1}, but bush-fire affected samples submitted to the Australian Wine Research Institute have shown levels in excess of 70 μg L^{-1} (Grainger, 2009).

Studies, tests and commercial applications in attempting to remove smoke taints by RO have been recently taking place. The first study to examine reducing smoke taint in wines using RO and solid phase adsorption was undertaken by Fudge *et al.* (2011). Three samples of only one variety (Pinot Noir) was used for the evaluation; the study concluded that the process reduced the level of volatile phenols derived from smoke, and improved the aromas and taste. However, the taints were found to return after a period of time and the authors put this down to hydrolysis of glycoconjugate precursors which the treatment process did not remove. The process of reduction by RO would be helped by rapid measures of smoke related compounds (e.g. UV 275 nm phenolics measures) as they go through the membrane filtration process to assist in determining when appropriate reductions have been achieved, and an improved understanding of the full range of smoke taint compounds is also required to assist in the design of specific membranes to selectively filter these compounds. Guaiacol and some other smoke related compounds are relatively small molecules so they can be removed, but other larger compounds that are not removed may dominate and become more evident in the aroma. Although there is potential for using this technique when the smoke taint is low, in highly tainted wines it would take many passes through the system to reduce levels to the threshold concentrations of smoke taint compounds. There is also concern that some fruit characters may be removed and the smoke characters come to the fore.

14.5 Wine stabilisation and pH adjustment

Membrane electrodialysis technique is widely used in large wineries for tartrate stabilisation and pH adjustment in wines as follows.

14.5.1 *Tartrate stabilisation*

Tartrate crystals in the bottle or glass are *not* a wine fault, and are very often visible in wines of the very highest quality (Grainger, 2009). However, to the consumer they can look alarming like pieces of broken glass, and most large and some small scale producers undertake one or other process to try and prevent crystal formation. The crystals are usually potassium bitartrate (potassium hydrogen tartrate) or occasionally calcium tartrate.

The conventional way of stabilising tartrates is by cold stabilisation or contact process. Using cold stabilisation the wine must be fined to negate the protective effect on micro-crystals, and is then chilled to just above its freezing point (e.g. −4°C for a wine containing 12% abv (9.471 g 100 mL^{-1}), or −8°C in the case of fortified wines containing 16–22% abv (12.628–17.364 g mL^{-1}), and is held in insulated tanks at this temperature for at least 8 days and often considerably more. The process is inefficient, and expensive in terms of capital outlay and the running costs of refrigeration units (Bird, 2010). A more effective, quicker and cheaper system is that of contact process, which involves chilling wine to 0°C and adding 4 g L^{-1} of wine of finely ground crystals of potassium bitartrate, with a vigorous stirring to keep in suspension or, alternatively, pumping of wine through the crystal bed. This method usually referred to as 'the contact process' will take 5 days or so. Ion exchange, replacing potassium and calcium ions in wine with sodium ions is, perhaps from a health point of view, undesirable and until recently banned in the EU, but the process is now permitted under Commission Regulation (EC) No. 606/2009 (EU, 2009b, 2009c). The addition of metatartaric acid or carboxymethylcellulose is permitted in the EU up to 100 mg L^{-1}, but they are only effective at protecting from tartrate precipitation for a limited amount of time.

A far more effective method of tartrate stabilisation is by the use of electrodialysis, which uses selective membranes allowing the passages of potassium, calcium and tartrate ions under the influence of an electrical charge. The machine operation can be monitored by continuous measurement of the wine being treated. The removal of tartrates is fast (a single cross-flow pass), efficient and reliable, and is adjusted to the characteristics of the wine being processed, and minimal pre-treatment of the wine is required. It is generally accepted that there is no effect on the organoleptic characteristics of wine thus treated. The speed of the process makes the system particularly attractive to producers who wish to get a wine, such as a new season's Sauvignon Blanc, to market early. There is also less wine loss, perhaps 1 % compared with up to 3.5% with cold stabilisation (Thurston, 2008). As the process takes place without refrigeration, there is a considerable reduction in energy costs, up to 95% compared with the 8+ days for cold stabilisation detailed above (Thurston, 2008). Fok (2008) reported using electrodialysis process to stabilise the wine where 8 watt hours of electricity per (US) gallon compared with 22 watt hours for wine cold stabilised in insulated tanks, and 1 200 watt hours if the tanks are un-insulated. Although Hestekin (2011) claims that the process is commonly used in the wine industry, the capital cost of equipment is high, making the machines unaffordable for most small wineries, although in some regions mobile machines and operators can be hired.

14.5.2 *pH adjustment*

The pH of must or wine can be reduced by using bipolar membrane electrodialysis. The process removes potassium (K) and replaces with hydrogen (H). The pH is reduced, but the titratable acidity remains unchanged, as does the level of sugar and phenolic compounds, and the case of wine, alcohol. The process achieves very accurate pH reductions, a typical drop after a single pass being by ~0.1 to 0.2 pH units, the legal maximum according to *International Oenological Codex* (OIV, 2012b) is 0.3 pH units. Brightness, colour, aromas and flavours are unaffected, which is not the case when tartaric acid is added, a process that is commonplace (especially for white wines) produced in New World countries and the hotter parts of southern Europe. As an extra benefit, the by-product of the process is suitable for using in wineries as a (most desirably chlorine free) cleaning agent.

14.6 Conclusions and future developments

Despite, or perhaps because of, the developments in the last 60 years of wine technology in both the vineyard and winery the move today, particularly among smaller quality conscious producers – the so-called 'boutique wineries' – is very much towards wines produced with minimum intervention, which are perceived by the media and the consumer to be more natural. Although the individual effects might be small, in every stage of handling there is the potential for a loss of quality (Bird, 2010). Although few would doubt that the use of modern technology has improved the average quality and consistency of wines, it is generally accepted that its use *per se* is no guarantee of making good wine.

Many growers have converted to organic or even biodynamic practices in the vineyard, and the 'natural' wine movement is rapidly gaining momentum as illustrated by the holding of two 'natural' wine fairs in London in 2012. To the wine lover, wine is the result of a unique combination of vineyard climate (including meso- and micro-climate), soil, aspect, the combination of which may be termed 'terroir', the weather in a particular year, grape variety(ies) and sensitive production techniques in both the vineyard and winery. However, bulk wine production is rapidly moving into a parallel world of 'the new oenology' as detailed by Fischer (2006). Here, grapes are a merely a commodity, a source for valuable sugars, acids, flavours, tannins, pigments. In this world grape processing obtains a defined must in a reproducible manner. Wine corrections are made by RO or spinning cone columns to achieve specifications. Ethanol, volatile acidity, flavour, off-flavour, tannins and coloured pigments may be fractionalised and recombined to produce a wine in the required style (Fischer, 2006).

The capital costs of membrane equipment are high, well beyond the means of many boutique wineries, particularly as the equipment is laying idle for the vast majority of the time. However, for the premier cru classés of Bordeaux – Lafite Rothschild, Margaux, Mouton Rothschild, and Haut Brion released the 2010 vintage wines ex-chateau at prices of €600 per bottle, Latour at €650, Cheval Blanc at €900 and Ausone at €1140, any techniques and equipment that help their wine achieve the highest perceived quality status is clearly worth the cost. Interestingly when the present author has visited the chais of several cru classés, the chief winemaker has readily shown the latest purchases of grape selection equipment, flotation pumps and nitrogen generators, but the membrane equipment, particularly the RO machine, have always remained well hidden. Indeed, many cru classés and other top wineries in both the Old and New Worlds have stated to the present author that they do not use any membrane equipment in the winery. Perhaps the development of cheaper and longer lasting membranes make processing equipment more attractive to smaller and less well-off wineries, and the appearance of mobile RO and electrodialysis machines is beginning to make the technology available to those to whom capital outlay is the main concern.

Perhaps the most exiting area for development of membrane technologies will be the arena of taint removal. Smoke taint and Brett have been discussed above, but other taints can seriously impact on the quality of wines and, in the case of some taints, render the product totally unsalable. Without doubt the most serious of these is contamination by one of the haloanisoles:

- 2,4,6-trichloroanisole (2,4,6-TCA) – perception threshold 1.5–3 ng L^{-1}
- 2,3,4,6-tetrachloroanisole – perception threshold 10–15 ng L^{-1}, but as low as 5 ng L^{-1} in sparkling wines
- tribromoanisole (TBA) – perception threshold 3.4 ng L^{-1}
- pentachloroanisole – perception threshold >50 μg L^{-1}.

Although contamination with haloanisoles is usually associated with 'cork' taint, cellars and the equipment therein, including oak barrels, bungs, filters, pallets, and particularly in the case of TBA, treated timbers in wooden structure in buildings can be contaminated and this may migrate into the wine. TCA is not removed by traditional filtration, but already progress is being made towards effective removal with membrane technology.

So-called ladybug taint is a growing and serious problem in some wine regions. The inclusion of *Harmonia axyridis* or *Coccinella septempunctata* with harvested grapes can give rise to generally undesirable green pepper flavours in wine as a result of 2-isopropyl-3-methoxypyrazine, which has a sensory threshold of 320 pg L^{-1}. Of all fining treatments, only charcoal is effective at removal of this taint, but unfortunately it also removes desirable aromas and flavours. This is certainly an area where developments in membrane technology may provide an urgently needed solution.

Advances have been made in the knowledge of membrane filtration fouling resulting in considerable technological improvements. However, fouling remains an issue to be resolved, and El Rayes *et al*. (2011) noted that the large-scale development of cross-flow MF is still hampered by the barriers induced by membrane fouling – it might be said to be one of the blocks to more widespread use of membranes in the wine industry. Furthermore, environmental, health and safety pressures on the use of traditional filter materials are increasing. As previously stated diatomaceous earth is a carcinogen, and can be a hazard to operative's breathing and also presents issues regarding landfill of residue. Diatomaceous earth contains crystalline silica, and its inhalation of which has been classified as carcinogenic by the World Health Organisation – International Agency for Research on Cancer (WHO, 1997). Operatives should wear respirators. Diatomaceous earth is a depleted substrate and there can be little doubt that its use as a depth filter in the wine industry will diminish in the near future.

The pace of advancement in technology leaves legislation trying to catch up. The International Organisation of Vine and Wine (OIV) details conditions and limitations of usage of winemaking techniques in the *International Code of Oenological Practices* (OIV, 2012a). The International Oenological Codex (OIV, 2012b) gathers descriptions of the main chemical, organic and gas products used in the making and keeping of wines. For wines produced, or marketed in the European Union, the EU Regulations detail permitted practices – these are now in line with the International Code of Oenological Practices. The current regulations are Council Regulation No. 479/2008 (EU, 2008), the framework regulation is Commission Regulation No. 606/2009 (EU, 2009b, 2009c), as subsequently amended by Commission Regulation No. 53/2011 (EU, 2011). Whilst derogations are given for experiments, unless a practice is in the regulations no wine for commercial sale may be made using such a practice. It can be argued, however, that such restrictions delay and limit future developments in the oenological use of membrane technology.

References

Agence Nationale de La Recherche (2009) Quality wines with reduced alcohol content (VDQA). ANR-05-PNRA-011.

Anderson, K. & Nelgen, S. (2011) *Global Wine Markets, 1961 to 2009: a Statistical Compendium*, University of Adelaide Press, Adelaide.

Arriagada-Carrazana, J.P., Sáez-Navarrete, C. & Bordeu, E. (2005) Membrane filtration effects on aromatic and phenolic quality of Cabernet Sauvignon wines. *Journal of Food Engineering*, **68**, 363–368.

Bird, D. (2010) *Understanding Wine Technology*, 3rd edition, DBQA Publishing, Newark.

Boulton, R.B., Singleton, B.F., Bisson, L.L. & Kunkee, R.E. (1999) *Principles and Practices of Winemaking*, Aspen Publishers Inc., New York

Calderón, F., Morata, A., Uthurry, C. & Suárez, J.A. (2004) Aplicaciones de la ultrafiltración en la industria enological – Ultimos avances tecnológicos. *Tecnología del Vino* **16**, 49–54.

Catarino, M., & Mendes, A. (2011) Dealcoholizing wine by membrane separation processes. *Innovative Food Science & Emerging Technologies*, **12**, 330–337.

Cutler, L. (2007) Industry Round Table: Filtration – Winemakers discuss the truth, fiction and consequence of filtration in winemaking. (http://www.winebusiness.com/wbm/?go=getArticle&dataId=49276)

De Pinho, M.N. (2010) Membrane processes in must and wine industries. *Membrane Technology* (eds. K.-V. Peinemann, S. Pereira Nunes and L. Giorno), Vol. 3 – Membranes for Food Applications, pp. 105–118, Wiley-VCH, Weinheim.

El Rayes, Y., Albasi, C., Bacchin, P., Taillandier, P., Raynal, J., Mietton-Peuchot, M. & Devatine, A. (2011) Cross-flow microfiltration applied to oenology: A review. *Journal of Membrane Science*, **382**, 1–19.

EU (1999) Council Regulation (EC) No. 1493/1999 on the common organisation of the market in wine. *Official Journal of the European Communities*, *L* **179**, 1–101.

EU (2008) Council Regulation (EC) No. 479/2008 on the common organisation of the market wine, amending Regulations (EC) No. 1493/1999, (EC) No. 1782/2003, (EC) No. 1290/2005, (EC) No. 3/2008 and repealing Regulations (EEC) No. 2392/86 and (EC) No. 1493/1999. *Official Journal of the European Communities*, **L 148**, 1–61.

EU (2009a) Council Regulation (EC) No. 491/2009 amending Regulation (EC) No 1234/2007 establishing a common organisation of agricultural markets and on specific provisions for certain agricultural products (Single CMO Regulation). *Official Journal of the European Communities*, **L 154**, 1–82.

EU (2009b) Commission Regulation (EC) No. 606/2009 laying down certain detailed rules for implementing Council Regulation (EC) No. 479/2008 as regards the categories of grapevine products, oenological practices and the applicable restrictions. *Official Journal of the European Communities*, **L 193**, 1–68.

EU (2009c) Commission regulation (EC) No 606/2009 laying down certain detailed rules for implementing Council Regulation (EC) No. 479/2008 as regards the categories of grapevine products, oenological practices and the applicable restrictions. *Official Journal of the European Communities*, **L 193**, 1–59.

EU (2011) Commission regulation (EU) No. 53/2011 amending Regulation (EC) No. 606/2009 laying down certain detailed rules for implementing Council Regulation (EC) No. 479/2008 as regards the categories of grapevine products, oenological practices and the applicable restrictions. *Official Journal of the European Communities*, **L 19**, 1–6.

Fischer, U. (2006) Application of new technologies in the field of international oenology. *XIX Giornata Internazionale Vitivinicola*, 20 May 2006, Siena. (http://www.enoteca-italiana.it/ftp/attivita/Enology%20Siena.pdf)

Fok, P.E. (2008) PG & E studies electrodialysis for cold stability. (http://www.practicalwinery.com/sepoct08/page1.htm)

Fudge, A.L., Ristic, R., Wollan, D. & Wilkinson, K.L. (2011) Amelioration of smoke taint in wine by reverse osmosis and solid phase adsorption. *Australian Journal of Grape and Wine Research*, **17**, S41-S48.

Godden, P. & Gishen, M. (2005) Trends in the composition of Australian wine. *Wine Industry Journal*, **20**, 21–46.

Grainger, K. (2009) *Wine Quality Tasting and Selection*, John Wiley & Sons, Chichester.

Grainger, K. & Tattersall, H. (2005) *Wine Production Vine to Bottle*, Blackwell Publishing, Chichester.

Guimberteau, G., Gaillard, M., & Wajsfelner, R. (1989) Observations récentes sur l'utilisation de l'osmose inverse en vinification. *Connaissance de la Vigne et du Vin*, **23**, 95–118.

Hernández, E., Raventós, M., Auleda, J.M & Ibarz, A. (2009) Freeze concentration of Mustin a pilot plant falling film concentrator. *Innovative Food Science and Emerging Technologies*, **11**, 130–136.

Hestekin, J., Ho, T. & Potts, T. (2010) Electrodialysis in the food industry. *Membrane Technology* (eds. K.-V. Peinemann, S. Pereira Nunes and L. Giorno), Vol. 3 – Membranes for Food Applications, pp. 75–104, Wiley-VCH, Weinheim.

Jackson, R.S. (2008) *Wine Science (Principles and Applications)*, 3rd edn., Academic Press – Elsevier, San Diego.

Koehler, J.A., Ulbricht, M., & Belfort, G. (2000) Intermolecular forces between a protein and a hydrophilic modified polysulfone film with relevance to filtration. *Langmuir*, **16**, 10419–10427.

Liguori, L., Attanasio, G., Albanese, D. & di Matteo, M. (2010) Aglianico wine dealcoholization tests. *20th European Symposium on Computer Aided Process Engineering* – ESCAPE20, pp. 325–330, Elsevier, Amsterdam.

Lipnizki, F. (2010) Cross-flow membrane applications in the food industry, *Membrane Technology* (eds. K.-V. Peinemann, S. Pereira Nunes and L. Giorno), Vol. 3 – Membranes for Food Applications, pp. 1–24 Wiley-VCH, Weinheim.

Massot, A., Mietton-Peuchot, M., Peuchot, C. & Milisic, V. (2008) Nanofiltration and reverse osmosis in winemaking. *Desalination*, **231**, 283–289.

Mietton-Peuchot, M. (2010) New applications for membrane technologies in enology, *Membrane Technology* (eds. K.-V. Peinemann, S. Pereira Nunes and L. Giorno), Vol. 3 – Membranes for Food Applications, pp. 110127, Wiley-VCH, Weinheim.

Mietton-Peuchot, M., Milisic, V. & Noilet, P. (2002) Grape must concentration by using reverse osmosis – Comparison with chaptalization. *Desalination*, **148**, 125–129.

OIV (2012a) *International Code of Oenological Practices*, Organisation Internationale de la Vigne et du Vin, OIV, Paris.

OIV (2012b) *International Oenological Codex*, Organisation Internationale de la Vigne et du Vin, OIV, Paris.

Peynaud, E. & Allard, J.J. (1970) Concentration des moûts de raisin par osmose inverse. *Compte-rendus Académie d'Agriculture*, **56**(18), 1476–1478.

Rankine, B. (1989) *Making Good Wine*, 1stedn., The Macmillan Company of Australia, Sydney.

Rankine, B., (2004) *Making Good Wine*, revised Macmillan edn., Pan Macmillan Australia Pty Limited, Sydney.

Rektor, A., Pap, N., Kókai, Z., Szabo, R., Vatai, G. & Békássy-Molnár, E. (2004) Application of membrane filtration methods for must processing and preservation. *Desalination*, **162**, 271–277.

Ribéreau-Gayon, J., Peynaud, E., Ribéreau-Gayon, P. & Sudraud, P. (1976) *Sciences et Techniques du Vin*, Vol. III – Vinification – Transformation du Vin, Dunod, Paris.

Ribéreau-Gayon, P., Glories, Y, Maujean, A. & Dubourdieu, D. (2006a) *Handbook of Enology* Vol. 2 – *The Chemistry of Wine: Stabilization and Treatments*, John Wiley & Sons, Chichester.

Ribéreau-Gayon, P., Dubourdieu, D., Donèche, B. & Lonvaud, A. (2006b) *Handbook of Enology* Vol. 1 – The Microbiology of Wine and Vinifications, John Wiley & Sons, Chichester. Note to editor, no page numbers are given as many chapters have been referred to.

Schneider, I. (2011) The influence of colloidal treatment media on wine filterability. *Der Winzer*, March, 16–17.

Starbard, N. (2008) *Beverage Industry Microfiltration*, Wiley-Blackwell, Iowa.

Thurston, C., (2008) Electrodialysis catches on for acid reduction – Selective tartrate removal system uses less 'juice' to stabilize wine and can save wineries thousands of dollars in energy costs. *Wine Business Monthly*, August 15. Available from http://www.winebusiness.com/wbm/?go=getArticle&dataId=59205.

Ugate, P., Agosin, E., Bordeu, E. & Villalobos, J.I. (2005) Reduction in 4-ethylphenol and 4-ethylguaiacol concentration in red wines using reverse osmosis and adsorption. *American Journal of Enology and Viticulture*, **56**, 30–36.

Ulbricht, M., Ansorge, W., Danielzik, I., Kőnig, M. & Schuster, O. (2009) Fouling in microfiltration of wine – The influence of the membrane polymer on adsorption of polyphenols and polysaccharides. *Separation and Purification Technology*, **68**, 335–342.

Urkiage, A., de las Fuentes, L., Acilu, M. & Uriarte, J. (2002) Membrane comparison for wine clarification by microfiltration. *Desalination*, **148**, 115–120.

Vernhet, A. & Moutounet, M. (2002) Fouling of organic microfiltration membranes by wine constituents: importance, relative impact of wine polysaccharides and polyphenols and incidence of membrane properties. *Journal of Membrane Science*, **201**, 103–122.

Vernhet, A. & Moutounet, M. (2003) Contribution to the understanding of fouling build-up during microfiltration of wines. *Journal of Membrane Science*, **211**, 357–370.

Vernhet, A., Pellerin, M.P., Belleville, J., Planque, J. & Moutounet, M. (1999) Relative impact of major wine polysaccharides on the performances of an organic microfiltration membrane. *American Journal of Enology and Viticulture*, **50**, 51–56.

Whiting, J. & Krstic, M. (2007) *Understanding the Sensitivity to Timing and Management Options to Mitigate the Negative Impact of Bushfire Smoke on Grape and Wine Quality – Scoping Study*, Department of Primary Industries, Victoria.

WHO (1997) *IARC Mongraphs on the Evaluation of Carcinogenic Risks to Humans*, International Agency for Research on Cancer. *Silica*, **68**, 3–8.

15 Application of Membrane Technology in Vinegar

F. López

15.1 Introduction

Vinegar is a liquid food obtained by the double fermentation (alcoholic and acetic) of raw materials of agricultural origin containing sugars and/or starch substances. In the first fermentation stage, the sugars are transformed into ethanol by yeasts, whereas in the second fermentation stage the ethanol is oxidised to acetic acid by bacteria.

A variety of raw materials can be used in the production of vinegar, such as grapes, rice, apples, berries, grains, whey and honey (Solieri & Giudici, 2009). Legislation on vinegar varies in different countries; for example, in Europe the content of acetic acid in the vinegar from wine and fruit must be at least 6 g 100 mL^{-1} and 5 g 100 mL^{-1}, respectively, whereas in the USA it must be at least 4 g 100 mL^{-1}. There are also legal differences in the content of residual ethanol in vinegar between different areas of production (Solieri & Giudici, 2009).

From a technological point of view, there are two well-established methods for vinegar production, i.e. traditional (slow) and submerged (quick) (Tesfaye *et al*., 2002). The former is the so-called Orleans method, which consists of a surface culture fermentation in which the acetic acid bacteria are placed on the air–liquid interface in direct contact with the oxygen of atmospheric air. The bacteria are mainly present on the surface of the acidifying liquid and, therefore, it is also considered to be a static method. Nowadays, this method is used to produce traditional and selected vinegars, and a high acetic degree can only be obtained after a very long period of time. Therefore, production time and costs are higher. This system allows simultaneous acetification and ageing. The second method involves a classical bioreactor operating at a high aeration flow rate with the bacteria flooding the culture medium. With this method, the acetification rate is notably higher, and the production costs are lower. The quality of the vinegar, however, is also lower.

Although the two acetification methods are different, as are the raw material and the type of vinegar produced, the general process for producing vinegar is similar. The process involves preparing the alcohol substrate to be used as a culture medium, carrying out its biological oxidation and applying the post-fermentation treatments to obtain the commercial vinegar.

Membrane Processing: Dairy and Beverage Applications, First Edition. A. Y. Tamime.

15.2 Process of vinegar making

Vinegar manufacturers' focus on transforming ethanol into acetic acid, and they generally use the process shown in Fig. 15.1. The first stage of the process consists of preparing the alcoholic substrate to be used as the culture medium for acetification. This treatment usually removes the yeasts and solids in suspension that might remain in the alcohol substrate, and is normally carried out by decanting in the reception tanks. However, producers of large volumes may use centrifugation and/or filtration.

Once the alcohol substrate has been transformed into vinegar, it may be more or less clear and may contain bacteria and suspended solids. Then the vinegar is treated so that it reaches the consumer with the best appearance possible as far as colour and cleanliness are concerned. The quality of the vinegar may also be affected by chemical phenomena, such as the actions of micro-organisms, yeasts and vinegar eels.

Sometimes vinegars need to be treated by clarification and stabilisation, and the traditional method consists of adding adjuvant, such as bentonite and polyvinyl polypyrrolidone, to facilitate the process and to remove the oxidised brown effect produced in white vinegars. To obtain a specific product standard and maintain its characteristics and quality, it is a common practice in the vinegar industry to decolourise a fraction of the vinegar and blend it with the bulk of the vinegar that has not decolourised (López *et al*., 2003). Traditionally, activated carbon has been used to decolourise vinegar. The method used consists of mixing the activated carbon with the vinegar, stirring and then separating the activated carbon by filtration or settling. This process is a semi-continuous operation and, generally, small producers separate activated carbon by settling, which means that considerable amounts of vinegar are lost and solid residues are generated. The vinegar is then usually filtered by a diatomaceous earth filter.

Another important operation in vinegar processing is the dilution of vinegar to a commercial concentration (e.g. 6 g 100 g mL^{-1} in Europe for wine vinegars). To prevent the residual alcohol that remains in the vinegar from acetifying or the acetic acid from

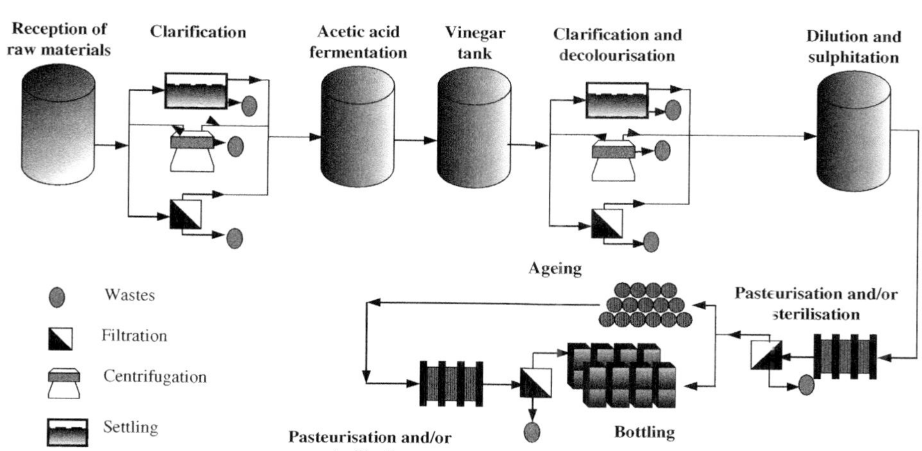

Fig. 15.1 Flow diagram showing the various stages during the manufacture of vinegar.

oxidising to carbon dioxide and water, sulphur dioxide needs to be added to the vinegar (Ormaechea-Landa, 1991). After this operation, the vinegar can be pasteurised and/or sterilised by heating or membrane filtration, and the product may be bottled or aged in wooden barrels. If the latter approach of processing is practised, the aged vinegar may need to be filtered again before bottling.

In vinegar production, filtration is often used where diatomaceous earth filters are normally used. This technique has the disadvantage that it generates solid waste and, in some cases, the vinegar may need to be sterilised to remove bacteria or vinegar eels – small nematodes that occur in some types of wine vinegar aged in wooden casks (for example, sherry vinegars). Filtration with diatomaceous earth filters requires a considerable amount of manpower and a certain loss of vinegar. Therefore, the use of membrane technologies, especially tangential flow filtration, improves the vinegar production process, reduces the environmental impact and makes cold sterilisation possible, thus reducing the vinegar treatment to a minimum, preserving the vinegar's own qualities, and minimising production costs.

15.3 Membrane technology in the production of vinegar

Lately, membrane science and technology has led to an improvement in the processes of the food industry, and has made it possible to design, rationalise and optimise innovative products. Membrane technology has improved the quality of the final product, decreased the thermal and environmental impact, and minimised the need for space and energy consumption, which means that it can be applied to all fields of production in the food and wine industry (Drioli & Romano, 2001) and in vinegar clarification (García-García *et al.*, 2009).

Many reported studies focus on using membranes on a laboratory- or pilot-scale in food industries producing fruit juice, beer, wine, milk, etc. Membrane technologies, such as the cold sterilisation method, have also been applied to food products, e.g. milk (Girard & Fukumoto, 2000), beer (Drioli & Romano, 2001) and pineapple juice (Carneiro *et al.*, 2002). Membrane application in the vinegar industry makes it possible to avoid a pasteurisation stage, which is still useful in this production area (Ormaechea-Landa, 1991). Cross-flow microfiltration (MF) is useful because clarification and sterilisation can be carried out simultaneously (Carneiro *et al.*, 2002).

Very few studies have been made of the application of membrane technology to vinegar; nevertheless, the application of MF of vinegar has been reported by the main membrane manufacturers. In industrial production, the filter modules used in cross-flow MF are hollow-fibre membranes or spiral-wound membranes manufactured with organic materials, such as polysulphone or regenerated cellulose, and ceramic tubular membranes manufactured with materials, such as zirconia or alumina. The pore sizes that are most commonly used are of 0.2, 0.45 and 0.65 μm, but the current trend is to use 0.2 μm.

López *et al*. (2005) made a study of the clarification of vinegars (white, rosé and red) by cross-flow MF on an industrial scale, and the main conclusions were as follows:

- The turbidity for the three kinds of vinegars was reduced considerably; for filtered vinegars it was less than 0.5 Nephelometric Turbidity Unit (NTU).
- The reduction of total solids in suspension was complete.

- These results show that MF can be used to simultaneously clarify and cold sterilise vinegar.
- The effect of MF on colour and polyphenol content was acceptable. The reduction in colour, expressed as a decrease in modified colour intensity, was practically negligible for white vinegars, 11% for rosé vinegar and 37% for red vinegar.
- The reduction in polyphenol content was less than 15% in all vinegars. It was greatest for red vinegar, followed by rosé and finally white vinegar. The polyphenolic profile of the filtered vinegars is quite similar to that of the initial vinegar. The reduction in benzoic and hydroxycinnamic acids was less than 10% for all vinegars. However, the reduction in procyanidin B2 was total for rosé and red vinegars.

The production of highly concentrated vinegar would be advantageous for the vinegar industry because it would decrease storage capacity and filtration volumes. However, in order to obtain high concentrations of acetic acid, parameters such as alcohol and acetic acid concentrations in the medium need to be controlled because the growth of acetic acid bacteria is directly associated with these variables. At the present, batch technology, which is the most widely used method of manufacture in the vinegar industry, does not allow these parameters to be optimised. The main problem is the number of viable acetic acid bacteria because they stop increasing above 40 g L^{-1} acetic acid, and are also inhibited by ethanol (Berraud, 2000). The inhibition by alcohol is solved using the fed-batch system, in which the ethanol is maintained at a particular concentration by selective feeding. The inconvenience of highly concentrated acetic acid can be solved by a fed-batch operation with two consecutive fermenters or a bioreactor coupled to a cell recirculation system (a filter that separates cells from the culture medium).

Another way of obtaining high concentration vinegar is to crystallise water from vinegar using freeze concentration. An alternative to concentrating vinegar is electrodialysis (i.e. using membrane technology), which can be used to concentrate vinegar, and Chukwu & Cheryan (1996) concentrated vinegar (9–12 g 100 mL^{-1} acetic acid) to produce triple-strength vinegar (30–33 g 100 mL^{-1} acetate).

Membrane bioreactors (MBRs) are an attractive technology, which can improve productivity. Various studies have been made on the vinegar industry. Of these, the first (Mehaia & Cheryan, 1991) studied the fermentation of date extracts to ethanol and vinegar in batch and continuous-membrane reactors. The production of ethanol with the membrane recycle bioreactor was more than eight times higher than with batch process fermentation. The acetic acid was produced from the fermented date juice with *Acetobacter aceti*. In a membrane recycle bioreactor, productivity was 10.8 g L^{-1} h^{-1} at an acetic acid concentration of 45 g L^{-1}. This is more than 20 times higher than that of a batch process.

Nishiwaki & Dunn (2005) analysed acetic acid production in a continuous two-stage bioreactor with cell recycling. In this method of production, both the filtrate and the bleed broth from the first stage are supplied to the second fermenter. At high substrate conversions, this configuration is found to provide higher acetic acid productivity than either a previous configuration in which only the first stage bleed broth is fed to the second stage or a single recycle chemostat at the same bleed ratios.

In vinegar decolourisation, nanofiltration (NF) can be an alternative technique of processing rather than the traditional system using activated carbon. Güell & López (2001) decoloured vinegar with a small pilot-plant that operates with a titanium dioxide membrane fused on a porous ceramic tube with a nominal molecular cut-off of 1 kDa. The decolourisation efficiency was 88.2% and 35.5% for the red and white vinegars, respectively. It has also been observed that the membrane treatment affects the fixed acidity and the dry matter content of the product, but leaves the other parameters almost unaffected. This study suggests that NF can be used successfully to reduce colour in vinegar.

15.4 Conclusions

The application of membrane technology in the vinegar industry focuses mainly on clarification and cold sterilisation using cross-flow MF. Nevertheless, other techniques, such as NF can feasibly be used to decolourise vinegar by removing brown compounds from oxidised vinegar and to decolourise vinegar to be used in vinaigrettes and canned products as an alternative technique of processing to activated carbon.

Membrane technology can also be used in MBRs with the primary objective of improving productivity. Nevertheless, this system is not currently applied on an industrial-scale.

References

Berraud, C. (2000) Production of highly concentrated vinegar in fed-batch culture. *Biotechnology Letters*, **22**, 451–454.

Carneiro, L., Sa, I.D., Gomes, F.D., Matta, V.M., & Cabral, L.M.C. (2002) Cold sterilization and clarification of pineapple juice by tangential microfiltration. *Desalination*, **148**, 93–98.

Chukwu, U. & Cheryan, M. (1996) Concentration of vinegar by electrodialysis. *Journal of Food Science*, **61**, 1223–1226.

Drioli, E., & Romano, M. (2001) Progress and new perspectives on integrated membrane operations for sustainable industrial growth. *Industrial Engineering and Chemistry Research*, **40**, 1277–1300.

García-García, I., Santos-Dueñas, I.M., Jiménez-Ot, C., Jiménez-Hornero, J.E. & Bonilla-Venceslada, J.L. (2009) Vinegar engineering. *Vinegars of the World* (eds. L. Solieri and P. Giudici), pp. 97–120, Springer-Verlag, Milan.

Girard, B. & Fukumoto, L. R. (2000) Membrane processing of fruit juices and beverages: A review. *Critical Reviews in Food Science and Nutrition*, **40**, 91–157.

Güell, C. & López, F. (2001) Vinegar decolorization using nanofiltration. *Engineering with Membranes* (ed. S. Luque & J.R. Álvarez), pp. II-85–II-88, Servicio Publicaciones Universidad de Oviedo, Oviedo.

López, F., Medina, F., Prodanov, M. & Güell, C. (2003) Oxidation of activated carbon: Application to vinegar decolorization. *Journal of Colloid and Interface Science*, **257**, 173–178

López, F., Pescador, P., Güell, C., Morales, M.L., García-Parrilla, M.C. & Troncoso, A.M. (2005) Industrial vinegar clarification by cross-flow microfiltration: Effect on colour and polyphenol content. *Journal of Food Engineering*, **68**, 133–136

Mehaia, M.A. & Cheryan, M. (1991) Fermentation of date extracts to ethanol and vinegar in batch and continuous membrane reactors. *Enzyme and Microbial Technology*, **13**, 257–261.

Nishiwaki, A. & Dunn, I.J. (2005) Analysis of acetic acid productivity in a continuous two-stage bioreactor with cell recycling. *Journal of Chemical Technology and Biotechnology*, **80**, 271–275.

Ormaechea-Landa, A.R. (1991) Proceso industrial de elaboración del vinagre. *El vinagre de vino* (eds. C. Llaguno and M.C. Polo), pp. 49–68, C.S.I.C., Madrid.

Solieri, L. & Giudici, P. (2009) Vinegars of the World. *Vinegars of the World* (eds. L. Solieri and P. Giudici), pp. 1–16, Springer-Verlag, Milan.

Tesfaye W., Morales M.L., García-Parrilla M.C. & Troncoso A.M. (2002) Wine Vinegar: technology, authenticity and quality evaluation. *Trends in Food Science and Technology*, **13**, 12–21.

Index

Membrane Processing: Dairy and Beverage Applications, First Edition. A. Y. Tamime.

Printed and bound by CPI Group (UK) Ltd, Croydon, CR0 4YY
15/09/2026
14962398-0001